CONFORMAL ARRAY ANTENNA THEORY AND DESIGN

IEEE PRESS SERIES ON ELECTROMAGNETIC WAVE THEORY

The IEEE Press Series on Electromagnetic Wave Theory consists of new titles as well as reprints ahd revisions of recognized classics in electromagnetic waves and applications which maintain long-term archival significance.

**BOOKS IN THE IEEE PRESS SERIES
ON ELECTROMAGNETIC WAVE THEORY**

Christopoulos, C., *The Transmission-Line Modeling Methods, TLM*
Clemmow, P. C., *The Plane Wave Spectrum Representation of Electromagnetic Fields*
Collin, R. B., *Field Theory of Guided Waves,* Second Edition
Dudley, D. G., *Mathematical Foundations for Microwave Engineering,* Second Edition
Elliot, R. S., *Antenna Theory and Design,* Revised Edition
Elliot, R. S., *Electromagnetics: History Theory, and Applications*
Felsen, L. B., *and Marcuvitz, N., Radiation and Scattering of Waves*
Harrington, R. F, *Field Computation of Moment Methods*
Harrington, R. F, *Time Harmonic Electromagnetic Fields*
Hansen et al., *Plane-Wave Theory of Time-Domain Fields. Near-Field Scanning Applications*
Ishimaru, A., *Wave Propagation and Scattering in Random Media*
Jones, D. S., *Methods in Electromagnetic Wave Propagation,* Second Edition
Lindell, I. V., *Methods for Electromagnetic Field Analysis*
Lindell, I. V, *Differential Forms in Electromagnetics*
Peterson et al., *Computational Methods for Electromagnetics*
Tai, C. T., *Generalized Vector and Dyadic Analysis. Applied Mathematics in Field Theory*
Tai, C. T., *Dyadic Green Functions in Electromagnetic Theory,* Second Edition
Van Bladel, J., *Singular Electromagnetic Fields and Sources*
Volakis et al., *Finite Element Method. for Electromagnetics: Antennas, Microwave Circuits, and Scattering Applications*
Wait, J., *Electromagnetic Waves in Stratified Media*

CONFORMAL ARRAY ANTENNA THEORY AND DESIGN

Lars Josefsson
Chalmers University of Technology, Sweden

Patrik Persson
Royal Institute of Technology, Sweden

IEEE Antennas and Propagation Society, *Sponsor*

IEEE Press

A WILEY-INTERSCIENCE PUBLICATION

Published by John Wiley & Sons, Inc., Hoboken, New Jersey.
Published simultaneously in Canada.

For general information on our other products and services or for technical support, please contact our Customer Care Department within the United States at (800) 762-2974, outside the United States at (317) 572-3993 or fax (317) 572-4002.

Wiley also publishes its books in a variety of electronic formats. Some content that appears in print may not be available in electronic format. For information about Wiley products, visit our web site at www.wiley.com.

Library of Congress Cataloging-in-Publication Data is available.

ISBN-13 978-0-471-46584-3
ISBN-10 0-471-46584-4

Printed in the United States of America.

10 9 8 7 6 5 4 3 2 1

CONTENTS

PREFACE

This book is the result of close cooperation between industry and academia, notably Ericsson and two universities in Sweden: Chalmers University of Technology (CTH) in Göteborg and The Royal Institute of Technology (KTH) in Stockholm. In 1987, a student at CTH presented her PhD thesis on the topic of conformal antennas. It was then considered an interesting but difficult technology with no immediate application. About 10 years later, many conformal array applications were seriously considered or in development. At that time, we became involved in several conformal R&D programs that also included experimental hardware for model verification. Our intention was to compile results from these efforts into one, thick internal report. However, with the support and encouragement from many colleagues, we set out on the much more demanding route to write a book on the subject.

Many standard textbooks on antennas include short sections on conformal array antennas, but usually only simple reference cases are treated. The mutual coupling (which is an important parameter) is often just briefly mentioned. Examples of array characteristics with the mutual coupling included are rare. Thus, we believe this book fills a gap in the existing literature.

Our purpose is to present the fundamental principles behind conformal antennas, as well as hands-on information necessary for the analysis and design of conformal antenna arrays. Graphical illustrations are used extensively, both for calculated and measured results, including results not published before. We describe theoretical methods for analysis and design, and include explicit formulas where applicable. From a practical point of view, mechanical aspects, beam-forming techniques, and packaging of conformal array antennas are included. Furthermore, scattering properties are discussed, which are of interest in stealth applications, for example. Lists of references are provided at the end of each chapter for further studies. Thus, we hope that the book will become a useful tool for the practicing antenna and systems engineer in understanding and working with these interesting antennas.

Each chapter starts with some introductory material, that is, the basic concepts that are essential to get an understanding of the more advanced aspects. The first three chapters present an overview of conformal array principles and applications, including the theory for circular arrays and phase mode concepts, and discussions of various shapes of conformal arrays.

In Chapters 4 and 5, theoretical methods for analysis and design are described, including explicit formulas; for example, for geodesics on more general surfaces than the canonical circular cylinder and sphere. Doubly curved surfaces and dielectric covered surfaces using high-frequency methods are also included. Two canonical examples are also discussed in detail, thus assisting the reader in his/her own conformal antenna analysis.

Chapters 6 and 7 deal with radiating elements on singly curved and doubly curved surfaces. The focus is on mutual coupling characteristics and element radiation properties. Element types include waveguide-fed apertures and microstrip patches. For both types, measured data supports the calculated results.

Chapters 8 and 9 treat conformal array antenna characteristics—radiation, impedance and polarization—as well as mechanical and packaging aspects. Feeding systems and beam-scanning principles are also included.

Chapter 10 discusses various synthesis methods, with some examples. Also, aspects such as optimizing the shape, distribution of elements, polarization, and bandwidth are included.

The final chapter deals with methods for the analysis of scattering (radar cross section) from conformal array antennas; in particular, waveguide-fed aperture elements with and without a dielectric coating. We include also a discussion on the problem of reducing the radar cross section without decreasing the antenna performance.

While written with engineering applications in mind, this book can also serve as a text for graduate courses in advanced antennas and antenna systems.

ACKNOWLEDGMENTS

Among the numerous colleagues and friends who have supported our work with their advice, encouragement, and contributions, we can mention only a few. In particular, we want to thank Dr. Björn Thors, Royal Institute of Technology/Ericsson AB, Sweden, for his contributions on scattering and radiation characteristics, and for valuable discussions. Thanks also go to Dr. Zvonimir Sipus, University of Zagreb, Croatia, for contributing results for antennas on singly and doubly curved surfaces. Dr. Hans Steyskal, U.S. Air Force Research Lab, helped with valuable comments, especially regarding array theory, beam forming, and synthesis techniques. Other contributors include Dr. Torleif Martin, Swedish Defence Research Agency (FOI); Prof. Lars Pettersson, Swedish Defence Research Agency (FOI); Dr. Silvia Raffaelli, Ericsson AB, Sweden; and Techn. Lic. Maria Lanne, Ericsson Microwave Systems AB/Chalmers University of Technology, Sweden. Special thanks go to Prof. Kjell Rosquist, Stockholm University, Sweden, for his guidance into the world of geodesics.

We are grateful for the permission to use results from several recent studies on conformal antennas in Sweden, among them work sponsored by Ericsson Microwave Systems AB, Ericsson AB, The Foundation for Strategic Research (SSF), and The Swedish Defense Material Administration (FMV).

Last but not least, we express our appreciation and gratitude to our families for their encouragement, understanding, and patience during the writing of this book.

ABBREVIATIONS AND ACRONYMS

A/D	analog/digital
AF	array factor
BI	boundary integral
BiCG	biconjugate gradient method
BOR	body of revolution
CAD	computer aided design
CP	contour path; also circular polarization, conformal path
DBF	digital beam forming
DE	differential equation
DOA	direction of arrival
EFIE	electric field integral equation
EGL	endfire grating lobe
EM	electromagnetic
EMP	electromagnetic pulse
ESM	electronic support measures
EWCA	European Workshop on Conformal Antennas
FDTD	finite-element time domain
FE	finite element
FE-BI	finite-element boundary integral
FBF	fast beam forming
FEM	finite-element method
FFT	fast Fourier transform
FMF	fast mode former
FSS	frequency-selective structure
GCM	geodesic constant method
GCS	geodesic coordinate system
GaAs	gallium arsenide
GHOR	general hyperboloid of revolution
GO	geometrical optics
GPOR	general paraboloid of revolution
GTD	geometric theory of diffraction
h/D	characteristic dimension of a paraboloid, height over diameter
HF	high frequency
HP	horizontal polarization
IBC	impedance boundary condition
IC	integrated circuit
IE	integral equation
IEP	isolated element pattern

IF	intermediate frequency
LEO	low earth orbit
LMS	least mean square
LNA	low-noise amplifier
LP	linear polarization
LPI	low probability of intercept
MEMS	microelectromechanical system
MMIC	monolithic microwave integrated circuit
MOM, MoM	method of moments
MPIE	mixed-potential integral equation
MTI	moving-target indication
MUSIC	multiple-signal classification
PEC	perfect electric conductor
PO	physical optics
PTD	physical theory of diffraction
Q	quality factor
RCS	radar cross section
RF	radio frequency
RX	receive
SAR	synthetic-aperture radar
SBR	shoot and bounce ray (method)
SDP	steepest decent path
SMI	sample matrix inversion
SNIR	signal-to-noise plus interference ratio
SPNT	single-pole N-throw switch
STAP	space–time adaptive processing
TACAN	tactical air navigation
TD	time domain
TD-PO	time domain physical optics
TD-UTD	time domain uniform theory of diffraction
TE	transverse electric
TEM	transverse electromagnetic
TM	transverse magnetic
TX	transmit
UCA	uniform circular array
ULA	uniform linear array
UPML	uniaxial perfectly matched layer
UTD	uniform theory of diffraction
VP	vertical polarization
VPD	variable power divider

1

INTRODUCTION

1.1 THE DEFINITION OF A CONFORMAL ANTENNA

A conformal antenna is an antenna that conforms to something; in our case, it conforms to a prescribed shape. The shape can be some part of an airplane, high-speed train, or other vehicle. The purpose is to build the antenna so that it becomes integrated with the structure and does not cause extra drag. The purpose can also be that the antenna integration makes the antenna less disturbing, less visible to the human eye; for instance, in an urban environment. A typical additional requirement in modern defense systems is that the antenna not backscatter microwave radiation when illuminated by, for example, an enemy radar transmitter (i.e., it has stealth properties).

The IEEE Standard Definition of Terms for Antennas (IEEE Std 145-1993) gives the following definition:

> **2.74 conformal antenna [conformal array].** An antenna [an array] that conforms to a surface whose shape is determined by considerations other than electromagnetic; for example, aerodynamic or hydrodynamic.
>
> **2.75 conformal array.** *See:* **conformal antenna.**

Strictly speaking, the definition includes also planar arrays if the planar "shape is determined by considerations other than electromagnetic." This is, however, not common practice. Usually, a conformal antenna is cylindrical, spherical, or some other shape, with the radiating elements mounted on or integrated into the smoothly curved surface. Many

Conformal Array Antenna Theory and Design. By Lars Josefsson and Patrik Persson variations exist, though, like approximating the smooth surface by several planar facets. This may be a practical solution in order to simplify the packaging of radiators together with active and passive feeding arrangements.

1.2 WHY CONFORMAL ANTENNAS?

A modern aircraft has many antennas protruding from its structure, for navigation, various communication systems, instrument landing systems, radar altimeter, and so on. There can be as many as 20 different antennas or more (up to 70 antennas on a typical military aircraft has been quoted [Schneider et al. 2001]), causing considerable drag and increased fuel consumption. Integrating these antennas into the aircraft skin is highly desirable [Wingert & Howard 1996]. Preferably, some of the antenna functions should be combined in the same unit if the design can be made broadband enough. The need for conformal antennas is even more pronounced for the large-sized apertures that are necessary for functions like satellite communication and military airborne surveillance radars.

A typical conformal experimental array for leading-wing-edge integration is shown in Figure 1.2. The X-band array is conformal with the approximately elliptical cross section shape of the leading edge of an aircraft wing [Kanno et al. 1996]. Figure 1.3 shows an even more realistically wing-shaped C-band array (cf. [Steyskal 2002]).

Array antennas with radiating elements on the surface of a cylinder, sphere, or cone, and so on, without the shape being dictated by, for example, aerodynamic or similar reasons, are usually also called conformal arrays. The antennas may have their shape determined by a particular electromagnetic requirement such as antenna beam shape and/or angular coverage. To call them conformal array antennas is not strictly according to the IEEE definition cited above, but we follow what is common practice today.

A cylindrical or circular array of elements has a potential of 360° coverage, either with an omnidirectional beam, multiple beams, or a narrow beam that can be steered over

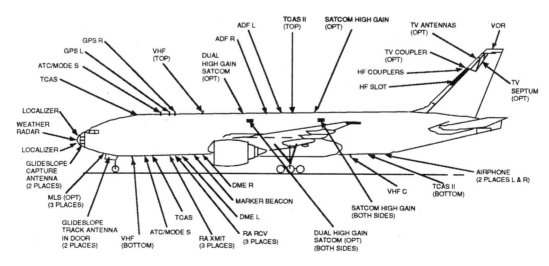

Figure 1.1. At least 20–30 antennas protrude from the skin of a modern aircraft. (From [Hopkins et al. 1997], reprinted by permission of the American Institute of Aeronautics and Astronautics, Inc.)

Figure 1.2. Conformal array antenna for aircraft wing integration [Kanno et al. 1996].

Figure 1.3. A microstrip array conformal to a wing profile in the test chamber. See also color insert, Figure 1. (Courtesy of Air Force Research Lab./Antenna Technology Branch, Hanscom AFB, USA.)

360°. A typical application could be as a base station antenna in a mobile communication system. Today, the common solution is three separate antennas, each covering a 120° sector. Instead, one cylindrical array could be used, resulting in a much more compact installation and less cost.

Another example of shape being dictated by coverage is shown in Figure 1.4. This is a satellite-borne conical array (and, hence, the drag problem is certainly not an issue here).

The arguments for and against conformal arrays can be discussed at length. The applications and requirements are quite variable, leading to different conclusions. In spite of this, and to encourage further discussion, we present a summary based on reflections by Guy [1999], Guy et al. [1999], Watkins [2001], and others in Table 1.1.

1.3   HISTORY

The field of phased array antennas was a very active area of research in the years from WW II up to about 1975. During this period, much pioneering work was done also for conformal arrays. However, electronically scanned, phased array antennas did not find widespread use until the necessary means for feeding and steering the array became available. Integrated circuit (IC) technology, including monolithic microwave integrated circuits (MMIC), filled this gap, providing reliable technical solutions with a potential for

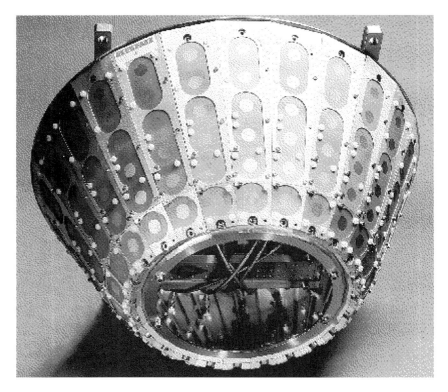

Figure 1.4. A conical conformal array for data communication from a satellite [Vourch et al. 1998, Caille et al. 2002]. See also color insert, Figure 5.

Table 1.1. Planar versus conformal array antennas

Parameter	Planar array	Conformal array
Technology	Mature	Not fully established
Analysis tools	Available	In development
Beam control	Phase only usually sufficient, fixed amplitude	Amplitude and phase, more complicated
Polarization	Single can be used (dual often desired)	Polarization control required, especially if doubly curved
Gain	Drops with increased scan	Controlled, depends on shape
Frequency bandwidth	Typically 20%	Wider than planar is possible
Angular coverage	Limited to roughly ±60°	Very wide, half sphere
RCS	Large specular RCS	Lower than planar
Installation on platform	Planar shape limits due to swept volume	Structurally integrated, leaves extra space. No drag
Radome	Aberration effects	No conventional radome, no boresight error
Packaging of electronics	Known multilayer solutions	Size restriction if large curvature, facets possible

low cost, even for very complex array antennas. An important factor was also the development of digital processors that can handle the enormously increased rate of information provided by phased array systems. Digital processing techniques made phased array antenna systems cost effective, that is, they provided the customers value for the money spent.

This being true for phased arrays in general, it also holds for conformal array antennas. However, in the area of conformal arrays, electromagnetic models and design know-how needed extra development. During the last 10 to 20 years, numerical techniques, electromagnetic analysis methods, and the understanding of antennas on curved surfaces have improved. Important progress has been made in high-frequency techniques, including analysis of surface wave diffraction and modeling of radiating sources on curved surfaces.

The origin of conformal arrays can be traced at least back to the 1930s when a system of dipole elements arranged on a circle, thus forming a circular array, was analyzed by Chireix [1936]. Later, in the 1950s, several publications on the subject were presented; see, for example, [Knudsen 1953a,b]. The circular array was attractive because of its rotational symmetry. Proper phasing can create a directional beam, which can be scanned 360° in azimuth. The applications were in broadcasting, communication, and later also navigation and direction finding. An advanced, more recent application using a large circular array is the French RIAS experimental radar system [Dorey et al. 1989, Colin 1996].

During the Second World War, HF circular arrays were developed for radio signal intelligence gathering and direction finding in Germany. These so-called Wullenweber[1] arrays (code word for the development project) were quite large with a diameter of about 100 meters. After the war, an experimental Wullenweber array was developed at the University of Illinois (see Figure 1.5). This array had 120 radiating elements in front of a reflecting screen. The diameter was about 300 m; note the size of the buildings in the center [Gething 1966]. Many similar systems were built in other countries during the Cold War.

---

[1]Named for J. Wullenweber, 1488–1537, Lord Mayor of Lübeck

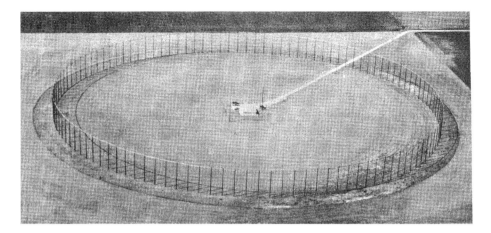

Figure. 1.5. The experimental 300 m diameter Wullenweber antenna at the University of Illinois. (Courtesy P. J. D. Gething, "High-Frequency Direction Finding," *Proceedings of IEE,* January 1966, p. 54.)

Some of these huge antennas may still be operating. See also [IRE PGAP Newsletter Vol. 3, December 1960].

During this period, new, efficient pattern synthesis methods and practical feeding and beam control schemes were investigated by several workers. For an overview see [Davies 1981, 1983]. A very useful approach in this work was based on the concept of phase modes. For the circular array, the excitation can be viewed as a periodic function in azimuth, with period 2π. The excitation can therefore be expressed as a Fourier series. Each term in this series is a phase mode, which can be generated in the practical situation by convenient networks, specifically the Butler matrix. A phase mode has constant amplitude but a linear phase progression from one radiating element to the next, totaling a multiple of 2π over the circumference. The phase mode concept proved to be an efficient tool in pattern synthesis, and will be described in more detail in Chapters 2 and 10. By measuring the phase modes on reception, the signal direction of arrival (DOA) can be determined [Rehnmark 1980]. Figure 1.6 shows a direction-finding application using this technique.

With omnidirectional elements, the full circle can be used. However, constructive addition of signals from both the front and the rear part of the circular array is not easily achieved, in particular not over an extended bandwidth. Most circular arrays therefore use directive radiating elements, pointing outward from the center. The Wullenweber antennas have usually a reflective element or screen behind each radiator, making them directive. Element directivity has been analyzed in relation to the phase mode concept and significant improvements compared to omnidirectional elements were demonstrated [Rahim et al. 1981].

In order to increase the directivity and narrow the beam in elevation, several circular arrays placed on top of each other can be used. A good example is the electronically scanned TACAN (tactical navigation) antenna [Christopher 1974, Shestag 1974]. The TACAN antenna can be placed on the ground, radiating a rotating phase-coded signal that helps aircraft to find their position in relation to, for example, an airfield.

Jim Wait did fundamental work on radiation from apertures in metallic circular cylinders; see [Wait 1959]. His work has been continued by many others employing either

Figure 1.6. Broadband circular array for signal-bearing measurements. See also color insert, Figure 7. (Courtesy of Anaren Inc., Syracuse, NY, USA.)

modal expansion techniques or high-frequency diffraction techniques [Hessel 1970, Pathak et al. 1980]. In particular, mutual coupling is included in the solutions. The methods will be described in Chapter 4.

Nose-mounted antennas in missiles or aircraft are protected by a pointed radome. Alternatively, the antenna elements could be put on the radome itself. This possibility has created an interest in conformal arrays on cones [Munger et al. 1974]. The progress in this field has been slow, however. Also, conformal spherical antennas have attracted interest. A well-known example is the dome radar antenna [Bearse 1975, Liebman et al. 1975]. This antenna has a passive-transmission-type lens of hemispherical shape. It is fed from its diameter plane by a planar-phase-steered array (Figure 1.7). The lens causes an extra

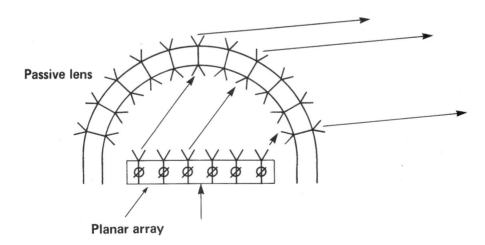

Figure 1.7. The dome array concept using a single planar array and a passive lens for hemispherical coverage.

deflection of the beam so that a scan of more than 90° is achieved. Such a wide coverage would normally require four planar arrays. However, there is a need for polarization control and the lens structure has some bandwidth limitations. It is an advantage that only one steered array is needed for hemispherical coverage, but detailed analysis [Kinsey 2000] indicates that the dome antenna does not offer any cost benefits over traditional solutions using planar arrays. According to Fowler [1998], the invention seems not to have been used in real applications.

Another array with more than hemispherical coverage, in this case for satellite communication from mobile units, is shown in Figure 1.8. This is an active faceted array with integrated transmit and receive electronics.

A great deal of important conformal work was done at the U.S. Naval Electronics Laboratory Center (NELC) in San Diego. The work included development of both cylindrical and conical arrays as well as feeding systems. Most of the activities in this field were closed around 1974. However, many technical results from this active period may be found in the *IEEE Transactions on Antennas and Propagation,* Special issue on conformal antennas (January 1974). Several workshops on conformal antennas were held in the United States, for example, in 1970 and 1975, but the proceedings may be hard to find. Most of the material was later published in scientific journals. At the 1996 Phased Array Symposium in Boston, several interesting conformal designs were presented in a Japanese session [Rai et al. 1996, Kanno et al. 1996]; see Figure 1.2.

An indication of a recent resurgence in the interest in conformal antennas is the series of Conformal Antenna Workshops, held in Europe every second year, starting from 1999. The first was held in Karlsruhe (Germany), the second in The Hague (The

Figure 1.8. A faceted active array antenna with six dual polarized dipole elements in each facet. See also Figure 8.4 and color insert, Figure 6. (Courtesy of Roke Manor Research Ltd., Roke Manor, Romsey, Hampshire, UK.)

Netherlands), the third in Bonn (Germany), and the fourth in Stockholm (Sweden) in 2005.

A paper in *Space/Aeronautics* magazine in 1967 [Thomas 1967] presented a very optimistic view of the development of conformal arrays for nose radar systems in aircraft; see Figure 1.9. Obviously, the development did not proceed quickly, mainly because of the limitations discussed previously. However, the conformal nose-mounted array has many advantages, especially an increased field of view compared to the traditional ±60° coverage of planar antennas.

A vision of a future "smart skin" conformal antenna is shown in Figure 1.10. This antenna constitutes a complete RF system, including not only the radiating elements but also feed networks, amplifiers, control electronics, power distribution, cooling, filters, and so on, all in a multilayer design that can be tailored to various structural shapes [Josefsson 1999, Baratault et al. 1993].

1.4 METAL RADOMES

What do radomes have to do with conformal array antennas? Radomes are usually thought of as dielectric shell structures protecting an antenna installation. If made of metal, a dense array of openings (slots) can provide the necessary transmission properties within a restricted range of frequencies. The result is a conformal frequency-selective structure (FSS). It is not an antenna, of course, but viewed from the outside it exhibits all the radiating characteristics of a curved antenna array of radiating elements, just like a conformal antenna. Hence, the (exterior) analysis problem of the structure has much in common with the analysis of conformal arrays. Pelton and Munk [1974] describe a conical metal radome that could be used in a high-speed aircraft or missile application.

A doubly curved FSS acting as a frequency and polarization filter is shown in Figure 1.11. Here we have a spherical array with two layers of rectangular slots in a copper sheet on a dielectric carrier [Stanek and Johansson 1995].

1.5 SONAR ARRAYS

The activities related to sonar arrays are often overlooked by the antenna community. These acoustic arrays used for underwater sensors are analogous to radar or communica-

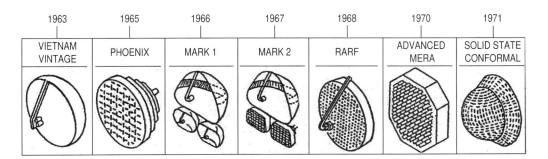

1963	1965	1966	1967	1968	1970	1971
VIETNAM VINTAGE	PHOENIX	MARK 1	MARK 2	RARF	ADVANCED MERA	SOLID STATE CONFORMAL

Figure 1.9. Predicted nose radar development as of 1967 [Thomas 1967].

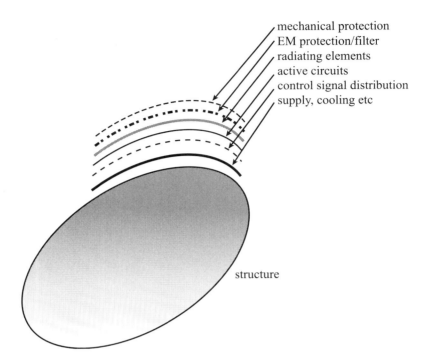

mechanical protection
EM protection/filter
radiating elements
active circuits
control signal distribution
supply, cooling etc

structure

Figure 1.10. Vision of a smart-skin antenna.

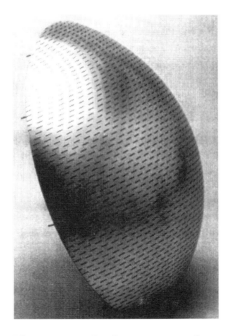

Figure 1.11. A spherical frequency-selective structure of resonant slots. See also color insert, Figure 8. (Courtesy of Ericsson Microwave Systems AB, Göteborg, Sweden.)

Figure 1.12. A cylindrical passive hydrophone array. See also color insert, Figure 10. (Courtesy of Atlas Elektronik GmbH, Submarine Systems, Bremen, Germany.)

tion arrays. The techniques for signal processing and beam forming are similar [Ziehm 1964, Stergiopoulos and Dhanantwari 1998, Gaulladet and de Moustier 2000]. The wave propagation is radically different, however, with sonic waves propagating almost six orders of magnitude more slowly than electromagnetic waves. The time scale is therefore radically different. The wavelengths used are about the same, and acoustic sensor arrays have an almost "electrical" appearance (Figure 1.12).

REFERENCES

Proceedings of First European Workshop on Conformal Antennas, Karlsruhe, Germany, 29 October 1999.

Proceedings of Second European Workshop on Conformal Antennas, The Hague, The Netherlands, 24–25 April 2001.

Proceedings of Third European Workshop on Conformal Antennas, Bonn, Germany, 22–23 October 2003.

Proceedings of Fourth European Workshop on Conformal Antennas, Stockholm, Sweden, 23–24 May 2005.

"Special Issue on Conformal Arrays," *IEEE Transactions on Antennas and Propagation,* Vol. AP-22, No. 1, January 1974.

Baratault P., Gautier F., and Albarel G. (1993), "Évolution des Antennes pour Radars Aéroportés. De la Parabole aus Peaux Actives," *Rev. Techn., Thomson-CSF,* pp. 749–793, September.

Bearse S. V. (1975), "Planar Array Looks Through Lens to Provide Hemispherical Coverage," *Microwaves,* July, pp. 9–10.

Caille G., Vourch E., Martin M. J., Mosig J. R., and Martin Polegre A. (2002), "Conformal Array Antenna for Observation Platforms in Low Earth Orbit," *IEEE Antennas and Propagation Magazine,* Vol. 44, No. 3, pp. 103–104, June.

Chireix H. (1936), "Antennes à Rayonnement Zénithal Réduit," *L'Onde Electrique,* Vol. 15, pp. 440–456.

Christopher E. J. (1974), "Electronically Scanned TACAN Antenna," *IEEE Transactions on Antennas and Propagation,* Vol. AP-22, No. 1, pp. 12–16, January.

Colin J.-M. (1996), "Phased Array Radars in France: Present and Future," *IEEE International Symposium on Phased Array Systems and Technology,* 15–18 October, pp. 458–462.

Davies D. E. N. (1981), "Circular Arrays: Their Properties and Potential Applications," in *IEE Proceedings of 2nd International Conference on Antennas and Propagation,* April, pp. 1–10.

Davies D. E. N. (1983), "Circular Arrays," in Rudge et al. (eds.): *The Handbook of Antenna Design,* Vol. 2, Peter Peregrinus Ltd., London.

Dorey J., Garnier G., and Auvray G. (1989), "RIAS, Synthetic Impulse and Antenna Radar," *Proc. International Conference on Radar,* Paris, 24–28 April, pp. 556–562.

Fowler C. A. (1998), "Old Radar Types Never Die; They Just Phased Array or . . . 55 Years of Trying to Avoid Mechanical Scan," *IEEE AES Systems Magazine,* pp. 24A–24L, September.

Gallaudet, T. C. and de Moustier C. P. (2000), "On Optimal Shading for Arrays of Irregularly-Spaced or Noncoplanar Elements," *IEEE Journal of Oceanic Engineering,* Vol. 25, No. 4, pp. 553–567, October.

Gething P. J. D. (1966), "High-Frequency Direction Finding," *Proceedings of IEE,* Vol. 113, No. 1, pp. 49–61, January.

Guy R. F. (1999), "Spherical Coverage from Planar, Conformal and Volumetric Arrays," *IEE Conference on Antennas and Propagation,* pp. 287–290, 30 March–1 April.

Guy R. F. E., Lewis R. A., and Tittensor P. J. (1999), "Conformal Phased Arrays," in *First European Workshop on Conformal Antennas,* Karlsruhe, Germany, 29 October.

Hessel A. (1970), "Mutual Coupling Effects in Circular Arrays on Cylindrical Surfaces—Aperture Design Implications and Analysis," in *Proceedings of Phased Array Symposium,* Polytechnic Institute of Brooklyn.

Hopkins M. A., Tuss J. M., Lockyer A. J., Alt K., Kinslow R., and Kudva J. N. (1997), "Smart Skin Conformal Load-bearing Antenna and Other Smart Structures Developments," in *Proceedings of American Institute of Aeronautics and Astronautics (AIAA), Structures, Structural Dynamics and Materials Conf.,* Vol. 1, pp. 521–530.

Josefsson L. (1999), "Smart Skins for the Future," in *RVK 99,* Karlskrona, Sweden, June, pp. 682–685.

Kanno M., Hashimura, T., Katada, T., Sato, M., Fukutani, K., and Suzuki, A., (1996), "Digital Beam Forming for Conformal Active Array Antenna," in *IEEE International Symposium on Phased Array Systems and Technology,* 15–18 October, pp. 37–40.

Kinsey R. R. (2000), "An Objective Comparison of the Dome Antenna and a Conventional Four-Face Planar Array," in *Proceedings of 2000 Antenna Applications Symposium,* Allerton Park, Illinois, USA, 20-22 September, pp. 360–372.

Knudsen H. L. (1953a), *Bidrag till teorien for Antennsystemer med hel eller delvis Rotationssymmetri,* Ph.D. Thesis, Teknisk Forlag, Copenhagen (in Danish).

Knudsen H. L. (1953b), "The Field Radiated by a Ring Quasi-Array of an Infinite Number of Tangential or Radial Dipoles," *Proceedings of IRE,* Vol. 41, No. 6, pp. 781–789, June.

Liebman P. M., Schwartzman L., and Hylas A. E. (1975), "Dome Radar—A New Phased Array System," in *Proceedings of IEEE International Radar Conference,* Washington, D.C., pp. 349–353.

Munger A. D., Vaughn G., Provencher J. H., and Gladman B. R. (1974), "Conical Array Studies," *IEEE Transactions on Antennas and Propagation,* Vol. AP-22, No. 1, pp. 35–43, January.

Pathak P. H., Burnside, W. D., and Marhefka, R. J. (1980), "A Uniform GTD Analysis of the Diffraction of Electromagnetic Waves by a Smooth Convex Surface," *IEEE Transactions on Antennas and Propagation,* Vol. AP-28, No. 5, pp. 631–642, September.

Pelton E. L. and Munk B. A. (1974), "A Streamlined Metallic Radome," *IEEE Transactions on Antennas and Propagation,* Vol. AP-22, No. 6, pp. 799–803, November.

Rahim T., Guy J. R. F., and Davies D. E. N. (1981), "A Wideband UHF Circular Array," in *IEE Proceedings of 2nd International Conference on Antennas and Propagation,* pp. 447–450, part 1, April.

Rai E., Nishimoto, S., Katada, T., and Watanabe, H. (1996), "Historical Overview of Phased Array Antenna for Defense Application in Japan," in *IEEE International Symposium on Phased Array Systems and Technology,* 15–18 October, pp. 217–221.

Rehnmark S. (1980), "Instantaneous Bearing Discriminator with Omnidirectional Coverage and High Accuracy," in *IEEE MTT-S International Symposium Digest,* May, pp. 120–122.

Schneider S. W., Bozada C., Dettmer R., and Tenbarge J. (2001), "Enabling Technologies for Future Structurally Integrated Conformal Apertures," in *IEEE AP-S International Symposium Digest,* Boston, 8–13 July, pp. 330–333.

Shestag L. N. (1974), "A Cylindrical Array for the TACAN System," *IEEE Transactions on Antennas and Propagation,* Vol. AP-22, No. 1, pp. 17–25, January.

Stanek T. and Johansson F. S. (1995), "Analysis and Design of a Hemispherical Metallic Radome," in *Proceedings of Workshop on EM Structures,* Nottingham, UK, September.

Stergiopoulos S. and Dhanantwari A. C. (1998), "Implementation of Adaptive Processing in Integrated Active-Passive Sonars with Multi-Dimensional Arrays," in *IEEE Symposium on Advances in Digital Filtering and Signal Processing.* Victoria, Canada, 5–6 June, pp. 62–66.

Steyskal H. (2002), "Pattern Synthesis for a Conformal Wing Array," in *IEEE Aerospace Conference Proceedings 2002,* Vol. 2, pp. 2-819–2-824.

Thomas P. G. (1967), "Multifunction Airborne Radar," *Space/Aeronautics,* February, pp. 74–85.

Wait J. R. (1959), *Electromagnetic Radiation from Cylindrical Structures,* Pergamon Press, London.

Watkins C. D. (2001), "WEAO Collaborative Research on Conformal Antennas and Military Radar Applications," in *Proceedings of Second European Workshop on Conformal Antennas,* The Hague, The Netherlands, 24–25 April.

Wingert D. A. and Howard B. M. (1996), "Potential Impact of Smart Electromagnetic Antennas on Aircraft Performance and Design," in *NATO Workshop on Smart Electromagnetic Antenna Structures,* Brussels, November, pp. 1.1–1.10.

Vourch E., Caille C., Martin M. J., Mosig J. R., Martin A., and Iversen P. O. (1998), "Conformal Array Antenna for LEO Observation Platforms," in *IEEE AP-S International Symposium Digest,* pp. 20–23.

Ziehm, G. (1964), "Optimum Directional Pattern Synthesis of Circular Arrays," *The Radio and Electronic Engineer,* pp. 341–355, November.

2

CIRCULAR ARRAY THEORY

In this chapter, we present an overview of the basics of circular array theory. This will serve as a background to the chapters on conformal array characteristics, design, and synthesis. We will also discuss the concept of phase modes, which is a useful tool for the understanding of the performance of circular arrays. It is also used in pattern synthesis. Omnidirectional patterns will be studied in some detail, especially the problem of reducing the amplitude ripple. The effect of directional elements on pattern bandwidth and the suppression of undesired spectral harmonics will also be investigated. Mutual coupling effects will not be introduced until later chapters, however.

2.1 INTRODUCTION

The circular array antenna can be seen as an elementary building block of conformal array antennas with rotational symmetry, just as the linear array is a building block of planar array antennas. Fundamental characteristics of planar arrays can be learned from studies of linear array antennas. Similarly, by studying circular arrays we can understand some basic aspects of conformal array antennas, especially cylindrical and conical arrays and other shapes with rotational symmetry. Furthermore, the radiation from an elliptical array can be derived from an equivalent circular array [Lo and Hsuan 1965].

The term "ring array" is sometimes used instead of "circular array" to distinguish it from a circular area filled with radiating elements, which is a circular *planar* array antenna [Figure 2.1 (a)]. We use the term circular array here, however, to mean an array of radiators distributed with equal spacing along the periphery of a circle, as in Figure 2.1 (b).

Conformal Array Antenna Theory and Design. By Lars Josefsson and Patrik Persson
© 2006 Institute of Electrical and Electronics Engineers, Inc.

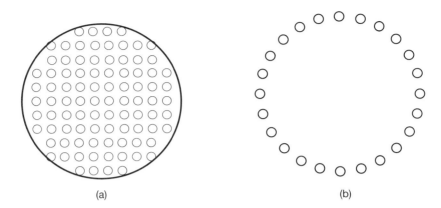

<u>Figure 2.1.</u> A circular planar array (a), and a circular (ring) array (b).

Circular array antennas with uniformly spaced radiating elements and with equal amplitude and phase excitation have long been used for the purpose of obtaining good omnidirectional patterns in the plane of the array (usually the horizontal plane). Later applications include phase-steered directional beams and arrays with several simultaneous beams, including broadband circular arrays. A linear variation of phase along the array circumference was analyzed in the 1930s by Chireix in France [Chireix 1936]. He demonstrated that a reduction in the elevation beamwidth and, consequently, a concentration of the radiation in the horizontal plane could be achieved with this phasing arrangement, while still preserving an omnidirectional amplitude pattern in the azimuth. This was expected to reduce fading in communication systems, since the interference between a ground wave and signals reflected from the ionosphere was reduced, resulting in an "antifading antenna." In fact, Chireix had introduced a *phase mode*. Phase mode theory was later used as an efficient tool for the understanding of the behavior of circular arrays and for pattern synthesis [Taylor 1952, Longstaff et al. 1967]. H. L. Knudsen made important contributions to circular array theory [Knudsen 1953, 1956], and several authors have studied the quality of the omnidirectional patterns in terms of amplitude ripple. The dependence on the number of radiators, spacing, and element directivity was analyzed by Chu [1959]. The use of directional elements has been shown to considerably improve the circular array pattern bandwidth compared to using isotropic elements [Rahim and Davies 1982]. Many more references dealing with the development of circular arrays could be mentioned; the reader is referred to several overview papers, in particular [Davies 1981, 1983, Mailloux 1994, Hansen 1998].

2.2 FUNDAMENTALS

2.2.1 Linear Arrays

Although our purpose is to study circular arrays, we will start with a short introduction to linear arrays [Rudge et al. 1983, Kummer 1992, Mailloux 1994]. Several aspects of circular array theory build on the previously developed theory for linear arrays. It is also informative to point out similarities and differences between the linear and circular cases. Much of linear array theory can be applied to planar arrays, just as the circular array case can be applied to cylindrical arrays.

The basic linear array has a number of discrete radiating elements, equally oriented and equally spaced along a straight line. Each element is viewed as an electric or magnetic current source, which gives rise to a radiated field, the solution to Maxwell's equations [Harrington 1961, Chapter 3]. We assume that the normalized radiation function for one such element, n, is $\overline{EL}_n(\bar{r})$, where $\bar{r}$ is the direction and distance from a reference point in the element to the field point. The reference point has the global coordinates $\bar{R}_n$, and the complex excitation of the element is V_n. Thus, the radiated field from this element, including polarization, is given by $V_n \overline{EL}_n(\bar{r})$. Summing over all elements, we obtain the radiated field at the field point P given by the coordinates $\bar{r}$ (see Figure 2.2):

$$\overline{E}(\bar{r}) = \sum_n V_n \overline{EL}_n(\bar{r} - \bar{R}_n)e^{-jk|\bar{r}-\bar{R}_n|} \tag{2.1}$$

Here k is the propagation constant, $k = 2\pi/\lambda$.

In the far field (r large), the dependence of the amplitude on the distance will be approximately the same for all radiators (typically e^{-jkr}/r). Since the elements were assumed to be identical and identically oriented, we can bring the element radiating function, except the relative phases, outside the summation in Eq. (2.1) and write for the far field:

$$\overline{E}(\bar{r}) = \overline{EL}(\bar{r})\sum_n V_n e^{-jk|\bar{r}-\bar{R}_n|} \tag{2.2}$$

The expression outside the summation is the *element factor*. It determines the polarization of the field, whereas the rest is the scalar *array factor*. For the latter, the spacing between elements is important as well as the element excitations.

We can simplify even further for the case with the field point at a large distance (see Figure 2.3.) The common phase e^{-jkr} is usually of no importance. Since we are mainly concerned with the angular dependence, we may also discard the amplitude dependence $1/r$ and write the radiation function in the following form:

$$\overline{E}(\theta, \phi) = \overline{EL}(\theta, \phi)\sum_n V_n e^{j\bar{k}\cdot\bar{R}_n} = \overline{EL}(\theta, \phi)AF(\theta, \phi) \tag{2.3}$$

where (θ, ϕ) are the angular coordinates and AF is the array factor. In linear and planar arrays, most of the interest is often focused on the array factor, since the element factor can

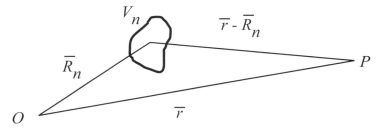

Figure 2.2. The geometry related to one radiator with excitation V_n.

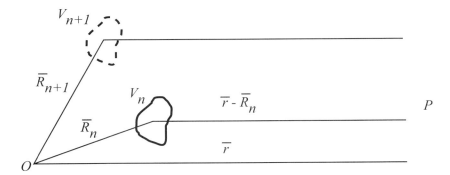

Figure 2.3. Geometry for the array and far field point P.

be considered a slowly varying function that does not impact very much on the shape of the radiation pattern, sidelobes, and so on.[1]

Let us now look specifically at the characteristics of the linear array in Figure 2.4. The array is assumed to have N elements (of which four are shown in Figure 2.4). We assume that the elements are omnidirectional. Thus, we have only a ϕ dependence with the array factor:

$$AF(\phi) = \sum_{0}^{N-1} V_n e^{jknd \sin \phi} \qquad (2.4)$$

By putting $\psi = kd \sin \phi$ in Eq. (2.4), we can write

$$AF(\psi) = \sum V_n e^{jn\psi} \qquad (2.5)$$

or, with $w = e^{j\psi}$, an even simpler form is obtained:

$$AF(w) = \sum V_n w^n \qquad (2.6)$$

This last expression is identified as a polynomial [Schelkunoff 1943] that can be characterized by its zeroes (a fact that has been found useful in pattern synthesis, see, e.g., [Orchard et al. 1985]).

An important special case is uniform excitation, in which all $V_n = 1$ (a uniform linear array or ULA). Equation (2.6) becomes a simple geometric series expression with the sum

$$AF(w) = w \frac{w^N - 1}{w - 1} \qquad (2.7)$$

where N is the total number of radiators. By putting the reference phase in the center of the array we obtain

$$AF(\phi) = \frac{\sin\left(\dfrac{Nkd}{2} \sin \phi\right)}{\sin\left(\dfrac{kd}{2} \sin \phi\right)} \qquad (2.8)$$

[1]This may be true according to the simplified theory above. However, mutual coupling can sometimes have a profound effect on the (embedded) element pattern and the resulting total array pattern.

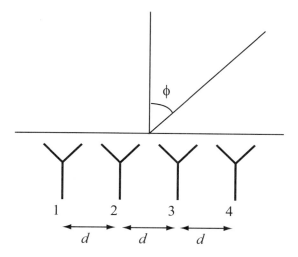

Figure 2.4. A linear array antenna with element spacings d.

Equation (2.8) is close to the form sin x/x, especially when the number of elements is large. The main beam maximum occurs at broadside, that is, at $\phi = 0$. The 3 dB beamwidth is approximately $\phi_3 = 0.88\ \lambda/Nd$ radians or $\phi_3 = 50\ \lambda/Nd$ degrees. The level of the first sidelobe is about 13.5 dB below the main beam maximum, or slightly higher if the number of array elements is small.

We mentioned the main beam maximum at $\phi = 0$. More correctly, the array factor has maxima at $kd \sin \phi = 2m\pi$, that is, at $\sin \phi = m(\lambda/d)$, where $m = 0, \pm1, \pm2, \ldots$. The secondary maxima, or *grating lobes,* appear in real space ($|\sin \phi| < 1$) if $d/\lambda > 1$. The grating lobes are replicas of the main beam in sine space. For no part of the grating lobe (considering the width of the grating lobe) within $\phi = \pm90°$, the requirement becomes $d/\lambda \leq 1 - 1/N$. In many practical cases, the element pattern will suppress the level of grating lobes at large angles from the broadside direction.

The angular spacing between the main beam and the first grating lobe, as measured in sine space, is λ/d. In order to keep the grating lobe maximum outside real space, often called "visible space," when the beam is phase steered, the element spacing requirement is

$$d/\lambda \leq 1/(1 + |\sin \phi_s|) \qquad (2.9)$$

where ϕ_s is the maximum scan angle. For typical scan angles of $\pm70°$, the requirement becomes almost $d/\lambda \leq 0.5$.

2.2.2 Circular Arrays

We will now turn to circular arrays, and in particular look at the uniform circular array (UCA). This generic array has elements equally spaced along the periphery of a circle (Figure 2.5), and the elements are excited with equal phase and amplitude.

We will start by an analysis of the radiation in the plane that contains the array, which we refer to as the azimuth (or horizontal) plane. Local coordinates on the circle are (R, φ), the radius and angle, respectively [see Figure 2.5 (a)]. Far-field coordinates are indicated by (r, θ, ϕ) [Figure 2.5 (b)]. As in the linear array case, we will neglect mutual coupling

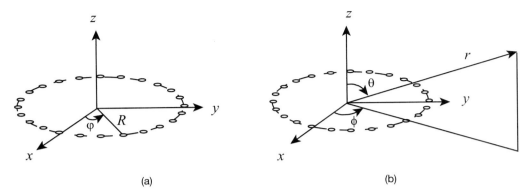

Figure 2.5. Local (a) and far-field (b) coordinates for the circular array.

effects, that is, we consider element excitations (voltages or currents) as given or speci-fied including the effects of mutual coupling. The mutual coupling problem in conformal arrays is treated extensively in Chapters 6 and 7. Limiting the analysis to the azimuth plane only makes the circular array problem a two-dimensional problem, that is, it has bearing on other two-dimensional problems like cylindrical array structures. However, variations with the elevation angle θ, away from the azimuthal plane, will also be dis-cussed.

Let us first make some elementary observations of the differences between linear and circular arrays. For the linear array in Figure 2.6 (a), the radiating elements are identical, with equal spacing d, and they all point in the same direction. The far-field radiation func-tion in the azimuth plane is, as we already know from Equations (2.3, 2.4),

$$E(\phi) = EL(\phi) \sum_n V_n e^{jknd \sin \phi} \qquad (2.10)$$

where, as before, V_n is the excitation amplitude of element n and k is the propagation con-stant, $k = 2\pi/\lambda$. The element factor $EL(\phi)$ is common to all elements and was therefore brought outside the summation sign in this equation. The radiation function is the product of the element factor and the array factor.[2]

The corresponding far-field expression for the circular array (Figure 2.6 (b)), in the az-imuth plane is

$$E(\phi) = \sum_n V_n EL(\phi - n\Delta\varphi) e^{jkR \cos(\phi - n\Delta\varphi)} \qquad (2.11)$$

where the phase has been referenced to the center of the circle. The identical elements are spaced $R\Delta\varphi$ along the circle, each element pointing in the radial direction. The ele-ment function can, therefore, in general not be brought outside the summation, since it is a function of the element position; there is no common element factor. Consequently, we can, in general, not define an array factor as in linear and planar arrays, unless the elements are isotropic, that is, have isotropic (omnidirectional) radiation at least in the horizontal plane. A typical example of the latter is an array of vertical dipoles with their axes perpendicular to the array plane. A further difference compared to linear and pla-

[2]We have dropped vector signs (overbar) to simplify notation.

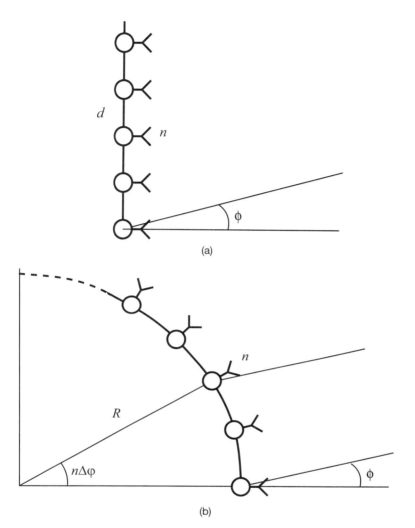

<u>Figure 2.6.</u> Linear array (a); part of a circular array (b).

nar arrays is found in the slightly more complicated phase expression because of the array curvature.

Equation (2.10) for the linear case is essentially a Fourier relationship between excitation and pattern functions. Much of the experience from Fourier analysis can be directly applied to linear (and planar) arrays, and there are several analogies with well-known time/frequency spectral relationships. Unfortunately, this knowledge is not so easily applied to the analysis of circular and conformal arrays. However, Fourier theory is still a valuable tool for analysis and synthesis of circular arrays, as we shall see later in the next section.

With the linear array, we can point a focused beam in a direction ϕ_0 by applying a linear phase shift $\psi(n)$ to the elements along the array:

$$\psi(n) = -knd \sin \phi_0 \tag{2.12}$$

With the circular array, we can similarly impose phase values to each element so that they add up coherently in the direction ϕ_0. We get the proper phase excitation (beam cophasal excitation) for each element n by choosing

$$\psi(n) = -kR\cos(\phi_0 - n\Delta\varphi) \tag{2.13}$$

Thus, the radiation function for the focused, circular case becomes

$$E(\phi) = \sum_n |V_n| EL(\phi - n\Delta\varphi) e^{jkR[\cos(\phi - n\Delta\varphi) - \cos(\phi_0 - n\Delta\varphi)]} \tag{2.14}$$

Sometimes, it is required to generate a beam with equal radiation in all directions in the azimuthal plane (an omnidirectional beam). Circular arrays are particularly suitable for this by virtue of their circular symmetry. A calculated result using Equation (2.11) for an eight-element array with isotropic elements and equal amplitude and phase, $V_n = 1$, is shown in Figure 2.7. The element spacing along the circle was selected to be 0.65 wavelengths. The parameter $kR = Nd/\lambda$ is 5.2. We notice a pattern ripple amounting to about 8 dB.

Another typical case is cophasal excitation in order to scan a focused beam to a specific direction. We take the same array as before, but apply a phase taper according to Equation (2.13) to steer the beam to 45°. The result is shown in Figure 2.8.

None of these two examples is particularly satisfying. The last example gave high sidelobes and the omnidirectional example had a rather large pattern ripple. We will in the

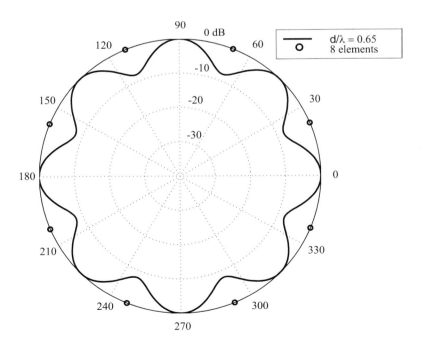

Figure 2.7. Radiation pattern with eight isotropic elements spaced at 0.65 wavelengths. Uniform excitation.

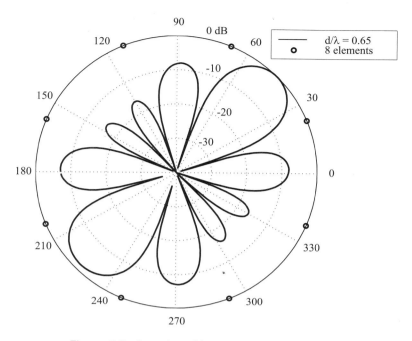

Figure 2.8. A cophasal beam scanned to +45°.

next sections look more closely at circular array behavior and investigate how performance can be optimized. Indeed, much better results can be obtained by proper choice of parameters.

2.3   PHASE MODE THEORY

2.3.1   Introduction

Let the excitation be given by $V(\varphi)$, representing a continuous excitation function. We will take a closer look at the discrete excitation later. The radiated far field in the azimuth plane can be written as

$$E(\phi) = \frac{1}{2\pi} \int_{-\pi}^{\pi} V(\varphi)EL(\phi - \varphi)e^{jkR\cos(\phi-\varphi)}d\varphi \tag{2.15}$$

The excitation function $V(\varphi)$ is obviously periodic (over 2π) so it can be expanded in a Fourier series. We use the complex Fourier series with the definition [Ersoy 1994]

$$f(t) = \sum_{m=-\infty}^{\infty} X(m/T)e^{2\pi jmt/T} \tag{2.16}$$

$$X(m/T) = X_m = \frac{1}{T} \int_{-T/2}^{T/2} f(t)e^{-2\pi jmt/T}dt \tag{2.17}$$

Applying this to the circular array with excitation $V(\varphi)$ and period $T = 2\pi$, we obtain

$$V(\varphi) = \sum_{-\infty}^{\infty} C_m e^{jm\varphi} \tag{2.18}$$

$$C_m = \frac{1}{2\pi} \int_{-\pi}^{\pi} V(\varphi) e^{-jm\varphi} d\varphi \tag{2.19}$$

Each coefficient C_m represents a phase mode, that is, $m = 0$ is the first mode with no phase variation along the circumference, the mth mode has m times 2π variation, and so on. The mode number m can take both positive and negative values.

Assuming isotropic elements (directional elements are treated in Section 2.3.3), we obtain the "array factor":

$$E(\phi) = \frac{1}{2\pi} \int_{-\pi}^{\pi} V(\varphi) e^{jkR \cos(\phi - \varphi)} d\varphi \tag{2.20}$$

Inserting Equation (2.18) and reversing the order of integration and summation yields

$$E(\phi) = \sum_{-\infty}^{\infty} \frac{1}{2\pi} C_m \int_{-\pi}^{\pi} e^{jm\varphi} e^{jkR \cos(\phi - \varphi)} d\varphi \tag{2.21}$$

Now, the far-field must also be a periodic function over 2π, so we can also expand the far field in a Fourier series:

$$E(\phi) = \sum_{-\infty}^{\infty} A_m e^{jm\phi} \tag{2.22}$$

Comparing Equations (2.21) and (2.22) we find that the far-field phase mode amplitude A_m is related to the excitation phase mode amplitude C_m according to the following expression:

$$A_m = C_m \frac{1}{2\pi} \int_{-\pi}^{\pi} e^{jm(\varphi - \phi)} e^{jkR \cos(\varphi - \phi)} d\varphi \tag{2.23}$$

Now, we have the integral form of the Bessel function of the first kind:

$$J_n(z) = \frac{1}{2\pi j^n} \int_{-\pi}^{\pi} e^{jn\theta} e^{jz \cos\theta} d\theta \tag{2.24}$$

and find the important result

$$A_m = j^m C_m J_m(kR) \tag{2.25}$$

The pattern function resulting from the mth excitation mode is written in Equation (2.26). Summation over all excitation modes gives the total pattern.

$$E_m(\phi) = j^m C_m J_m(kR) e^{jm\phi} \tag{2.26}$$

In Equation (2.26) the Bessel function $J_m(kR)$ of order m and argument kR can be seen as a scaling factor between the two domains (near field excitation and far field radiation). The Bessel function is continuous but exhibits large oscillations and has several zeroes (Figure 2.9). With array sizes and frequencies kR, such that the Bessel function J_m is close to zero, it will not be possible to excite the corresponding far field component E_m. This has important implications for pattern synthesis and pattern stability versus frequency; more about this later.

For a given argument kR, the higher-order Bessel functions decay rapidly with increasing order and, hence, higher excitation phase modes radiate poorly. How many higher-order modes need to be included before their contributions become negligible? An indication is given by Figure 2.10, showing the error when excluding a particular order as a function of the argument. The three lines correspond to 0.1%, 1%, and 10% error normalized to the value of $J_0(0)$. The rule of thumb is that Bessel orders up to the value of the argument shall be included [Davies 1981], that is, m should be chosen at least equal to kR. This value corresponds to a phase shift of 2π per wavelength along the periphery; compare this to endfire excitation in linear arrays. However, from Figure 2.10 we conclude that two to five additional modes (depending on the requirements) could be desirable [Ziehm 1964]. Similar arguments apply to array shapes other than circular; for instance, the spherical array discussed by MacPhie [1968].

Since according to Equation (2.26) the mode amplitudes are multiplied by $J_m(kR)$, the impact of a particular Bessel order depends very much on the excitation mode amplitude C_m (and vice versa). In fact, the magnitude of the Bessel function for a fixed argument changes quite a lot between orders, as illustrated in Figure 2.11. This graph was computed for two selected values of kR, representing element spacings of 0.4 λ and 0.8 λ, respectively, when the number of elements is 8.

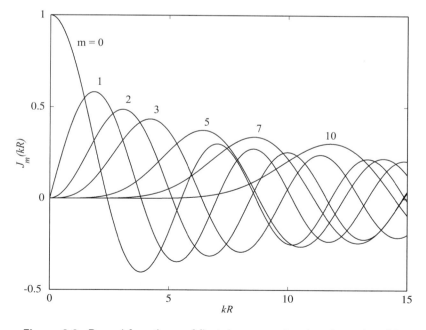

<u>Figure 2.9.</u> Bessel functions of first degree and orders from 1 to 10.

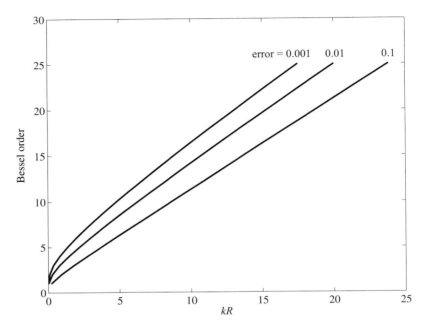

Figure 2.10. The required Bessel function order (required mode number) for given error limits.

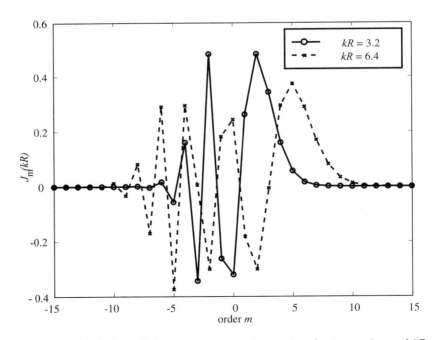

Figure 2.11. Variation of $J_m(kR)$ versus mode number for two values of kR.

A graph calculated for several combinations of orders from −10 to +10 and arguments ranging from 3 to 10 is shown in Figure 2.12. The graph shows the magnitude of the Bessel function plotted against the order normalized to the argument. We can understand the envelope of all the curves in the figure to represent a "window function" between the excitation and radiation domains. Modes inside this window are "visible," hence, radiating, whereas modes outside this window are not radiating. This is analogous to the concept of visible/invisible space in linear array theory, although the division between the two regions in the circular case is less distinct.

2.3.2 Discrete Elements

The discrete array excitation is a sampled version of the continuous case. Feeding the elements with a linearly increasing phase gives rise to a corresponding phase mode excitation. However, in the discrete case we also obtain harmonics to the fundamental phase mode. Depending on the value of $J_m(kR)$, these harmonics may result in additional radiated modes besides the fundamental mode, causing distortions in the radiation pattern. For this reason, the harmonics are sometimes called "distortion modes" [Davies 1981].

We will start by discussing the effects of discretizing the array by assuming isotropic elements. The effect of directional, discrete elements is treated in Section 2.3.3.

A constant, continuous function has a Fourier expansion with just a single spectral component C_0 (the "DC component"). Sampling the function at N equispaced points (which would correspond to N discrete elements) results in a line spectrum (for this and other cases, see Table 2.1). The lines are separated by N/T, where T is the (periodic) extent of the source (in our case 2π). Mathematically, this result is obtained from Equation (2.17) by writing

$$X(q/T) = X_q = \frac{1}{T}\int_{-T/2}^{T/2} \sum_{n=1}^{N} \delta(t - t_n)e^{-j2\pi qt/T}dt \qquad (2.27)$$

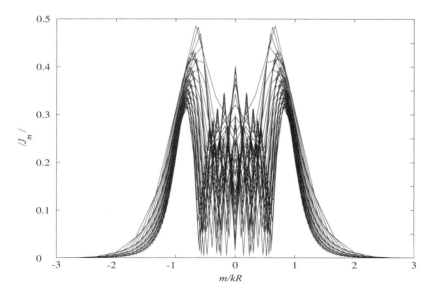

Figure 2.12. Bessel windowing function versus normalized order m/kR.

where t_n is the location of the source element number n and $\delta(t)$ is the Dirac delta function:

$$
\begin{cases}
\delta(t) = 0 & t \neq 0 \\
\displaystyle\int_{-\varepsilon}^{\varepsilon} \delta(t)\,dt = 1, & \varepsilon > 0
\end{cases}
\tag{2.28}
$$

We use q to represent the spectral line number, whereas m is reserved for the phase mode order. We obtain

$$
X_q = \frac{1}{T} e^{j\pi q(1+\frac{1}{N})} \sum_{n=1}^{N} e^{-j2\pi q n/N} = \begin{cases} \pm 1/T & \text{when } q = 0, \pm N, \pm 2N \ldots \\ 0 & \text{otherwise} \end{cases}
\tag{2.29}
$$

The exponent outside the summation sign dictates the phase, resulting in alternating signs for the (real) spectral lines.

It is easy to show that a linear phase progression in the source domain transforms into a shift of the spectrum; compare this to distortion-less filters [Collin 1996]. With a phase slope of m times 2π over the source period T, the spectrum shifts m/T. Finally, a general excitation generates a full line spectrum, repeated with a period of N/T. The cases just discussed are summarized in Table 2.1.

The first two cases in Table 2.1 are continuous sources that may be represented by just one phase mode. The corresponding radiation patterns [cf. Eq. (2.26)], are perfectly omnidirectional with no ripple. In the third case, the source is discrete and we have infinitely many modes in the spectrum. However, as seen from the behavior of the Bessel function (Figure 2.12), only orders slightly beyond $\pm kR$ need to be included. The Bessel function acts as a spatial filter, determining which modes will radiate (belong to the "visible space"). Now, at best only the central part of the spectrum should be passed through this spatial filter, that is, modes up to $\pm N/2$. Since the Bessel argument kR equals Nd/λ, we will have this case if d/λ is less than roughly 0.3–0.4. With larger spacings, the higher spectral orders, if excited, will also radiate. Since in many practical applications d/λ exceeds 0.4, we often need to include the first two ambiguous spectral components at $\pm N/T$. The interference between the fundamental and the two ambiguous components is the source of pattern ripple. It is also easy to understand that the spectrum shift, resulting from an excitation phase slope, changes the Bessel weighting with possible significant modification of the pattern shape. With higher modes, the ambiguous spectrum comes closer to and eventually moves into the "visible space," causing pattern distortion.

In the omnidirectional example (Figure 2.7), the pattern was computed from a direct summation of the N element contributions [Eq. (2.11)]. Using phase mode theory, we get an identical result from a summation of the three relevant phase modes, modes 0 and $\pm N$, as demonstrated in Figure 2.13.

It may be worth pointing out in summary that the rightmost column of Table 2.1 represents the actual phase mode spectra, extending to $\pm$ infinity. When a discrete array is fed with one phase mode only, the resulting excitation spectrum will, in general, contain also harmonics or distortion modes, as we have illustrated here. To minimize the impact of these distortion modes, small kR values are required; that is, for a given diameter array, small element separations. However, it is sometimes still possible to find array solutions that are acceptable even when distortion modes are present.

Matrix-feeding systems can be used to excite a number of phase modes simultane-

Table 2.1. Mode spectra for various array excitations.

Case	Source domain	Spectrum
Continuous	Constant ⊓ $-T/2$ $T/2$	Single line $|$ 0
Continuous, with phase slope	Constant + phase slope $m\pi$ $-T/2$ $T/2$ $-m\pi$	Single line, shifted $|$ ⋮ m/T
Discrete, constant	N point sources $-T/2$ $T/2$	Line spectrum π 0 π $|$ $|$ $|$ $-N/T$ 0 N/T
Discrete, constant with phase slope	N point sources + phase slope $m\pi$ $-T/2$ $T/2$ $-m\pi$	Shifted line spectrum $|$ ⋮ $|$ ⋮ $|$ ⋮ $-N/T$ 0 N/T
General case	N point sources, sampled excitation $-T/2$ $T/2$	General spectrum $-N/T$ 0 N/T

ously, each phase mode corresponding to one beam port of the matrix. Ideally, the corresponding patterns would all be (amplitude) omnidirectional, but with different phase progressions in the azimuth. This means that the angular bearing of a received signal can be determined by comparing the phases of the different mode excitations. Typical feeding arrangements to accomplish this and other arrangements are described in more detail in Chapter 9. Another aspect of using several phase modes is found in pattern synthesis applications. A discussion of conformal antenna pattern synthesis is given in Chapter 10.

2.3.3 Directional Elements

When the radiating elements are directive, the array performance in terms of bandwidth and pattern stability is usually improved compared to arrays with isotropic elements [Rahim and Davies 1982]. Also, with directive elements the phase mode concept helps in understanding the mechanisms.

The general expression for the radiation from a circular array [Eq. (2.15)] is repeated here:

$$E(\phi) = \frac{1}{2\pi} \int_{-\pi}^{\pi} V(\varphi) EL(\phi - \varphi) e^{jkR\cos(\phi - \varphi)} d\varphi \qquad (2.30)$$

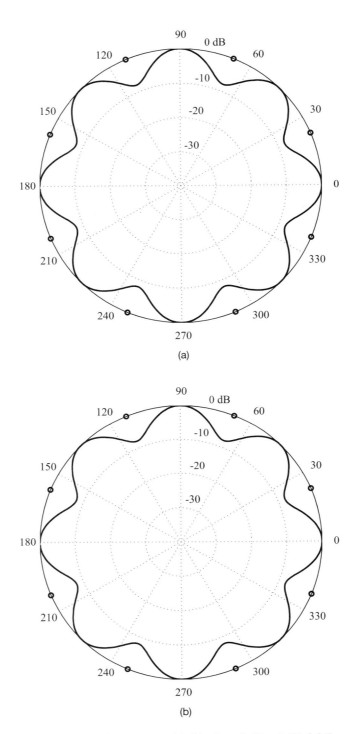

Figure 2.13. Omnidirectional patterns with $N = 8$ and $d/\lambda = 0.65$. (a) Computed from a summation over all elements; (b) directly from a summation of three phase modes (cf. Figure 2.7).

The element radiation function is periodic over 2π and can, hence, also be expanded in a Fourier series:

$$EL(\alpha) = \sum_{p=-\infty}^{\infty} D_p e^{jp\alpha} \tag{2.31}$$

Inserting this in Equation (2.30) yields

$$E(\phi) = \frac{1}{2\pi} \int_{-\pi}^{\pi} V(\varphi) \sum_{p=-\infty}^{\infty} D_p e^{jp(\phi-\varphi)} e^{jkR\cos(\phi-\varphi)} d\varphi \tag{2.32}$$

Now, since we can expand the excitation in phase modes, $V(\varphi) = \Sigma C_m e^{jm\varphi}$, we can obtain the result for each phase mode m after changing the order of integration and summation:

$$E_m(\phi) = C_m e^{jm\phi} \sum_{p=-\infty}^{\infty} D_p \frac{1}{2\pi} \int_{-\pi}^{\pi} e^{j(m-p)(\varphi-\phi)} e^{jkR\cos(\varphi-\phi)} d\varphi \tag{2.33}$$

or

$$E_m(\phi) = \left[C_m \sum_{p=-\infty}^{\infty} D_p j^{m-p} J_{m-p}(kR) \right] e^{jm\phi} = A_m e^{jm\phi} \tag{2.34}$$

This is essentially the product of (the spectral forms of) the excitation function and the element radiation function, respectively, a result that could be anticipated from Equation (2.30), which is a convolution between excitation and radiation. Summation over all modes m gives the total radiation. For each phase mode, we have to sum the contributions from all components D_p of the radiating element expansion. [cf. isotropic elements, Eq. (2.26)].

It is of interest to investigate the impact of various realistic element patterns. A particular element has its characteristic Fourier expansion (D_p coefficients) that influence the array mode amplitudes A_m, including their frequency dependence and so on. An isotropic element has, of course, only a DC term in the expansion of Equation (2.33). Other pattern shapes of interest include a $(\cos\phi)$ function or $(1 + \cos\phi)$ or something similar. In most practical cases, the element pattern is a slowly varying function dominated by low spectral orders.

The problem with nulls of the Bessel function mapping improves radically with directional elements. Consider, for example, the element pattern $(1 + \cos\phi)$ [Rahim and Davies 1982], which has only three spectral components, $p = 0$ and $p = \pm1$. Thus, in Equation (2.34) we will sum terms like $J_m + 0.5j(J_{m+1} - J_{m-1}) = J_m - jJ'_m$, where J'_m is the first derivative of the Bessel function. This summation has no nulls within a wide region of interest, substantially eliminating the Bessel null problem. Therefore, the mapping from excitation to radiation function remains stable over a quite large kR range, enabling broadband array performance.

Figure 2.14 shows typical element pattern spectral representations for the following cases: isotropic pattern, $(1 + \cos\phi)$ pattern, $(\cos\phi)$ pattern, and the E-plane pattern computed for an axial slot in a conducting cylinder. The spectral magnitudes are shown (absolute value). The slot pattern is quite similar to the $(1 + \cos\phi)$ pattern apart from some

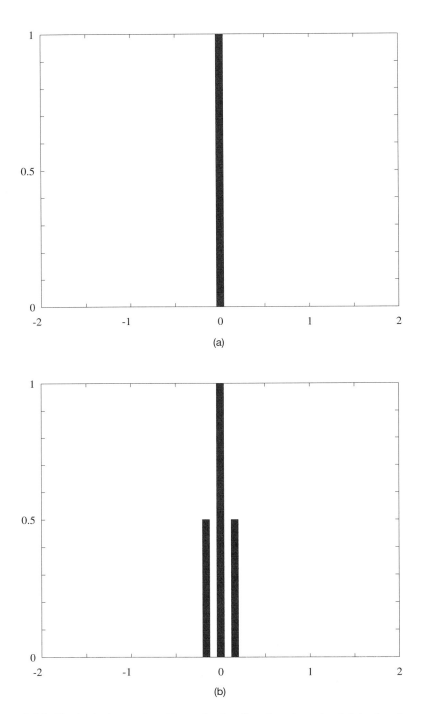

<u>Figure 2.14.</u> Typical element patterns in the Fourier domain: (a) isotropic, and (b) (1 + cos ϕ). (*continued*)

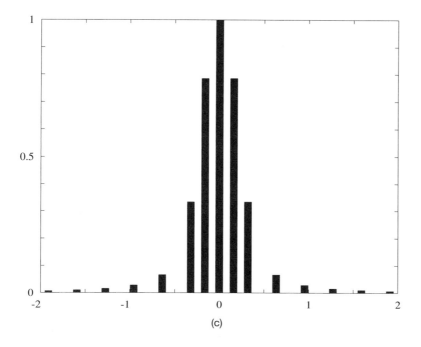

(c)

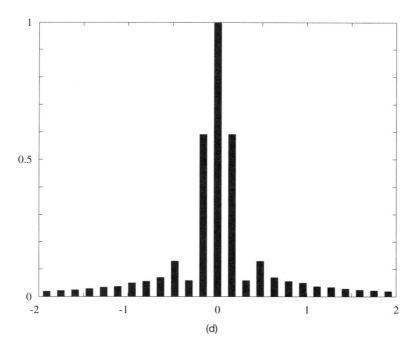

(d)

<u>Figure 2.14 (*continued*).</u> Typical element patterns in the Fourier domain: (c) (cos ϕ), and (d) slot element in cylinder.

low-amplitude, higher-order components, and can therefore be expected to perform well in circular arrays over large bandwidths. The similarity between the two in the nontransformed domain (that is, as a function of the azimuth angle) is demonstrated in Figure 2.15. The derivation of the slot pattern is presented in Chapter 6. The (cos ϕ) case is similar to a microstrip patch *H*-plane pattern.

The transfer function from excitation mode to radiation mode, *mode stability* [Davies 1983], is demonstrated in Figure 2.16 for several element types and mode orders. The spectra in Figure 2.14, including the first ±10 components were used. For isotropic elements, the mode stability is given directly by the Bessel function (Equation 2.25), whereas the general case is given by Equation (2.34). We see that all directional elements show promise for broadband performance. The case with $m = 5$ illustrates also that effective radiation requires the array to be big enough in order to avoid superdirectivity ($kR = Nd/\lambda <$ 4; one could, e.g., use 10 elements and $d/\lambda < 0.4$).

2.4 THE RIPPLE PROBLEM IN OMNIDIRECTIONAL PATTERNS

A common application of circular arrays is to generate an omnidirectional radiation pattern in the azimuthal plane. An example of this using isotropic radiators was presented in Figure 2.7, where a ripple amplitude of about 8 dB was obtained. We will here present several additional design cases, and study how the ripple depends on the choice of design parameters.

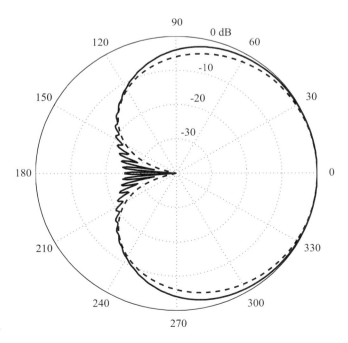

Figure 2.15. Slot pattern (solid line) and 1 + cos ϕ pattern (dashed) compared in the angular domain.

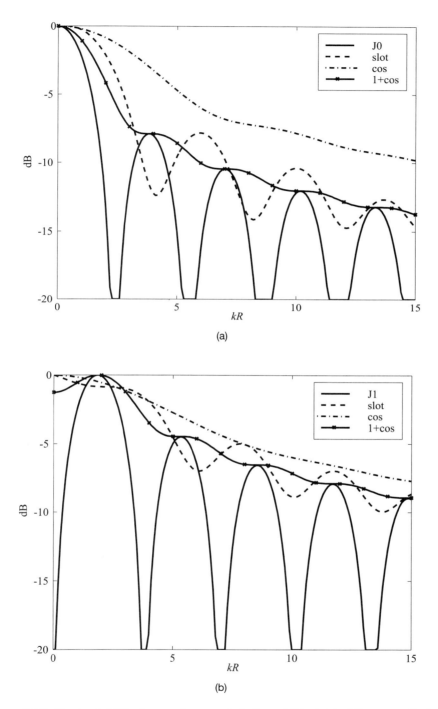

(a)

(b)

Figure 2.16. Mode stability for modes $m = 0$, 1, 2, and 5 in (a) to (d), respectively. El-ement types: isotropic (solid line), slot (dashed), $\cos \phi$ (dash-dotted), and $1 + \cos \phi$ (solid line with crosses).

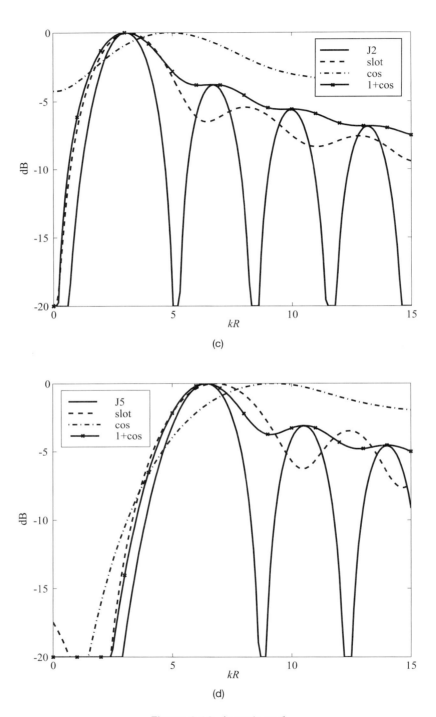

(c)

(d)

Figure 2.16. (*continued*)

2.4.1 Isotropic Radiators

The radiated phase mode $m = 0$ represents an ideal omnidirectional pattern without amplitude ripple. In a practical case, we have discrete radiating elements, however, and, therefore, higher-order modes will be excited in addition to the fundamental mode. The interference between the modes causes a ripple in the resulting far-field pattern. Thus, it is desirable to have just one mode, the fundamental $m = 0$ mode, to dominate. For this mode, the $J_0(kR)$ Bessel function obviously plays an important role. Its first three zeroes and the corresponding element spacings for $N = 4$, 8, and 16 elements are presented in Table 2.2. Note that $kR = Nd/\lambda$.

As an example, $d/\lambda = 0.5406$ with $N = 16$ is a combination that cannot produce a $m = 0$ mode in the far field. This combination is therefore not suitable for the generation of an omnidirectional pattern. The pattern would be determined by the distortion modes.

Figure 2.17 shows the ripple amplitude in decibels (lowest dip compared to maximum value) plotted versus element spacing for 4, 8, and 16 elements in the array. Isotropic radiators were assumed. From this figure, we find that a spacing less than 0.5 wavelengths is required with four elements in order to obtain a ripple less than 5 dB. About 8 dB ripple is obtained with eight elements and 0.65 wavelengths spacing, as demonstrated in Figure 2.7, whereas 16 elements with the same spacing results in very low ripple. A large spacing greater than one wavelength is more or less useless.

Figure 2.18 shows a detailed examination of the ripple behavior for element spacings in the region from 0.5 to 0.6 wavelengths. The sharp dips in the curves correspond to the third null in the J_0 function; compare Table 2.2.

Further examples are shown in Figure 2.19, where contour lines for 1, 3, and 10 dB ripple amplitudes are plotted in a graph with the number of elements and the element spacing on the two axes. As seen, the dependence on these parameters is neither simple nor easy to predict. In the figure, the locations of the first nulls of the Bessel function J_0 are indicated with dash-dotted lines. The highest ripple typically occurs along these lines, as expected.

It has been observed [Chu 1959, Knudsen 1956] that odd numbers of elements sometimes produce less ripple than even numbers. Figure 2.20, which is extracted from Figure 2.19 for $d/\lambda = 0.8$, verifies that this often is the case.

In order to minimize the ripple, the harmonics of the fundamental mode must not be allowed to radiate. Figure 2.10 gave an indication of which modes are important—mode numbers up to $kR + 5$ will radiate. Since for the fundamental mode excitation the first harmonics appear at modes $\pm N$, we find an approximate criterion for no radiation of distortion modes:

$$N > 5/(1 - d/\lambda) \qquad (2.35)$$

Table 2.2. Zeroes of $J_0(kR)$ for various combinations of N and d/λ.

Zero Number	J_0 argument	$N = 4$	$N = 8$	$N = 16$
			d/λ	
1	2.405	0.601	0.3	0.15
2	5.520	1.38	0.69	0.345
3	8.654	2.16	1.08	0.5406

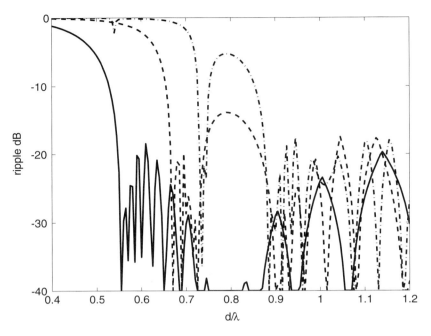

Figure 2.17. The ripple amplitude for circular arrays with isotropic elements; four elements (solid line), eight elements (dashed line), and 16 elements (dash–dotted line).

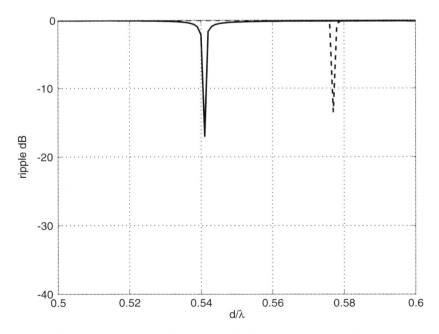

Figure 2.18. Ripple resonance effect with 15 (dashed line) and 16 elements (solid line). Isotropic radiators.

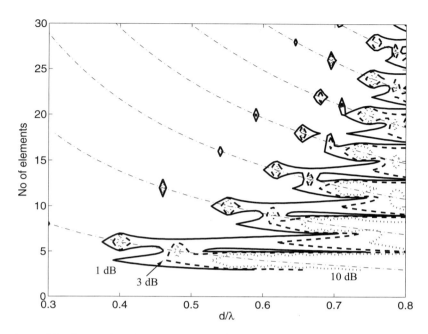

Figure 2.19. Ripple amplitude contour lines (1 dB solid, 3 dB dashed, and 10 dB dotted) for circular ring arrays with isotropic elements and $m = 0$ mode excitation. Along the dash–dotted lines, $J_0(kR) = 0$.

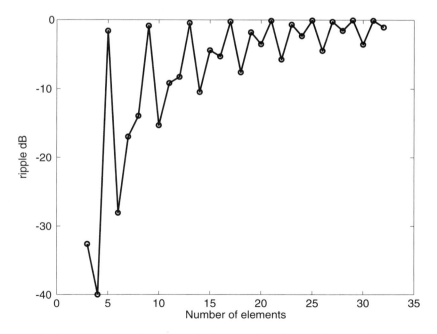

Figure 2.20. Omnidirectional pattern ripple as a function of the number of elements when $d/\lambda = 0.8$.

This explains why more elements are required as the element spacing increases. As Figure 2.12 indicates, there is a gradual cutoff of higher modes, so the number of elements required depends on how large a ripple one can tolerate.

2.4.2 Higher-Order Phase Modes

Each individual excitation mode, not just the zero mode, has an omnidirectional (amplitude) far field pattern (but with an azimuthal phase variation). However, with higher excitation modes the radiation from distortion modes tends to increase, causing increased pattern ripple. We can see from Figure 2.20 that particularly good performance (about 1.5 dB of ripple) is obtained with five isotropic elements, $d/\lambda = 0.8$, and $m = 0$. The same array but with $m = 1$ gives 5 dB of ripple, and $m = 2$ gives 25 dB of ripple. In general, low phase modes are to be preferred. Figure 2.21 is similar to Figure 2.19, but for $m = 1$ excitation. The dash–dotted lines represent $J_1 = 0$ in this case.

Thus, even if a feeding network is designed to generate a particular phase mode in a circular array antenna, other modes will also be generated. These harmonics can be suppressed if the element spacing is small: $kR < N/2$. Higher modes can also be caused by mutual coupling among the radiating elements [Jones and Griffiths 1988]. The phase mode amplitude is furthermore dependent on the matching of the phase mode impedance; more about this in Chapter 8.

2.4.3 Directional Radiators

In our discussion of circular arrays with isotropic elements, we have seen that the array parameters (radius and number of elements) must be chosen carefully in order to avoid

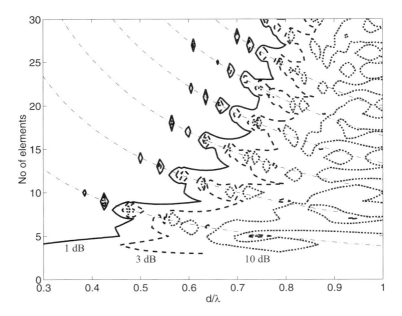

Figure 2.21. Ripple amplitude contour lines (1 dB solid, 3 dB dashed, and 10 dB dotted) for ring arrays with isotropic elements and $m = 1$ mode excitation. Along the dash-dotted lines, $J_1(kR) = 0$.

excessive ripple and/or canceling of the radiation from certain modes. This implies also that the array is narrowband. Physical insight into the problem can be gained by considering the interference between the contributions from the different parts of the array. Figure 2.22 shows the calculated near field amplitude and a cut along a diameter of an array with 16 isotropic radiators. The array acts as a resonator with significant standing wave amplitude inside the array. This effect can be minimized by using directional elements that have beams pointing away from the center of the array.

A simulation with cosine-shaped element patterns using the omnidirectional mode $m = 0$ is shown in Figure 2.23. The element pattern is

$$EL(\phi) = \begin{cases} \cos \phi & |\phi| < \pi/2 \\ 0 & |\phi| > \pi/2 \end{cases} \tag{2.36}$$

Similarly, Figure 2.24 shows the case with $(1 + \cos \phi)$ element patterns.

These examples show that a considerable improvement in ripple performance is obtained if elements with a directive beam pointing radially away from the center of the array are used. The case presented in Figure 2.17 for isotropic elements is repeated in Figure 2.25, but now with $\cos \phi$ element patterns.

2.5 ELEVATION PATTERN

Leaving the azimuth plane, we can write the radiation function for arbitrary directions (θ, ϕ) and isotropic elements:

$$E(\theta, \phi) = \frac{1}{2\pi} \int_{-\pi}^{\pi} V(\varphi) e^{jk\bar{R}(\varphi)\cdot\hat{r}(\theta,\phi)} d\varphi = \frac{1}{2\pi} \int_{-\pi}^{\pi} V(\varphi) e^{jkR \sin \theta \cos(\phi-\varphi)} d\varphi \tag{2.37}$$

Thus, R is replaced by $R \sin \theta$ in the previous azimuth plane expressions. Comparing with Equation (2.26), we obtain the pattern from the mth phase mode excitation and isotropic elements:

$$E_m(\phi, \theta) = j^m C_m J_m(kR \sin \theta) e^{jm\phi} \tag{2.38}$$

The fundamental $m = 0$ mode with all elements radiating in phase has its maximum radiation for $\theta = 0$ (zenith direction). With higher-order modes, more power is concentrated in the horizontal plane, as was discovered by Chireix [1936]. Since kR is replaced by $kR \sin \theta$, the critical dependence on the Bessel function and the pattern stability with frequency becomes an issue also with regard to the elevation angle. The variation with azimuth angle ϕ will depend on the elevation angle and there can be nulls at certain elevations; the amplitude of each phase mode depends on the elevation angle. However, the use of directional elements improves performance also in the elevation plane [Rahim and Davies 1982]. The elevation beamwidth for a particular phase mode is obtained directly from the corresponding Bessel function, as seen from Equation (2.38).

2.6 FOCUSED BEAM PATTERN

The pattern function for a circular array with cophasal excitation given by Equation (2.14) is repeated here:

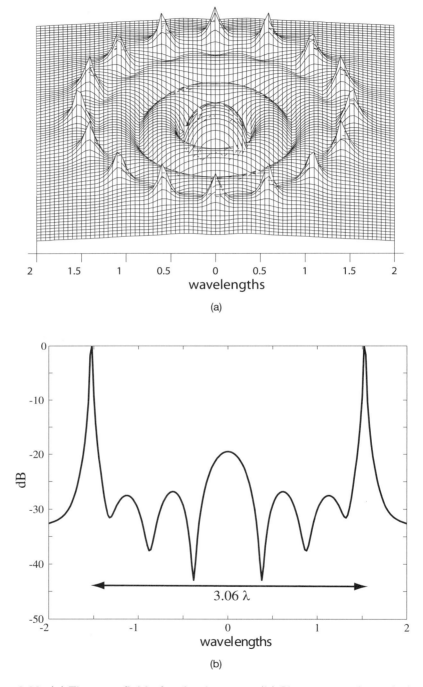

Figure 2.22. (a) The near field of a circular array. (b) Shows a cut through the array center. 16 isotropic elements spaced at 0.6 wavelengths. The array diameter $2R$ is 3.06 λ.

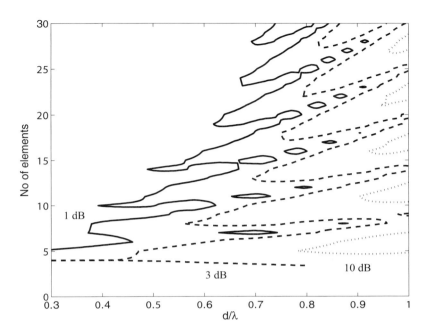

Figure 2.23. Ripple amplitude contour lines (1 dB solid, 3 dB dashed, and 10 dB dotted) for ring arrays with $m = 0$ mode excitation and $\cos \phi$ element patterns.

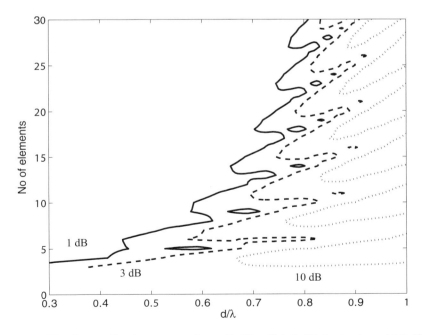

Figure 2.24. Ripple amplitude contour lines (1 dB solid, 3 dB dashed, and 10 dB dotted) for ring arrays with $m = 0$ mode excitation and $1 + \cos \phi$ element patterns.

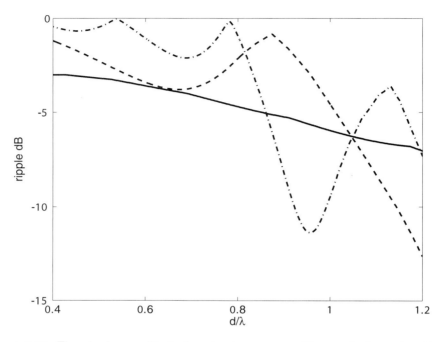

<u>Figure 2.25.</u> The ripple amplitude for circular arrays with cos ϕ element patterns. Compare Figure 2.17. Four elements = solid line, eight elements = dashed line, and 16 elements = dash–dotted line.

$$E(\phi) = \sum_n |V_n|\, EL(\phi - n\Delta\varphi)e^{jkR[\cos(\phi - n\Delta\varphi) - \cos(\phi_0 - n\Delta\varphi)]} \qquad (2.39)$$

The radiating elements are phased to produce a beam in the direction $\phi = \phi_0$ in the azimuth plane. Simplifying to the continuous case with isotropic radiators, $|V_n| = 1$, and letting $\phi_0 = 0$ we obtain the pattern function

$$E(\phi) = \frac{1}{2\pi} \int_0^{2\pi} e^{jkR[\cos(\phi - \varphi) - \cos\varphi]} d\varphi \qquad (2.40)$$

which after some manipulations may be written

$$E(\phi) = J_0\left(2\, kR \sin \frac{\phi}{2}\right) \qquad (2.41)$$

We compare this ideal Bessel pattern with that of a discrete cophasal array of the same size (in this example, $kR = 9.6$) in Figure 2.26. The two patterns are almost identical in the main beam and the near sidelobe regions. The first sidelobe level is about –8 dB, compared with about –13.5 dB for a uniform line source array. A continuous line source antenna would have a sin x/x pattern instead of the Bessel function obtained for the circular ring source.

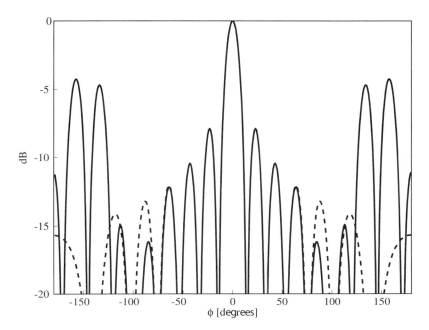

Figure 2.26. Radiation patterns for a continuous circular ring source pattern (dashed line) and a corresponding discrete array with 16 isotropic elements spaced at 0.6 λ (solid line).

The 3 dB beam width is approximately

$$\phi_{3dB} = 82.2\ \lambda/\text{Diameter, ring source} \tag{2.42}$$

$$\phi_{3dB} = 50.6\ \lambda/\text{Length, line source} \tag{2.43}$$

measured in degrees in both cases.

Returning to our example in Figure 2.8 for an eight-element array with 0.65 λ spacing, we plot in Figure 2.27 the same pattern in Cartesian coordinates and compare it with the "ideal" Bessel function pattern. A third curve (shown dashed) is a case with the same diameter but with 13 radiators at a denser spacing of 0.4 λ. As seen here, the dashed curve is very close to the ideal Bessel pattern (dash-dotted).

Consider again the Bessel window in Figure 2.12, from which we see that only components with roughly $m/kR \leq 1.5$ are radiated. In order to exclude undesired spectral harmonics, the window should only pass the unambiguous range up to at most $N/2$, and thus $N/2$ should be greater than approximately $1.5\ kR$, or $N > 15$. The example in Figure 2.27 with $N = 13$ almost gives the limiting Bessel function pattern shape.

Is there equivalence between circular and linear arrays? In this chapter we have pointed to several differences between the two cases. Still, there is one striking equivalence that we may note. For the linear array with $2N + 1$ isotropic radiators we have the pattern

$$E_{lin}(\phi) = \sum_{-N}^{N} V_n e^{jknd\ \sin\ \phi} \tag{2.44}$$

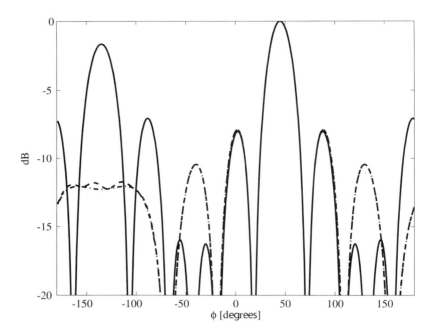

Figure 2.27. Patterns focused in the 45° direction with a $kR = 5.2$ array. Eight elements (solid line), 13 elements (dashed line), and Bessel pattern (dash–dotted line).

For the circular array with $2M + 1$ phase modes, we have the pattern

$$E_{circ}(\phi) = \sum_{-M}^{M} A_m e^{jm\phi} \tag{2.45}$$

Thus, by making each phase mode A_m equal to the corresponding element amplitude V_n, identical patterns result [Davies 1965]. In the linear array case, the pattern would be given as a function of $kd \sin \phi$, and in the circular array case as a function of the angle ϕ, but otherwise the pattern shapes, sidelobe levels, and so on would be identical. This equivalence will be further discussed in Chapter 10 on synthesis methods.

However, there is also one major difference worth pointing out. In the linear array case, the element excitations V_n directly enter the pattern function. In the circular case, it is the radiating phase modes A_m instead, and we know that they are under the influence of the Bessel function $J_m(kR)$ and not directly controlled by the element excitations.

REFERENCES

Chireix H. (1936), "Antennes à Rayonnement Zénithal Réduit," *L'Onde Electrique,* Vol. 15, pp. 440–456.

Chu Ta-Ching (1959), "On the Use of Uniform Circular Arrays to Obtain Omnidirectional Patterns," *IRE Transactions on Antennas and Propagation,* Vol. AP-7, No. 4, pp. 436–438, October.

Collin R. E. (1966), *Foundations for Microwave Engineering,* p. 132, McGraw-Hill, New York.

Davies D. E. N. (1965), " A Transformation Between the Phasing Techniques Required for Linear and Circular Aerial Arrays," *Proceedings of IEE,* Vol. 112, No. 11, pp. 2041–2045, November.

Davies D. E. N. (1981), "Circular Arrays: Their Properties and Potential Applications," in *IEE Proceedings of Second International Conference on Antennas and Propagation,* pp. 1–10, April.

Davies D. E. N. (1983), "Circular Arrays," in Rudge et al. (eds.): *The Handbook of Antenna Design,* Vol. 2, Peter Peregrinus Ltd., London.

Ersoy O. K. (1994), "A Comparative Review of Real and Complex Fourier-Related Transforms," *Proceedings of IEEE,* Vol. 82, No. 3, pp. 429–447, March.

Hansen R. C. (1998), *Phased Array Antennas,* Wiley.

Harrington R. F. (1961), *Time Harmonic Electromagnetic Fields,* Prentice-Hall.

Jones M. R. and Griffiths H. D. (1988), "Prediction of Circular Array Phase Mode Characteristics," *Electronics Letters,* Vol. 24, No. 13, pp. 811–812, 23 June.

Knudsen H. L. (1953), "The Field Radiated by a Ring Quasi-Array of an Infinite Number of Tangential or Radial Dipoles," *Proceedings of IRE,* Vol. 41, No. 6, pp. 781–789, June.

Knudsen H. L. (1956), "Radiation from Ring Quasi-Arrays," *IRE Transactions on Antennas and Propagation,* Vol. AP-4, No. 3, pp. 452–472, July.

Kummer W. H. (1992), "Basic Array Theory," *Proceedings of IEEE,* Vol. 80, No. 1, pp. 127–140, January.

Lo T. Y. and Hsuan H. C. (1965), "An Equivalence Theorem between Elliptical and Circular Arrays," *IEEE Transactions on Antennas and Propagation,* Vol. AP-13, pp. 247–256, March.

Longstaff I. D., Chow P. E. K., and Davies D. E. N. (1967),"Directional Properties of Circular Arrays," *Proceedings of IEE,* Vol.114, No. 6, pp. 713–718, June.

MacPhie R. H. (1968), "The Element Density of a Spherical Antenna Array," *IEEE Transactions on Antennas and Propagation,* Vol. AP-16, No. 1, pp. 125–127, January.

Mailloux R. J. (1994), Phased Array Antenna Handbook, Artech House.

Orchard H. J., Elliott, R. S., and Stern G. J. (1985), "Optimising the Synthesis of Shaped Beam Antenna Patterns," *Proceedings of IEE, Part H.,* Vol. 132, No. 1, pp. 63–68, February.

Rahim T., and Davies D. E. N. (1982), "Effect of Directional Elements on the Directional Response of Circular Antenna Arrays," *Proceedings of IEE,* Part H, Vol. 129, No. 1, pp. 18–22, February.

Rudge A. W., Milne K., Olver A. D., and Knight P. (eds.) (1983), *The Handbook of Antenna Design,* Vols. 1–2, Peter Peregrinus Ltd., London.

Schelkunoff S. A. (1943), "A Mathematical Theory of Linear Arrays," *Bell System Technical Journal,* Vol. 22, pp. 80–107.

Taylor T. T. (1952), "A Synthesis Method for Circular and Cylindrical Antennas Composed of Discrete Elements," *IRE Transactions on Antennas and Propagation,* pp. 251–261, August.

Ziehm, G. (1964), "Optimum Directional Pattern Synthesis of Circular Arrays," *The Radio and Electronic Engineer,* pp. 341–355, November.

3

THE SHAPES OF CONFORMAL ANTENNAS

In this chapter, we will investigate how antenna performance, in particular the gain and coverage, depends on the shape of the conformal antenna. We consider situations in which a wide angular region of space is covered by the antenna system, typically wider than can be realized with one planar array antenna. Solutions with several planar arrays, each pointing in a different direction, can also be used, and we will compare these with curved conformal solutions. Finally, the use of multifaceted arrays, which approximate a curved surface, will be discussed.

3.1 INTRODUCTION

A typical conformal array antenna has radiating elements evenly distributed on a smoothly curved surface. This is the ideal case in which even the radiator shape matches the curvature of the surface. Alternatively, the surface can be built up of many small, planar patches or facets, with one or a few radiators per facet. In this way, the surface approximates a curved surface and the radiator shapes match the plane facets. A third possibility is to use rather few large planar surfaces with several radiators per surface—a multisurface array. In the typical multisurface case, one surface is used at a time. The beam from one planar array is phase steered to a maximum value (edge of scan) at which the adjacent plane surface takes over. Thus, a continuous wide angular coverage can be obtained.

Conformal Array Antenna Theory and Design. By Lars Josefsson and Patrik Persson
© 2006 Institute of Electrical and Electronics Engineers, Inc.

A number of possible generic surfaces are shown in Figure 3.1. The cylinder in Figure 3.1 (a) and the six-sided prism (b) are suitable for two-dimensional coverage in one plane (360°). The half sphere (c) and the pyramid (d) with its four planar surfaces can both be candidates in the (three-dimensional) case when a full hemisphere is covered. Obviously, many more shapes can be considered.

The shape of a conformal antenna is often determined by mechanical or aerodynamic considerations. However, if we have the freedom to choose the shape to maximize antenna performance, how shall we optimize it? Moreover, how would a curved array compare with solutions with a few planar antennas? In the next two sections, we will analyze the performance of solutions for two-dimensional and three-dimensional cases that use a number of planar surfaces as well as curved surfaces. The performance comparison will be made based on simple geometrical criteria.

In a planar phased array antenna, the beam can usually not be steered more than about 60°–70° from the normal of the array [Mailloux 1994]. The limitation is due to beam broadening (reducing directivity) and element impedance variations due to mutual coupling, resulting in increased mismatch at large scan angles. The combined effect can be expressed as a reduction in the antenna gain (or equivalently, the effective antenna area, A_{eff}). The variation is approximately given by $(\cos\theta_s)^\alpha$, where θ_s is the scan angle from the normal to the surface and α is a parameter close to one. Thus the reduction varies roughly as the projection of the antenna surface in the scan direction (see Figure 3.2).

The parameter α, introduced to account for scan loss, has a value typically in the range 1.0 to 1.3. $\alpha = 1$ corresponds to the ideal case (no mismatch), whereas a value greater than 1.0 can be introduced to account for increasing mismatch loss as the beam is scanned away from the normal. A simple physical model can be obtained by representing the array antenna by a current sheet, equivalent to densely spaced elementary dipole elements [Wheeler 1965]. From this, we can derive a scan-dependent reflection coefficient. The current sheet has an active resistance varying as $\cos\theta_s$ in the E-plane, and as $1/\cos\theta_s$ in the H-plane. For both planes the resulting mismatch loss becomes

$$1 - |\Gamma(\theta_s)|^2 = 1 - \tan^4(\theta_s/2) \tag{3.1}$$

This can be compared with the assumed $(\cos\theta_s)^\alpha$ variation that includes both the projection effect and, with $\alpha > 1$, also mismatch. For up to about a 60° scan angle, the two models give roughly the same result if α is chosen to be 1.2. The total scan loss (projection and mismatch) for the two models is illustrated in Figure 3.3, where the range of α values is 1.0 to 1.3.

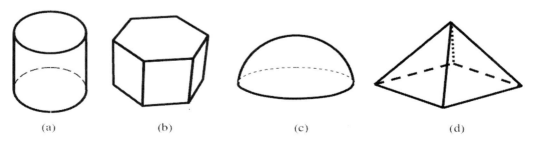

(a)   (b)   (c)   (d)

Figure 3.1. Curved surfaces (a, c) and multisurface (faceted) arrangements (b, d).

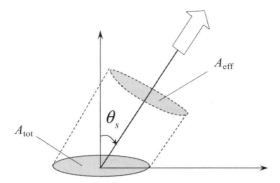

Figure 3.2. Effective (projected) area of a planar array antenna.

The effective area (excluding mismatch losses) of a curved, conformal array is the projection of the active part of the array in the beam direction, see Figure 3.4, where a half cylinder or a half sphere is assumed. The extent of the active area in the curved case (active sector) is one parameter that has to be selected in the design of the array system.

In principle, a curved array can be phase scanned just like planar arrays (cf. the multisurface alternative). This would imply keeping a fixed active sector and phase scanning the beam until the adjacent sector is made active. While this is certainly possible and is

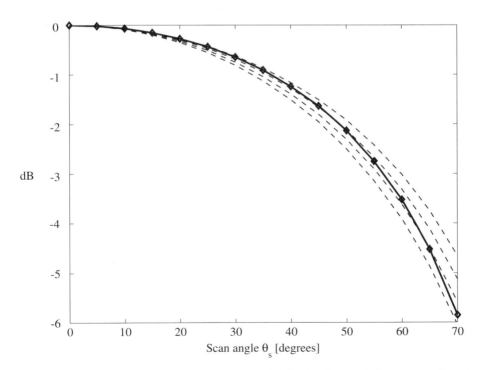

Figure 3.3. Total scan loss. Current sheet model (with diamonds) compared to the $(\cos \theta_s)^\alpha$ model, with $\alpha = 1.0$, 1.1, 1.2, and 1.3 (dashed curves, top to bottom).

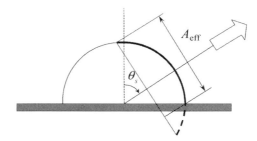

<u>Figure 3.4.</u> Effective (projected) area of a curved half-sphere or half-cylinder array antenna.

sometimes the preferred beam steering method, we have here assumed that the curved array is step scanned by shifting the active region in small increments of one element division at a time. This method of commutating is advantageous since it minimizes the gain variations. However, a complex switching network may be required. The technology is discussed in more detail in Chapter 9.

To obtain a cophasal distribution over the active region, the elements must be phase controlled. In particular, the edge elements are scanned (locally) to a large angle. Due to the shifting of the active region, all elements must usually be designed for this same large scan angle. The value of this element maximum look angle defines the extent of the active sector.

We use the following assumptions in comparing the different antenna shapes:

- The curved and faceted solutions will fit within the same given base area or volume.
- The radiating elements are assumed to be evenly distributed over the surface.
- To ensure a reasonable match of the radiating elements, they are used within a certain scan angle interval only (thus defining the active region of the array).
- For antennas consisting of several planar surfaces, only one planar array is active at a time. When an antenna element is not active, it is turned off.
- The gain (effective antenna area) is assumed to vary as $(\cos \theta_s)^\alpha$, where $\alpha > 1$ is used to account for additional mismatch scan loss.
- The number of elements required to cover the total antenna surface A_{tot} is used as a measure of cost.
- The different antenna shapes are compared based on their minimum effective antenna area A_{eff} within the coverage region. Dividing by the total number of elements gives a normalized performance/cost index.

3.2   360° COVERAGE

3.2.1   360° Coverage Using Planar Surfaces

We consider a two-dimensional case in which the requirement is to cover 360° in one plane. The antenna height (h) is kept constant and the antenna fits within a cylinder with radius R. The minimum number (n) of planar surfaces is obviously three, like a three-sided prism. In that case, each planar surface has to be steered ±60°. With more surfaces, the scan requirement is reduced, as shown in Table 3.1.

Table 3.1. Comparison of two-dimensional prism alternatives

Number of surfaces	Maximum scan angle	Shape
3	60°	
4	45°	
5	36°	
6	30°	
8	22.5^{o}	

As the maximum scan angle is reduced, the gain reduction with scanning becomes smaller. On the other hand, each plane surface becomes smaller when more surfaces are used. It is easily shown that four surfaces is approximately optimum in the sense that the minimum effective antenna area is maximized. Let the n-sided prism with height h be inscribed in a cylinder with radius R. Each planar surface area has an area

$$\text{Area} = 2hR \sin \frac{\pi}{n} \tag{3.2}$$

Since the required maximum scan angle is π/n, we get the minimum effective antenna area (at edge of scan, with $\alpha = 1$)

$$A_{\text{eff, min}} = hR \sin \frac{2\pi}{n} \tag{3.3}$$

This expression is maximized by $n = 4$.

Let the cost be represented by the total number of elements in the antenna system. We can then introduce a performance index by dividing the minimum effective area by the number of elements required. Now, the element spacing d/λ depends on the maximum scan angle θ_{max} according to

$$\frac{d}{\lambda} = \frac{1}{1 + \sin \theta_{\max}} \tag{3.4}$$

The element spacing along the prism axis (d_h) is not relevant for the two-dimensional case. Thus, the normalized "performance divided by cost" index q_p can be written

$$q_p = \frac{A_{\text{eff, min}}/\lambda d_h}{N} = \frac{\left(\cos \dfrac{\pi}{n}\right)^{\alpha}}{n\left(1 + \sin \dfrac{\pi}{n}\right)} \tag{3.5}$$

A plot of this function on a decibel scale is shown in Figure 3.5. It is seen that four or five surfaces perform about the same. If the array elements are well matched over the scan region ($\alpha = 1$) four surfaces may be preferred. With less well-matched elements, it pays to use more surfaces, thus reducing the maximum scan angle from each.

3.2.2   360° Coverage Using a Curved Surface

We will now see how a circular cylinder performs. The total antenna area for a circular cylinder with radius R and height h is simply

$$A_{\text{tot}} = 2\pi R h \tag{3.6}$$

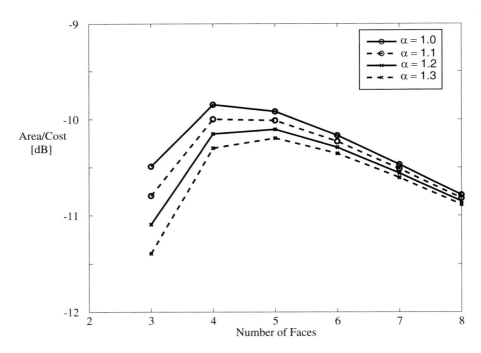

Figure 3.5. Normalized performance (effective area/cost) versus number of antenna faces for omnidirectional coverage (360°).

The effective area, that is, the projected area, will depend on the size of the angular sector $\Delta\phi$ of the cylindrical shape that is used (the active area). We obtain (cf. Figure 3.6)

$$A_{\text{eff}} = 2Rh \sin(\Delta\phi/2) \tag{3.7}$$

Equation (3.7) is valid if no degradation due to scan dependent mismatch loss is included (i.e., $\alpha = 1$). To include losses, we use $\alpha > 1$. The effective area for this more general case becomes

$$\frac{A_{\text{eff}}}{Rh} = 2\int_0^{\Delta\phi/2} (\cos\theta)^\alpha d\theta \tag{3.8}$$

$\Delta\phi/2$ is the maximum element look angle (maximum scan angle). In Figure 3.7, the effective area is shown as a function of the active sector angle $\Delta\phi$ for different values of α.

The maximum projected area that can be used is, of course, $2Rh$, but this would require the edge elements to be steered up to 90° from the local normal. This is probably not possible or at least not very effective. An active sector of about 120° is more reasonable, corresponding to a maximum element look angle of 60°.

The larger the sector, the smaller the element spacings that must be used [Eq. (3.4)], and that increases the cost. Our performance/cost index for the circular cylinder case illustrates this:

$$q_c = \frac{A_{\text{eff}}/\lambda d_h}{N} = \frac{\displaystyle\int_0^{\Delta\phi/2} (\cos\theta)^\alpha d\theta}{\pi\left(1 + \sin\dfrac{\Delta\phi}{2}\right)} \tag{3.9}$$

The performance index in decibels is shown in Figure 3.8 for the cylindrical case as a function of the angular size of the active sector and for different values of α. The perfor-

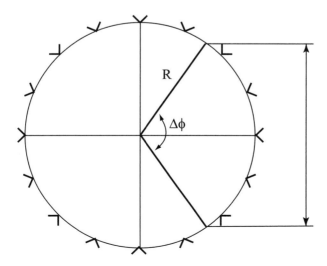

Figure 3.6. Active sector of a cylindrical array.

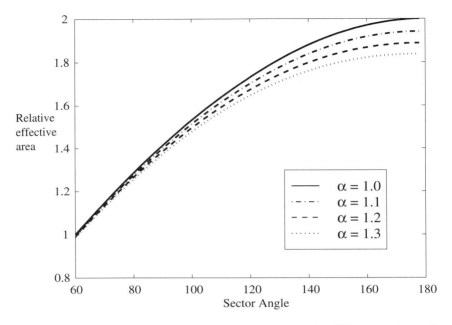

Figure 3.7. Relative effective area versus sector angle for different values of α.

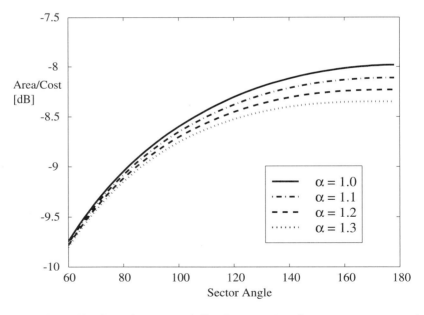

Figure 3.8. Normalized performance (effective area/cost) versus sector angle for omnidirectional coverage (360°).

mance for the solution using a number of planar surfaces was shown in Figure 3.5. From a study of these results, we draw the following conclusions:

- For planar surfaces, four or five surfaces are optimum.
- For cylinders, the larger the sector, the better, but going much beyond 120° is of limited value.
- The cylinder case is better than the planar alternative by about 1.5 dB, and almost 2 dB better if a large scan angle can be used effectively (α close to unity).

3.3 HEMISPHERICAL COVERAGE

3.3.1 Introduction

Now we consider the three-dimensional case in which the beam is steered over a full hemispherical volume, 2π steradians. For our discussion, we think of this as the upper hemisphere with scan angles θ_s from zenith to the horizon and 0°–360° in the azimuth. (For some applications, the coverage might be oriented differently, as in a nose-mounted radar in an aircraft, where the region would be the forward hemisphere instead.) The possible solutions include an arrangement of planar surfaces (pyramids), as discussed in Section 3.1, and several types of curved surfaces. The planar surfaces need not to be connected, since only one surface is used at a time. For installations on ships, for instance, two surfaces can be mounted separately to cover the forward region, and two surfaces positioned to cover the aft region.

The curved surface that first comes to mind is probably a half sphere, matching the shape of the required coverage. Many more surfaces can be considered, however, like conical, ellipsoidal, and paraboloidal surfaces. Furthermore, the shape within one class of surface depends on one or more parameters, such as eccentricity or focal distance.

As before, we base our comparison on the effective antenna area. To our previous list of assumptions for the analysis, we add the following:

- The antenna arrangements will fit within a base diameter of 2R.
- The active surface for a curved solution is defined by a maximum local scan angle of 60°.
- All curved alternatives are rotationally symmetric.
- The total antenna area is taken as a measure of cost.

It turns out that the optimum multisurface solution (for pyramidal shapes) has a maximum scan angle near 60°. We, therefore, impose a 60° scan limit also for the curved cases for simplicity. With this fixed, we do not need to introduce the α parameter (we let $\alpha = 1$) when comparing the different shapes.

In order to illustrate the concepts of active region and active elements, we take an antenna array with radiating elements on a surface with a paraboloidal shape as an example (see Figure 3.9). In this example, the antenna has a height-to-diameter ratio of $h/D = 0.5$. The radiating element positions are indicated by circles, suggesting circular waveguide apertures as radiators. The antenna has 177 elements, of which a certain number can be used (are active), depending on the scan direction.

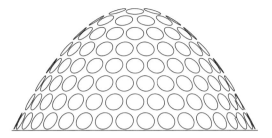

Figure 3.9. Conformal array antenna with the shape of a paraboloid.

Each element is pointing in the direction of the normal to the surface. The angle between the normal and the desired scan directions (element look direction) determines weather the element will be active (belongs to the active region) or not. As already discussed, we define the maximum useful look direction to be 60°. Figure 3.10 presents a view of the antenna as seen on an axis (looking down from the zenith) with a contour plot of cosines of the look directions over the surface. All values greater than or equal to cos 60° = 0.5 belong to the active region in this example, where a beam direction of 30° from zenith was assumed. The active elements (98 out of 177) for this case are shown in Figure 3.11 in the same view.

3.3.2 Hemispherical Coverage Using Planar Surfaces

This problem was discussed by Knittel [1965]. There must be at least three planar surfaces, which can be combined into a three-sided pyramid. The largest scan angle for

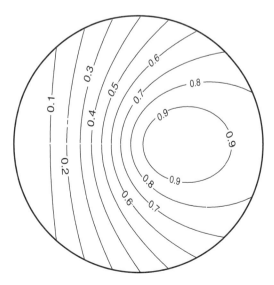

Figure 3.10. Contour plot showing cosine of the element look directions for 30° scan from the zenith.

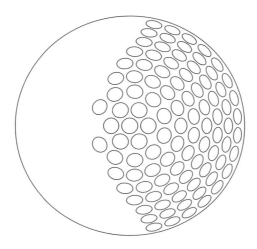

<u>Figure 3.11.</u> Active elements for 30° scan direction.

hemispherical coverage is then 63°. A four-sided pyramid requires a largest scan angle of 55°.

We obtain the ratio between the minimum effective area and the total area (performance-to-cost index) for pyramids:

$$\frac{A_{\text{eff}}}{A_{\text{tot}}} = \frac{\cos 63°}{3} = 0.151 \qquad \text{(three-sided)} \qquad (3.10)$$

$$\frac{A_{\text{eff}}}{A_{\text{tot}}} = \frac{\cos 55°}{4} = 0.143 \qquad \text{(four-sided)} \qquad (3.11)$$

Thus, the two cases perform about the same. The three-sided pyramid shows the largest antenna area in the most favorable direction (normal to the surface) but in the worst directions, the results differ very little. In fact, as pointed out by Knittel [1965], the four-sided pyramid requires fewer elements than the three-sided pyramid for equal minimum gain. This is not apparent from Equations (3.10)–(3.11), but closer investigation reveals that the three-sided pyramid performs slightly worse than assumed here because the maximum scan angle is larger by 8°, causing some extra mismatch loss not accounted for in our formulas. This makes the three-sided and four-sided pyramids more equal in performance. Additionally, the three-sided pyramid requires slightly denser element spacings, thus

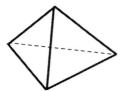 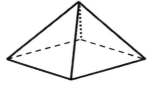

<u>Figure 3.12.</u> Three- and four-sided pyramids.

more elements. The result is that the four-sided pyramid turns out to be the most cost-effective of the two. However, both alternatives are inferior to the curved cases, as we shall see.

3.3.3 Half Sphere

The hemispherical surface provides an active area subtended by a conical angle of 120° based on our assumed maximum scan angle 60°. The effective area is $A_{\text{eff}} = \frac{3}{4}\pi R^2$ and could be expected to be independent of the scan direction since we are dealing with a spherical surface. However, for scan angles larger than a certain value, in our case 30° from the zenith, the effective area decreases and is halved at the 90° scan limit (cf. Figure 3.4).

For scan angles θ_s between 30 and 90° from the zenith, the effective area variation is given by

$$A_{\text{eff}} = (\pi - \psi)R^2 \sin^2 \theta_{\text{max}} + 0.5\, R^2 \sin 2\theta_{\text{max}} \cot \theta_s \sin \psi + R^2 \cos \theta_s (\chi - 0.5 \sin 2\chi) \quad (3.12)$$

where

$$\cos \psi = \cos \theta_s \cot \theta_{\text{max}}$$

$$\sin \chi = \sin \theta_{\text{max}} \sin \psi$$

In these expressions, θ_{max} is the maximum local scan angle. Dividing by the total area $2\pi R^2$, we get the performance-to-cost index shown in Figure 3.13 as a solid line.

The gradual reduction in performance from the 30° scan angle and onward is due to the cutoff of the lower part of the hemisphere (see Figure 3.4). One alternative is to extend the sphere downward to make the effective area constant with scan. However, this significantly increases the total area and, hence, the cost. The performance index (effective area/total area) reduces to 0.20 (the dashed line in Figure 3.13).

3.3.4 Cone

A conical shape can be used for hemispherical coverage provided it is not too pointed or too blunt. Since we imposed a maximum element look angle of 60°, the total cone angle θ_t [see Figure 3.14 (a)] must be in the interval from 60° to 120°.

The total cone surface is

$$A_{\text{tot}} = \frac{\pi R^2}{\sin \dfrac{\theta_t}{2}} \quad (3.13)$$

The active area has an azimuthal width of $2\phi_0$ [see Figure 3.14 (b)]. It depends on the scan angle θ_s from the zenith and the maximum element look angle θ_{max}:

$$\cos \phi_0 = \frac{\cos \theta_{\text{max}} - \theta_s \sin (\theta_t/2)}{\sin \theta_s \cos (\theta_t/2)} \quad (3.14)$$

The effective area is obtained by projection of the active area on to the beam direction θ_s from the zenith. Omitting the detailed derivation, we present the resulting expression:

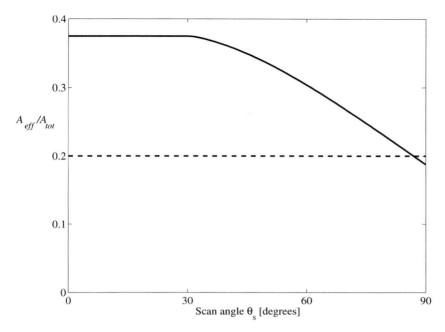

Figure 3.13. Hemisphere. Normalized performance (effective area/cost) versus scan angle from the zenith. The dashed line is for an extended sphere with constant effective area.

$$A_{\text{eff}} = R^2 \left[\frac{\sin \theta_s \sin \phi_0}{\tan (\theta_t/2)} + \phi_0 \cos \theta_s \right] \qquad (3.15)$$

and the performance index becomes

$$\frac{A_{\text{eff}}}{A_{\text{tot}}} = \frac{1}{\pi} [\sin \theta_s \sin \phi_0 \cos (\theta_t/2) + \phi_0 \cos \theta_s \sin (\theta_t/2)] \qquad (3.16)$$

where ϕ_0 is obtained from Equation (3.14).

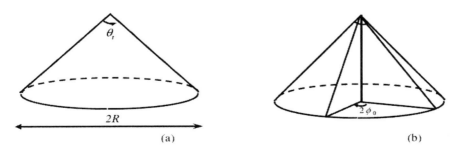

Figure 3.14. Cone parameters. (a) cone angle θ_t and (b) active sector $2\phi_0$.

The effective area changes rapidly with the scan angle. We show in Figure 3.15 how the performance index varies with the scan angle for five cones with different cone angles from 60° to 120°. In most cases, best performance is obtained near the zenith direction.

The average of A_{eff}/A_{tot} taken over the hemispherical scan region is maximum for the largest cone angle (120°) and about 20% less for the smallest cone angle (60°). If we instead allow a 70° maximum local scan angle, we can use a 140° cone with about 20% larger A_{eff}/A_{tot}.

It is of interest to note that truncated cones have a normalized performance just like ordinary cones. This is so because cutting off the tip reduces the effective area and the total area in the same proportions.

3.3.5 Ellipsoid

The equation for an ellipsoid can be written as

$$\frac{x^2}{a^2} + \frac{y^2}{b^2} + \frac{z^2}{c^2} = 1 \qquad (3.17)$$

The hyperboloid equation is similar:

$$\frac{z^2}{c^2} - \frac{x^2}{a^2} - \frac{y^2}{b^2} = 1 \qquad (3.18)$$

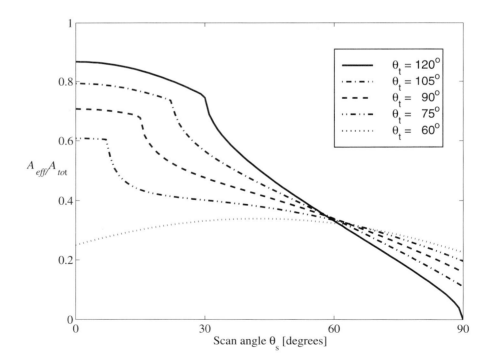

Figure 3.15. Cones with different cone angles. Normalized performance (effective area/cost) versus scan angle from zenith.

The asymptotes of hyperboloids are cones and, hence, pointed hyperboloids and cones produce similar results.

We consider rotationally symmetric bodies here with $a = b$ and z as the axis of rotation. For the special case of a half sphere that we previously discussed, we have $a = b = c = R$, R being the radius. By stretching the semisphere so that the height is larger than the radius, the coverage at large scan angles from the zenith can be improved at the expense of reduced performance toward the zenith. More radiating elements are needed, however, increasing the cost. If, instead, the sphere is made flatter, performance toward the zenith region is improved at the expense of reduced performance at large scan angles. Thus, by modifying the antenna shape the coverage can be tailored according to special requirements.

The normalized (effective area/cost) performance for ellipsoidal cases with different height-to-radius ratios is shown in Figure 3.16.

3.3.6 Paraboloid

The equation for a paraboloid of revolution can be written as

$$4f(z - h) = -(x^2 + y^2) \tag{3.19}$$

where f is the focal distance. In polar coordinates, with origin in the focal point (F), we have (cf. Figure 3.17)

$$r = \frac{2f}{1 + \cos \vartheta} \tag{3.20}$$

The polar angle to the rim of the antenna is given by

$$\tan \frac{\vartheta_{max}}{2} = \frac{4h}{D} = \frac{D}{4f} \tag{3.21}$$

The normal direction at the element location points an angle $\vartheta/2$ away from the z direction.

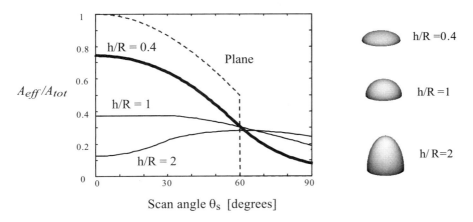

Figure 3.16. Normalized effective area for ellipsoidal shapes with different heights.

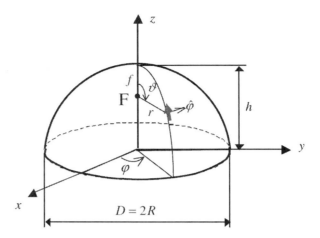

<u>Figure 3.17.</u> Polar coordinates with a radiating element indicated at position (r, ϑ).

In Cartesian coordinates, the paraboloidal surface can be written as

$$z = -h\left[\left(\frac{x}{R}\right)^2 + \left(\frac{y}{R}\right)^2\right]^\varepsilon \tag{3.22}$$

where we have introduced the shaping parameter ε [Czarnecki 2001]. This parameter allows modifications to the paraboloidal shape, where $\varepsilon = 1$ gives the normal paraboloid and $\varepsilon = 0.5$ gives the equation for a cone. Examples of possible cross sections (generating curves) are shown in Figure 3.18. Thus, the same equation can be used to characterize many typical shapes, from pointed to flat-topped surfaces. Similar variations can be applied to hyperboloids and ellipsoids.

The results for the paraboloid are included in the discussion in the next section, where different shapes are compared. The electromagnetic aspects, including mutual coupling and radiating patterns for paraboloidal array antennas, are discussed in detail in Chapters 7 and 8.

3.3.7 Comparing Shapes

The studied shapes differ in curvature and in the slope of the sides. This is important in the trade-off of performance versus scan angle. We have looked mainly at the minimum effective area within the scan region but, in reality, the requirements are often more detailed and different weights can be given to different regions of the coverage.

A summary of how different shapes make use of the available size ($2R$ base diameter) is shown in Figure 3.19, where we have also included a hyperbolic shape and the planar case for reference. All cases shown in this figure (except the pyramid and the plane) have $h/R = 1$. The hyperboloid has an eccentricity $e = 1.1$. We see that performance very much depends on the antenna shape. The pyramid makes poor use of the available base area (inscribed in the $2R$ diameter circle). The single plane surface performs well up to the scan limit of 60°, but covers only 50% of the required volume.

The same cases are shown again in Figure 3.20 with the effective area normalized with respect to the total area (= cost).

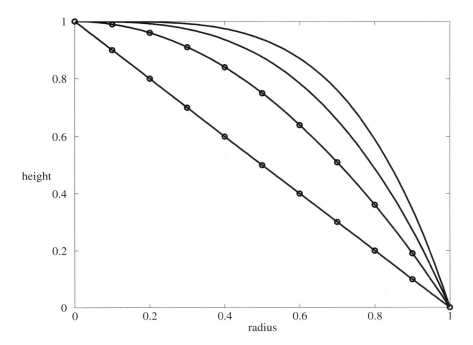

Figure 3.18. Paraboloidal shapes with $\varepsilon = 0.5$ (cone), 1.0 (paraboloid), 1.5, and 2 (paraboloid-type). The cone and the paraboloid are indicated by circles.

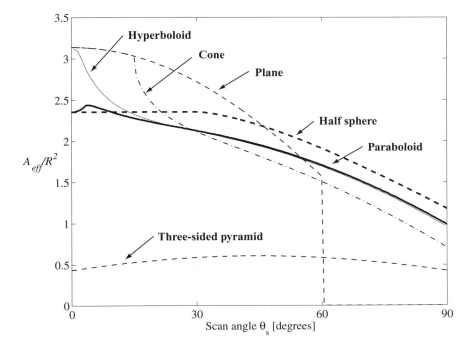

Figure 3.19. The effective area for several types of surfaces; $h/R = 1$.

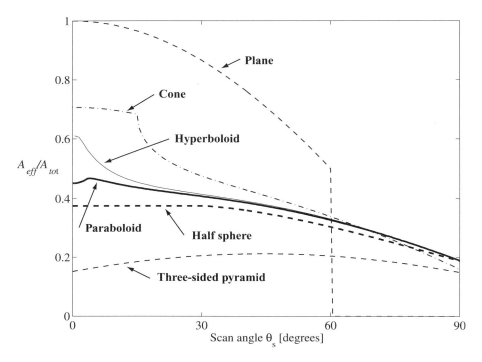

<u>Figure 3.20.</u> The normalized effective area versus scan angle from the zenith; $h/R = 1$.

Finally, we summarize effective area versus total area for several shapes in Figure 3.21. The maximum and minimum effective areas within the scan region are given. It is desirable to obtain a large effective area with a shape that has a small total area. The closed cylinder is an extreme case with a very large total area (both the top planar area and the curved sides are used here), resulting in high cost. It performs marginally better than the hyperboloid but has twice the total area. The half sphere is also relatively expensive. The variations with scan angle were illustrated in Figure 3.13. These examples are all for $h = R$. By changing the height, the results are modified; examples were given in Figures 3.15 and 3.16. Conical shapes perform favorably in this comparison. A further advantage of cones is the constant slope of the surface, possibly simplifying a physical realization.

It should be remembered that the presented results are based on a geometrical analysis. We have not introduced electromagnetic aspects, impedance variations, pattern performance, polarization effects, and so on. These aspects will be addressed in Chapter 8 and other places. It is clear that specific application requirements may lead to other shapes optimized not for maximum effective antenna area, as was assumed here, but for some other criterion. For instance, it has been found that a pyramidal arrangement with five to six surfaces (one pointing upward) can be optimal depending on what system requirements are most important [Knittel 1965]. It has also been shown that six surfaces can be the best alternative when the impact on system noise temperature is of main concern [Wu 1996].

However, based on our chosen criteria, we have found that curved solutions, in general, exhibit much better performance than solutions based on planar surfaces. This is true for both the two-dimensional and three-dimensional cases.

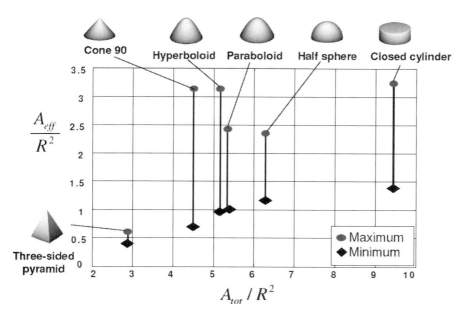

Figure 3.21. Maximum and minimum values of the effective antenna area versus the total antenna area for several conformal antenna shapes.

3.4 MULTIFACETED SURFACES

We conclude this chapter with some remarks on multisurface arrays. As long as a multi-faceted surface is a good approximation to its curved counterpart, the results presented for the curved shapes apply also to the faceted ones [Marcus et al. 2003]. An example of an effort to build a spherical surface from many planar, triangular subsurfaces is shown in Figure 3.22.

The sphere has the smallest surface compared to its volume of all canonical shapes. The *largest* surface compared to its volume is found in the tetrahedron (three-sided pyra-

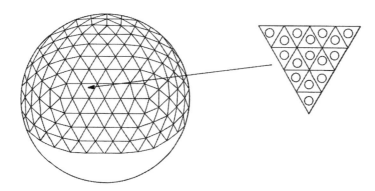

Figure 3.22. A spherical surface built from triangular planar facets [Bondyopadhyay 2002].

mid) [Miranda 2000]. Are there regular arrays of identical facets that closely approximate the sphere? Only few such shapes (Platonic solids) exist, among them the octahedron and icosahedron, with 8 and 20 equilateral triangles, respectively. The dodecahedron has 12 pentagons. It is, therefore, not possible to find a *regular* element layout on a spherical surface with a large number of facets or elements. However, the icosahedron presents a good starting point. See also page 309.

Horiguchi et al. [1985] describe a projection method starting from the regular triangles of an icosahedron. The triangle is subdivided into smaller triangles that are projected onto the sphere surrounding the icosahedron. These small triangles can then be further divided into even smaller triangles and projected again, and so on. The result is a periodic lattice with some variation in element distances. Alternatively, one could randomly disperse the positions (within certain limits) as described by Vallecchi and Gentili [2003].

A particular appeal of the (almost) spherical array, apart from the wide scanning capability, is the uniformity of the element surroundings; all elements see the same environment, which simplifies the design. Figure 3.23 shows a proposed multisurface array [Löffler et al. 2001] with many small planar surfaces, each with several radiators. It is an intermediate case between one element per facet and the pyramid in Figure 3.1. See also the faceted array in Figure 1.8.

The multisurface array in Figure 3.24 is a crude approximation of a half sphere array with 27 microstrip patch elements on a number of plane facets. Sufficient gain for a satellite communication and navigation ground segment system is obtained by switching between the various elements. This results in a relatively low-cost solution [Löffler et al. 1997, Wiesbeck et al. 2002]. Each radiating element has been designed for maximum gain, up to 10 dB, which is achieved using superstrate layers. An important observation is that high-gain elements must be separated enough so that the effective apertures do not overlap, which would cause reduced efficiency.

A proposed multifaceted solution for mobile base stations using dual ±45° polarizations is shown in Figure 3.25. The performance for this faceted design in terms of ripple and beam forming was calculated to be about the same as a curved design [Raffaelli and Johansson 2003, 2005].

The multifaceted surfaces are advantageous because they simplify the packaging of radiators, active electronics, and so on into integrated planar multitile modules [Tomasic et

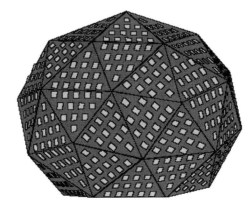

Figure 3.23. A multisurface array approximating a spherical shape [Löffler et al. 2001].

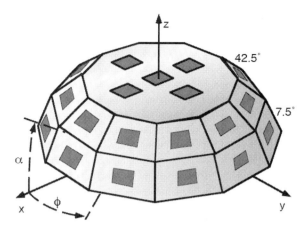

Figure 3.24. A multisurface solution approximating a half sphere [Wiesbeck et al. 2002].

al. 2003]. It has been shown that a few large inclined planar arrays can be combined to generate similar pattern performance as a corresponding curved array with uniform excitation, that is, about −13 dB sidelobe ratio [Hsiao and Cha 1974]. The nonsmooth shape, however, makes it difficult to meet more stringent radiation pattern requirements (unless a quite large number of elements and facets are used). A particular concern is the diffraction from the edges and/or gaps between facets. The mutual coupling and element pat-

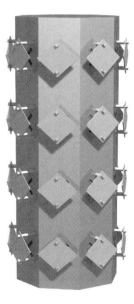

Figure 3.25. Faceted cylindrical microstrip patch array (CAD model). [Raffaelli and Johansson 2003]. See also color plate 11. (Courtesy of Ericsson AB, Göteborg, Sweden.)

terns on faceted and curved cylinders have been compared using UTD by Marcus et al. [2003]. The size of the facets was in this study at least a couple of wavelengths across. The results showed only small differences, but with an increased ripple for the faceted cases.

REFERENCES

Bondyopadhyay P. K. (2002), "New Phased Array Antenna Architecture for the X-band Radar," *IEEE AP-S International Symposium,* pp. 360–363, San Antonio.

Czarnecki D. (2001), *Study of Doubly Curved Antennas,* Master Thesis performed at Ericsson Microwave Systems in 2001 (in Swedish).

Horiguchi S., Ichizone T., and Mushiake Y. (1985), "Radiation Characteristics of Spherical Triangular Array Antenna," *IEEE Transactions on Antennas and Propagation,* Vol. AP-33, pp. 472–476, April.

Hsiao J. K. and Cha A. G. (1974), "Patterns and Polarizations of Simultaneously Excited Planar Arrays on a Conformal Surface," *IEEE Transactions on Antennas and Propagation,* Vol. AP-22, pp. 81–84, January.

Knittel G. H. (1965), "Choosing the Number of Faces of a Phased Array Antenna for Hemisphere Scan Coverage," *IEEE Transactions on Antennas and Propagation,* Vol. AP-13, pp. 878–882, November.

Löffler D., Rostan F., Xia J., and Wiesbeck W. (1997), "Semi-Spherical Array for Satellite-Communication Systems using High Gain Microstrip Patch Elements," in *Proceedings of International Conference on Electromagnetics in Advanced Applications (ICEAA 97),* Torino, Italy, 15–18 September, pp. 429–432.

Löffler D., Wiesbeck W., Eube M., Schad K.-B., and Ohnmacht E. (2001), "Low Cost Conformal Phased Array Antenna Using High Integrated SiGe-Technology," in *Proceedings of IEEE International Symposium AP-S,* pp. 334–337, June.

Mailloux R. J. (1994), *Phased Array Antenna Handbook,* Artech House.

Marcus C., Persson P., and Pettersson L. (2003), "Investigation of the Mutual Coupling and Radiation Pattern due to Sources on Faceted Cylinders," in *Proceedings of Antenn 03,* Kalmar, Sweden, 13–15 May, pp. 213–218.

Miranda L. (2000), *Sacred Geometry,* Wooden Books Ltd.

Raffaelli S. and Johansson M. (2003), "Conformal Array Antenna Demonstrator for WCDMA Applications," in *Proceedings of Antenn 03,* Kalmar, Sweden, 13–15 May, pp. 207–212.

Raffaelli S. and Johansson M. (2005), "Cylindrical Array Antenna Demonstrator: Simultaneous Pencil and Omni-directional Beams," in *Proceedings of 4th European Workshop on Conformal Antennas,* Stockholm, Sweden, 23–24 May, pp. 93–96.

Tomasic B., Turtle J., Liu S., Schmier R., Bharj S., and Oleski P. (2003), "The Geodesic Dome Phased Array Antenna for Satellite Control and Communication—Subarray Design, Development and Demonstration," in *IEEE International Symposium on Phased Array Systems and Technology 2003,* Boston, 14–17 October, pp. 411–416.

Vallecchi A. and Gentili G. B. (2003), "Broad Band Full Scan Coverage Polarization agile Spherical Conformal Array Antennas: Pseudo-Uniform vs. Pseudo-Random Element Arrangements," in *IEEE International Symposium on Phased Array Systems and Technology 2003,* Boston, 14–17 October.

Wheeler H. A. (1965), "Simple Relations Derived from a Phased Array Made of an Infinite Current Sheet," *IEEE Transactions on Antennas and Propagation,* Vol. AP-13, pp. 506–514, July.

Wiesbeck W., Younis M., and Löffler D. (2002), "Design and Measurement of Conformal Antennas," in *Proceedings of IEEE AP-S International Symposium,* San Antonio, pp. 84–87.

Wu T. K. (1996), "Phased Array Antenna for Tracking and Communication with LEO Satellites," in *Proceedings of IEEE International Symposium on Phased Array Systems and Technology,* 15–18 Oct, pp. 293–296.

4

METHODS OF ANALYSIS

In this chapter, we will describe some methods suitable for the analysis of conformal antennas. In general, analyzing conformal antennas is a difficult problem and solutions are often found only for specific geometries. Thus, there are many different methods available. We will discuss some of the most commonly used methods and their advantages and disadvantages. The methods will be described with an application point of view in mind. References will be given to a number of papers in which the methods have been used and where more details can be found. At the end, some final remarks are added, together with guidelines for selecting suitable methods.

4.1 INTRODUCTION

Conformal array antennas can have many shapes. The surface is often electrically large (in terms of wavelength) and it may be convex or concave or both. Furthermore, the surface can have edges or other discontinuities and a dielectric layer can cover the surface. Altogether, the surface can be very complex. Important steps in array analysis are to find the mutual coupling among the radiating elements and the (isolated and/or embedded) radiation patterns of the individual elements. A major part of the analysis is, therefore, to find the electromagnetic fields on the surface and the far field in the presence of this complex and, in general, arbitrarily shaped body.

The available methods (for any EM problem, in fact) are often divided into two categories: frequency domain methods and time domain methods. Frequency domain meth-

Conformal Array Antenna Theory and Design. By Lars Josefsson and Patrik Persson
© 2006 Institute of Electrical and Electronics Engineers, Inc.

ods are probably the most commonly used methods for analysis of antennas, including conformal arrays. This class is often subdivided into low-frequency methods and high-frequency methods. "High frequency" implies that the properties and size parameters of the geometry vary slowly with the frequency.

In this chapter, we will use a somewhat different classification of the methods: electrically small surfaces and electrically large surfaces. Both frequency and time domain methods are included in a logical way, making it easier to find a suitable method for a practical application. The distinction between electrically small and electrically large surfaces is not always clear—the two categories overlap to some extent. The assumption of an electrically large surface is often valid for large bodies when the frequency is in the microwave frequency band or higher. However, it is necessary to know several methods and find out when they are applicable. A hybrid approach (a combination of different methods) is often used in practice.

We will focus on the most commonly used methods but not describe the methods in great detail. Instead, we will try to point out what to take into account and describe the features of the methods, both good and bad. With this information, we hope the reader will be able to find a suitable method when facing a conformal antenna problem. However, the practical application of a certain method is not discussed; this is up to the user to find in the relevant references. Some help is provided by the many references that are listed, and also by two canonical examples that will be described in Section 4.5.

Before entering into the discussion of the various methods we present in Section 4.2 the typical electromagnetic problem that we have to solve, and we introduce also some parameters that will be used in the next several chapters.

Section 4.3 on electrically small surfaces deals with frequency domain methods such as modal solutions, integral equations (IE)/method of moments (MoM), and the finite element method (FEM). Time domain methods are also included here, including the most commonly used, the finite-difference time-domain method (FDTD). These methods become intractable for structures that are large in terms of the wavelength, since the associated matrices become very large. Thus, an upper frequency limit exists for these methods. The time domain methods are often used for analyzing antennas for large frequency bands, which, however, requires much computer memory (and time). The classical FDTD has a disadvantage when applied to conformal antennas: the staircase approximation of curved surfaces. Some attempts to improve the method will be discussed.

Section 4.4 discusses methods for electrically large surfaces. The methods considered here have a lower frequency limit. Examples of methods suitable for this class of surfaces are geometrical optics (GO), geometrical theory of diffraction (GTD), uniform theory of diffraction (UTD), physical optics (PO), and physical theory of diffraction (PTD).

In Section 4.5, we present some more details for two canonical problems. The two cases are a waveguide-fed aperture antenna and a microstrip-patch antenna located at an arbitrarily shaped surface. The approach is an integral equation formulation, solved with the method of moments. The reason for choosing this approach is that we can point out several interesting aspects, and at the same time retain a physical insight into the solution. With a full numerical approach, this insight is more difficult to illustrate.

Section 4.6 concludes this chapter with a short discussion and comparison of the various methods. We have included a table with guidelines and some additional comments. Hopefully, this will help the reader in choosing a proper method.

4.2 THE PROBLEM

In the previous chapters of this book, we have presented a background and general characteristics of conformal antennas. We will now proceed and analyze conformal antennas in more detail, that is, solve the associated electromagnetic problems. This includes finding the fields in space due to radiating sources (elements) on a curved surface. When this is done, the antenna array can be analyzed with respect to radiation pattern characteristics and mutual coupling among the elements, impedance, and so on. The generic problem is illustrated in Figure 4.1.

This is a boundary value problem in which we have to match a field representation outside S with the supposedly known field or current distribution at the antenna elements. When the solution has been found, we have obtained the Green's function ($\bar{\bar{G}}$) for the antenna problem. Thus, the field outside S is related to the radiating element current ($\bar{J}$) simply by the following general formula:

$$\bar{E} = \sum_{n=1}^{N} \int_{S_n} \bar{\bar{G}} \cdot \bar{J} dS \tag{4.1}$$

where the sum is taken over the number of elements considered. Unfortunately, the number of geometries where an explicit field representation outside S can be found is very limited. In many cases, asymptotic expressions are needed, and in some cases a numerical representation is the only possible solution.

When we have found a solution to the electromagnetic problem, we can start analyzing the conformal (array) antenna. For the radiation problem some approximations can be obtained when considering the far zone region. A brief overview of circular array antenna theory was presented in Section 2.2. The electromagnetic solution to the problem also gives us the tool to study the surface field or the near field, in order to find the mutual coupling among the radiating elements. This analysis is often difficult since no far field approximations are possible. However, the mutual coupling is very important and is needed for a full analysis of the antenna array characteristics. Calculating the mutual coupling between two radiators implies finding the surface field at the receiving element n due to the source element m, and then match the boundary conditions at the receiving element. For aperture antennas, this can typically be expressed as

$$Y_{nm} = \frac{1}{V_m V_n} \int_{S_n} \bar{E}_n \times \bar{H}_n(m) d\bar{S} \tag{4.2}$$

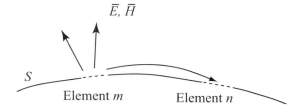

Figure 4.1. The problem: find the field outside the curved surface S due to one or more radiating elements on the surface S.

where $\overline{H}_n(m)$ is the surface magnetic field at aperture n due to the electric field $\overline{E}_m$ in aperture m. Hence, $\overline{E}_n$ denotes the electric field at the receiving aperture and V_m and V_n are the modal amplitudes (which equals 1 if normalized).

If we take two radiators like this from a large array with the other radiators short-circuited in case of apertures (or open-circuited for dipole type elements), we get essentially an element in the mutual admittance (impedance) matrix. From this matrix, we can derive the scattering matrix for the full array:

$$\mathbf{S} = (\mathbf{I} - \mathbf{Y})(\mathbf{I} + \mathbf{Y})^{-1} \qquad \text{or} \qquad \mathbf{S} = 2(\mathbf{I} + \mathbf{Y})^{-1} - \mathbf{I} \qquad (4.3)$$

The radiation properties and the mutual coupling among the elements are studied in great detail in Chapters 6–8. Another type of analysis is to study the scattering properties. In this case, the excitation is given by an incident wave. This problem will be considered in Chapter 11.

So far, we have not defined the type of radiating element used. The conformal array antenna characteristics depend to a great extent on the curvature of the surface, element lattice, and other array parameters, but also on the antenna element type. The choice of radiating element will to some extent determine how to solve the analysis problem. We will give two explicit examples in Section 4.5, where waveguide fed aperture antennas and microstrip patch antennas are considered. These two types of antenna elements are used in many conformal antenna applications.

4.3 ELECTRICALLY SMALL SURFACES

4.3.1 Introduction

In this section, methods for the analysis of electrically small surfaces are discussed. We will start by discussing the modal solution and also give some explicit results that are easy to apply for certain canonical surfaces. If more advanced shapes are of interest, other methods, often numerical methods, are more suitable, but they require more effort of the user.

A difficulty with many of the methods presented in this section is that they become numerically intractable as the dimensions increase. As already indicated, there is no exact boundary between electrically large and small surfaces. A radius of curvature of up to five wavelengths can often be analyzed by, for example, a modal solution method, but this depends on the available time for computations and available memory. One interesting approach is to use a hybrid method by which the attractive features of different methods are combined.

4.3.2 Modal Solutions

4.3.2.1 Introduction. The classical way of analyzing conformal antennas is to use separation of variables and express the solution in a modal representation, that is, in a vector harmonic series representation. This is possible only for special shapes, such as a perfectly conducting circular cylinder, elliptic cylinder, cone, or sphere. Many researchers studied the circular cylinder during the 1940s and 1950s; see, for example [Carter 1943, Pistolkors 1947, Silver and Saunders 1950, Levin 1951, Harrington and Lepage 1952, Knudsen 1959]. J. R. Wait [Wait 1959] summarized most of the early work on cylinders in his book *Electromagnetic Radiation from Cylindrical Structures.* The cone solution is found in, for example, [Bailin and Silver 1956, 1957, Pridemore-Brown and Stewart 1972, Hansen 1981a], and the sphere solution is found in, for example, [Karr 1951, Wait 1956b, Mushiake and Webster 1957, Belkina and Wainstein 1957]. See also [Zakharyev et al. 1970], in which the radiation patterns of spherical antennas are calculated with a modal solution. A simplification of the modal solution for structures with a rotational symmetry axis is to reduce the three-dimensional problem to a spectrum of two-dimensional problems by applying a Fourier transformation with respect to the symmetry axis, as described by R. F. Harrington [Harrington 1961].

However, if the geometry is more complex, a modal analysis is very difficult if not impossible to obtain. The modal solution is complicated, and it has poor convergence if the size of the geometry is increased (in terms of wavelength). Hence, the process of extracting numerical results can be difficult and time-consuming. Nevertheless, the modal solution is the basic building block for obtaining analytical expressions of the field in the presence of a curved surface. We will, therefore, derive explicit expressions for the fields in the presence of a circular cylinder, which will be used later in this book. This is done following the approach presented by Harrington [Harrington 1961].

4.3.2.2 *The Circular Cylinder.*

The circular cylinder is assumed to be infinitely long and perfectly conducting. In our example, the antenna element is an aperture antenna. We assume that the tangential electric field distribution (E_{tan}) on the surface is known, as indicated in Figure 4.2.

Since the electric field $\overline{E}$ and the magnetic field $\overline{H}$ outside the cylinder have zero divergence, they can be expressed in terms of the magnetic vector potential $\overline{A}$ and the electric

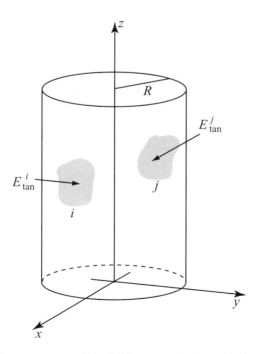

Figure 4.2. Two apertures (*i* and *j*) in a conducting circular cylinder.

vector potential $\overline{F}$. The components of the potentials satisfy the scalar wave equation. The solution is found by using the method of separation of variables. However, we will first reduce the problem to a two-dimensional problem by applying a Fourier transform with respect to the symmetry axis (here the cylinder or z axis). Thus, define $\widetilde{E}_{\tan}$ as a Fourier transform of the tangential components of $\overline{E}$ as

$$\widetilde{E}_{\tan}(n, k_z) = \frac{1}{2\pi} \int_0^{2\pi} \int_{-\infty}^{\infty} E_{\tan}(r, \varphi, z)e^{-j(n\varphi+k_zz)}dzd\varphi \tag{4.4}$$

with the inverse given by

$$E_{\tan}(r, \varphi, z) = \frac{1}{2\pi} \sum_{n=-\infty}^{\infty} \int_{-\infty}^{\infty} \widetilde{E}_{\tan}(n, k_z)e^{j(n\varphi+k_zz)}dk_z \tag{4.5}$$

The wave functions A_z and F_z are now constructed in a fashion similar to that of Equation (4.4), but expressed in terms of the second-order Hankel function $H_n^{(2)}(k_r r)$ representing outward-traveling waves. Using the boundary conditions, we can find field expressions that are valid outside the cylinder ($R \leq r < \infty$) in terms of the known Fourier transform of the tangential electric field at the apertures. The electric field and the corresponding magnetic field are given in Equations (4.6)–(4.9):

$$E_\varphi(r, \varphi, z) = \frac{1}{2\pi} \sum_{n=-\infty}^{\infty} \int_{-\infty}^{\infty} \left[g_n(k_z)k_r H_n^{\prime(2)}(k_r r) + \frac{nk_z}{j\omega\varepsilon r} f_n(k_z)H_n^{(2)}(k_r r) \right] e^{-jk_zz+jn\varphi}dk_z \tag{4.6}$$

$$E_z(r, \varphi, z) = \frac{1}{2\pi} \sum_{n=-\infty}^{\infty} \int_{-\infty}^{\infty} \frac{k_r^2}{j\omega\varepsilon} f_n(k_z)H_n^{(2)}(k_r r)e^{-jk_zz+jn\varphi}dk_z \tag{4.7}$$

$$H_\varphi(r, \varphi, z) = \frac{1}{2\pi} \sum_{n=-\infty}^{\infty} \int_{-\infty}^{\infty} \left[-f_n(k_z)k_r H_n^{\prime(2)}(k_r r) + \frac{nk_z}{j\omega\mu r} g_n(k_z)H_n^{(2)}(k_r r) \right] e^{-jk_zz+jn\varphi}dk_z \tag{4.8}$$

$$H_z(r, \varphi, z) = \frac{1}{2\pi} \sum_{n=-\infty}^{\infty} \int_{-\infty}^{\infty} g_n(k_z)H_n^{(2)}(k_r r)e^{-jk_zz+jn\varphi} \frac{k_r^2}{j\omega\mu}dk_z \tag{4.9}$$

where

$$f_n(k_z) = \frac{j\omega\varepsilon\widetilde{E}_z(n, k_z)}{k_r^2 H_n^{(2)}(k_r R)} \tag{4.10}$$

$$g_n(k_z) = \frac{1}{k_r H_n^{\prime(2)}(k_r R)} \left[\widetilde{E}_\varphi(n, k_z) - \frac{k_z}{k_r^2} \frac{n}{R} \widetilde{E}_s(n, k_z) \right] \tag{4.11}$$

and $k_r^2 = k^2 - k_z^2$.

The above solution has poor convergence if the radius of the cylinder is electrically large ($kR \gg 1$). This can be understood by recognizing the harmonic series as a series of standing waves around the cylinder. Therefore, the required number of terms in the summation over n is directly related to the size of the cylinder; many terms are necessary for an electrically large cylinder. According to Wait [1959], at least $2kR$ terms are needed in the summation for 2% accuracy. This problem can be reduced by deriving alternative formulations. In Wait's book, two different approaches are used. One option is to rewrite the solution as a residue series; see also [Stewart and Golden 1971] for the circular cylinder and [Golden et al. 1974] for the cone. However, there may be additional numerical problems in evaluating special functions such as the Bessel and Hankel functions for large orders and large arguments.

To overcome the difficulties for large surfaces, it is possible to use asymptotic expressions, both for special functions and integrals. This will be discussed in more detail in Section 4.4, where such techniques are considered.

We also have to solve the integral (in k_z) that represents the inversion to obtain the three-dimensional solution from the spectrum of two-dimensional solutions. In general, this must be done numerically, except for the far zone ($r \rightarrow \infty$), where the radiating field can be found by using an asymptotic formula for the k_z integral [Harrington 1961]. Thus, the quite simple far field expressions obtained from Equations (4.6)–(4.7) (when rewritten in spherical coordinates) are:

$$\begin{pmatrix} E_\theta \\ E_\varphi \end{pmatrix} = \begin{pmatrix} j\omega\mu \\ -jk \end{pmatrix} \frac{e^{-jkr}}{\pi r} \sin\theta \sum_{n=-\infty}^{\infty} e^{jn\varphi} j^{n+1} \begin{pmatrix} f_n(-k\cos\theta) \\ g_n(-k\cos\theta) \end{pmatrix} \qquad (4.12)$$

A similar asymptotic expression for the mutual admittance cannot be found, since we are then dealing with the near field. Instead, different techniques have been used to facilitate the k_z integration, for example, by separating the solution into its real and imaginary parts and changing the integration contour [Safavi-Naini and Lee 1976, Sohtell 1987]. See also [Wait 1959, Hansen 1981a] for more references on this topic.

If the above (aperture) antenna is coated with one or more dielectric layers, the procedure can be extended to include these layers, although the complexity of the solution increases as well as the numerical burden. Explicit formulas for a single layer can be found in, for example [Persson and Rojas 2003, Thors and Rojas 2003], and multiple layers are treated in, for example [Sipus et al. 1998]. A related problem is when the radiating element is a microstrip patch antenna. This type of antenna can also be analyzed with a modal solution, at least when studying the radiating properties. For rectangular patches, a transmission-line model can be used, in which the patch is modeled as two radiating slots with an approximately half-wavelength transmission line in between. The theory (for the planar case) can be found in many papers or textbooks; some examples are [Derneryd 1976, Sengupta 1984, Balanis 1997]. In [Munson 1974] and [Calander and Josefsson 2001], this approach was applied to conformal antennas. A better approximation, valid for any kind of patch geometry, is to view the area under the metallic radiator as a cavity bounded by metallic walls. Impedance boundary conditions at the wall can be used to take the losses in the cavity into account. The general theory can be found in, for example, [Lo et al. 1979, Richards et al. 1981, Carver and Mink 1981]. In [Wu and Kauffman 1983, Jakobsen 1984, Sohtell 1986, 1987, Krowne 1983, Luk et al. 1987] the cavity model was used for conformal microstrip patch antennas. These approximations are fast and give a fairly good alternative to the full-wave analysis needed for a more complete model of the microstrip patch antenna.

4.3.2.3 *A Unit Cell Approach.*   For arrays on circular cylinders (coated or not) an attractive formulation is obtained by the theory of periodic structures (Floquet's theorem). This approach is well proven for the planar case, in which a unit cell is defined in the (infinite) antenna array and the field is described using Floquet modes [Borgiotti 1968, Amitay et al. 1972]. The same procedure can in some special cases be employed for

arrays of equally spaced elements on curved symmetrical surfaces. One example is the circular cylinder (Figure 4.3), which can include dielectric layers; the sphere is another example.

When using the unit cell approach, the fields in all cells must be identical, except for a constant phase shift from cell to cell. For the cylindrical array, this is straightforward for the axial direction of the cylinder. However, special care must be taken in the circumferential direction. Along each ring of the array, the phase shift between the cells must be such that the complete ring produces a total phase change that is a multiple of 2π, that is, the phase shift between two adjacent cells is $2\pi n/N$, where N is the number of unit cells in one ring. This makes the array infinitely periodic in both dimensions, and the analysis is reduced to a study of one unit cell.

A few examples from the literature follow. An array of infinitely long longitudinal slots equispaced on a conducting cylinder was analyzed by Borgiotti and Balzano [1970]. A three-dimensional infinite periodic cylindrical array of waveguide elements located on a PEC cylinder was considered by Munger et al. [1971] and Borgiotti and Balzano [1972]. Balzano then extended the theory to a coated PEC cylinder in 1974 [Balzano 1974]. An axial dipole array was studied by Herper et al. [1985]. A more recent example is [Gerini and Zappelli 2004], in which a circular cylinder was analyzed using a multi-mode equivalent network approach.

4.3.3 Integral Equations and the Method of Moments

For arbitrarily shaped bodies, no exact analytical solution exists, as discussed in Section 4.3.2. Instead, numerical methods have to be used. One frequently used method is the method of moments (MoM) or moment method. Strictly speaking, the MoM is not an analysis method by itself. However, there is an underlying integral equation (IE) which has to be solved numerically, and the MoM is a powerful method to do this. The IE can, in general, be written as

$$\mathrm{L}_{IE}(w, w')x(w') = f(w) \tag{4.13}$$

where $\mathrm{L}_{IE}(w, w')$ is the integral operator, $x(w')$ is the unknown fields or sources, and $f(w')$ is the forcing function. The principle behind the MoM is to convert the integral equation

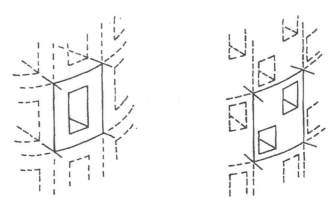

Figure 4.3. Examples of unit cells [Munger et al. 1971].

into a matrix equation. This is done by expanding the unknown quantity in appropriate basis (or expansion) functions with unknown coefficients. Projection on a set of testing (or weighting) functions gives a system of equations. As a result, the IE in Equation (4.13) is transformed into a matrix form as

$$\mathbf{A}\mathbf{x} = \mathbf{f} \tag{4.14}$$

with the solution formally given by $\mathbf{x} = \mathbf{A}^{-1}\mathbf{f}$. More details about the procedure can be found in any of the numerous references in this area; see, for example [Kantorovich and Krylov 1964, Vorovbyev 1965, Harrington 1968, Adams 1974, Poggio and Miller 1973, Miller and Burke 1992, Wang 1991].

The MoM procedure is general and applicable to conformal antennas located at arbitrarily shaped surfaces. Unfortunately, both calculating the elements in **A** and solving the matrix equation can be cumbersome if the number of unknowns is large. This is a general problem when using the MoM, but it can be much more pronounced for conformal antennas. If the appropriate Green's function cannot be found, as is the case in many conformal antenna applications, the free space Green's function must be used, resulting in an increased number of unknowns. At higher frequencies and/or for electrically large surfaces, the MoM becomes impractical in many cases.

Despite these complications, the MoM has found widespread use in the analysis of conformal antennas. It is often used in combination with other techniques such as spectral domain and asymptotic techniques in order to simplify the analysis. In the references cited below, several examples are found in which the MoM has been used, sometimes combined with other methods. Slots in (coated) PEC surfaces are considered in Peterson and Mittra [1989] and Sipus et al. [2001a–b]. Microstrip patch antennas have been studied extensively; see for example, [Habashy et al. 1990; Silva et al. 1991; Wong et al. 1993; Gottwald and Wiesbeck 1994; da Silva and Lacava 1995; Vecchi et al. 1997a–b; Bertuch et al. 1999; Raffaelli et al. 2001, 2005; Sipus et al. 1998; Sipus 2003; Svezhentsev and Vandenbosch 2001; Svezhentsev 2003; Jakobus and van Tonder 2003]. A perturbed microstrip antenna is considered in Thiel and Dreher [1999]; see also [Thiel and Dreher 2001a–c, 2002; Thiel 2002]. A conformal spiral antenna is studied in Nakayama et al. [1999].

Boulder Microwave Technologies was offering a commercial software for analysis of conformal microstrip-patch antennas, called Clementine [Hall and Wu 1996] in the late 1990s. This software is based on a mixed-potential integral equation (MPIE) model. However, to our knowledge, this software is no longer available.

4.3.4 Finite Difference Time Domain Methods (FDTD)

4.3.4.1 Introduction. Time domain methods have to date not been used extensively for the analysis of conformal antennas. One main reason is the difficulty in discretizing (meshing) curved surfaces. Discrete models require a lot of memory and computation time and there is a disadvantage in using the standard staircase approximation of a curved surface. However, progress in conformal antennas for broadband applications could motivate an increased interest in time domain methods compared to frequency domain methods.

This section concentrates on FDTD and techniques to avoid staircase approximation errors. There are other time domain methods such as TD integral methods [Bennet and Mieras 1981, Rao and Wilton 1991, Rynne 1991, Sun and Rusch 1994b], time domain uniform theory of diffraction (TD-UTD) [e.g., Rousseau and Pathak 1995a,b, 1996, Veruttipong 1990], and time domain physical optics (TD-PO) [Sun and Rusch 1994a]. However, most of these methods are used for transient and scattering studies. This is of interest regarding, for example, the effects of electromagnetic pulses (EMPs) on complex systems such as aircraft and spacecraft, but it will not be discussed here. The interested reader is referred to the references.

The standard FDTD algorithm [Yee 1966, Taflove 1995, 1998] is based on an orthogonal, regular Cartesian lattice. For nonplanar boundaries, a staircase approximation is usually adopted, since this is the only way to model curved surfaces in standard FDTD. To reduce the staircase approximation error, one solution may be to use a finer mesh to better resolve the boundaries. Unfortunately, as shown in [Cangellaris and Wright 1991], this does not necessarily mean that the solution will converge to the correct answer. As a consequence, a number of studies during the last 10–15 years have focused on developing FDTD-based algorithms using various types of conformal grids (see Figure 4.4). Below is a short summary of the most relevant conformal grid FDTD techniques that can be found in the literature.

4.3.4.2 Conformal or Contour-Path (CP) FDTD.   One possible solution to the staircase approximation problem is to use a locally distorted grid combined with a globally orthogonal grid [Jurgens et al. 1992, 1993; Railton 1993; Yu et al. 2000; Yu and Mittra 2000, 2001; Yu et al. 2001]. Thus, the space cells away from the boundary are the same unit cells as used in the Cartesian grid arrangement, whereas the cells adjacent to the boundary are deformed to conform to the surface. This method is known as conformal (C) or contour-path (CP) FDTD.

In the locally deformed cells at the boundaries, Maxwell's curl equations cannot be discretized using standard central differences. For these cells, the CP algorithm is used,

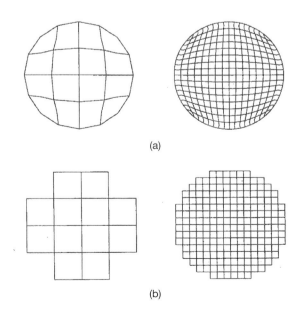

(a)

(b)

Figure 4.4. (*a*) Sparse and dense nonorthogonal grids for a cylindrical cavity. (*b*) Sparse and dense staircase grids for a cylindrical cavity. [Harms et al. 1992.]

which is based on Maxwell's equations in integral form. The details can be found in [Taflove 1995]. Figure 4.5 shows the deformed cells for a PEC surface.

In a modified version of the CP algorithm, see, for example [Yu and Mittra 2000], both the electric and magnetic fields inside the distorted cell are located at the same positions as those in the conventional FDTD scheme (see Figure 4.6). Now, the entire FDTD cell is used and not only the distorted part. Those parts of the contour that are located inside the PEC region are discarded. This increases the stability of the solution.

The advantage of CP FDTD is that it does not require extensive modifications of the standard (Cartesian) FDTD algorithm. The CP method need only be applied to cells near a material interface, whereas the traditional FDTD method is applied to the remaining cells. However, electrically large surfaces still require the same number of cells as the traditional FDTD; only the staircase approximation has been avoided. For non-PEC boundaries, the method must be slightly modified; the reader is refereed to for example, [Taflove 1995, Yu et al. 2000, 2001; Yu and Mittra, 2000, 2001]. See also [Byun et al. 1999], in which microstrip antennas have been analyzed using CFDTD. A three-dimensional CFDTD algorithm is presented in [Dey and Mittra 1994].

4.3.4.3 FDTD in Global Curvilinear Coordinates.

Another, more general, attempt to get rid of staircase approximations is to use the FDTD algorithm in generalized curvilinear coordinates. To handle generalized (nonorthogonal) coordinates, Maxwell's equations are formulated in tensor notation. It was Holland [Holland 1983; see also Mei et al. 1984] who first proposed the generalized FDTD method in curvilinear coordinates. A refined approach was described by Fusco [Fusco 1990] and applied to a two-dimensional case. An attempt to study a three-dimensional problem was also proposed by Fusco [Fusco et al. 1991]. Since the number of coordinate systems that can be represented analytically is small, Fusco and coworkers used a numerically defined mesh wherein the mesh point coordinates were given in an underlying Cartesian system. They applied the algorithm to a body of revolution and determined both the near and far fields. However,

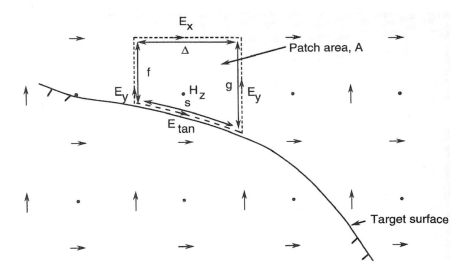

Figure 4.5. The CP FDTD methodology for a perfectly conducting surface [Jurgens et al. 1992].

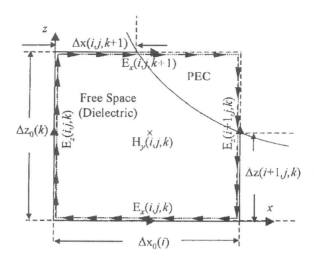

<u>Figure 4.6.</u> The contour path of the conformal FDTD in a deformed cell [Yu and Mittra 2000].

special treatments where needed in the vicinity of the rotation axis, since this axis represent a coordinate singularity. Later, a more efficient method was proposed [Lee et al. 1992]. This method has been used for analyzing curved microstrip patch antennas (see Figure 4.7).

As can be imagined, this method is very demanding if large, generally shaped surfaces are considered. It requires a large memory because of the large amount of geometrical information that must be stored. Because of the increased algorithmic complexity, it is also time-consuming. In [Kashiwa et al. 1993], the curved antenna shown in Figure 4.7 needed 20 MB of memory compared to 3 MB for the flat antenna case. The computation time for the curved antenna was 8.5 hours and 3.5 hours for the flat case. To the authors' knowledge, no major improvements have been made since then and the curvilinear FDTD algorithm in its present form may not be suitable for analyzing conformal antennas. However, an advantage of curvilinear FDTD could be that it interfaces more easily with the mesh generators in mechanical CAD systems.

4.3.4.4 *FDTD in Cylindrical Coordinates.*   FDTD in cylindrical coordinates is a special case of the curvilinear FDTD algorithm discussed above. In this area, several recent studies have been applied to conformal antennas. The discretization of Maxwell's equations in cylindrical coordinates can be found in, for example [Kunz and Luebbers 1993, Taflove 1998]. Figure 4.8 shows an FDTD unit cell in cylindrical coordinates. Since the z-axis represents a coordinate singularity, special care must be taken when updating the E_z component if the z-axis is included in the computational domain; for example, if a cylinder of finite length is considered. In this case, the E_z component can be calculated by integration of the H_φ component around the z-axis (Stokes's theorem).

Also, infinitely long PEC cylinders can be analyzed by this method if the tangential electric field is set to zero at a radius R_0, defining the lower boundary of the computational volume in the ρ-direction as shown in Figure 4.9. The upper and lower boundaries in the z-direction and the outer boundary in the ρ-direction must be terminated using a uni-

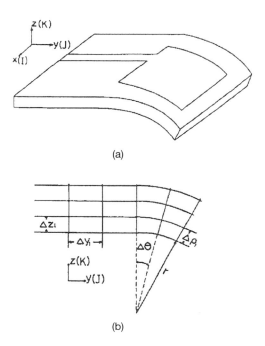

(a)

(b)

Figure 4.7. Typical grid when using FDTD in curvilinear coordinates [Kashiwa et al. 1994]. (a) Rectangular microstrip antenna on curved substrate. (b) Mesh configuration on *yz* cross section.

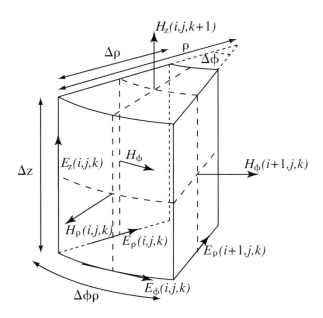

Figure 4.8. Field components placement in the FDTD unit cell in cylindrical coordinates, after [Marcano et al. 1998].

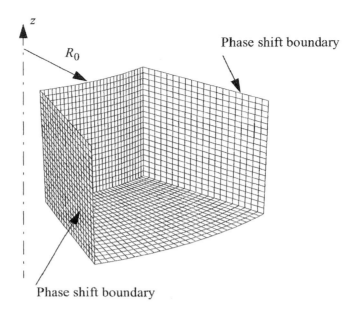

Figure 4.9. Schematic three-dimensional view of a sector computational volume when using cylindrical FDTD [Martin et al. 2003a].

axial perfectly matched layer (UPML) boundary condition [Teixeira and Chew 1997, Taflove 1998].

An important feature when using cylindrical coordinates is that a Fourier decomposition in the azimuthal direction can be used, regardless of whether a solution in the time domain or the frequency domain is considered. Hence, the solution can be written in the form

$$F(\rho, \varphi, z) = \sum_n F_n(\rho, z)e^{jn\varphi} \qquad (4.15)$$

where F is any of the six field components. Thus, Maxwell's equations can either be solved for each orthogonal mode n for a body of revolution (BOR method) or for a group of modes that can exist in a sector of the cylinder. The latter method is useful when the circular array antenna has elements equally spaced around the cylinder.

If the antenna (or scattering object) is N-periodic around the cylinder it is sufficient to divide the cylinder into N sectors and use a computational volume occupying one of these sectors with an angle $2\pi/N$ of the total cylindrical volume. Thus, the results for the entire cylindrical volume are obtained by running the FDTD program N times using phase shift boundary conditions at the vertical walls at $\varphi = \varphi_0$ and $\varphi = \varphi_0 + 2\pi/N$ (see Figure 4.9). The phase shift boundary conditions in the azimuth direction can be written as

$$E_t\left(\varphi_0 + \frac{2\pi}{N}\right) = E_t(\varphi_0)e^{j\alpha}$$

$$(4.16)$$

$$H_t(\varphi_0) = H_t\left(\varphi_0 + \frac{2\pi}{N}\right)e^{-j\alpha}$$

where E_t and H_t represent any components of the electric and magnetic fields, although in our case only the tangential components are used. The phase shift α in Equation (4.16) is given by

$$\alpha = 2\pi n/N, \qquad n = 0, 1, 2, \ldots, N - 1 \tag{4.17}$$

Since the fields are complex-valued (even though the simulation is performed in the time domain), they contain information of both positive and negative phase shifts. Consequently, it can be shown that it is sufficient with $N/2 + 1$ simulations, corresponding to phase shifts with $n = 0, 1, 2, \ldots, N/2$ in Equation (4.17).

This method has been used to study radiation properties and mutual coupling for elements on a circular cylinder with and without a dielectric layer [Martin et al. 2003a,b]. An interesting extension of the cylindrical FDTD is the hybrid cylindrical–Cartesian FDTD method approach [Craddock et al. 2001, Paul and Craddock 2001, Paul et al. 2003]. This method is shown to be well suited to the analysis of faceted conformal antennas. With this model, all surfaces in the array are modeled without staircase approximations. Furthermore, the cylindrical–Cartesian FDTD hybrid method requires the same amount of memory and computation times as an equivalent planar array.

4.3.5 Finite Element Method (FEM)

4.3.5.1 Introduction. FEM or the finite element method is a general method for solving boundary-value problems. This method has been applied to many different problems, not only electromagnetic (EM) problems. Thus, a large number of papers and books have been published on this topic; see, for example [Courant 1943, Desai and Abel 1972, Martin and Carey 1973, Norrie and de Vries 1973, Zienkiewicz and Taylor 1989, Jin 1993, Volakis et al. 1998].

The boundary problem to be solved is defined by a differential equation and the relevant boundary conditions. A typical problem is the scalar or vector wave equation with proper boundary conditions. For such problems, the differential equation (DE) can be written in general form as

$$L_{DE}\,\kappa = f \tag{4.18}$$

defined in a domain Ω. Here, L_{DE} is the differential operator, κ is the unknown quantity, and f is the excitation or forcing function. For a complete definition, the boundary conditions on the boundary Γ enclosing Ω must also be defined.

When solving this problem with FEM, there are four basic steps involved (see the references for a more detailed description). First, the domain must be subdivided or discretized into a finite number of subregions (or elements). With a sufficiently fine subdivision, very complex geometries (and materials) can be analyzed. The discretization will affect the computer storage requirements, the computation time, and the accuracy of the solution. Second, the solution within each element is approximated by a suitable function; often a first-order polynomial is used. In the third step, all elemental equations are brought together and the boundary conditions are imposed to get the final system of equations. If a proper discretization was made, the FEM formulation results in a banded matrix formulation. Finally, this system of equations can be solved in different ways, preferably by using a method that takes advantage of the banded matrix.

One disadvantage of FEM is that the Sommerfeld radiation condition is not incorporated automatically. Thus, the computational domain must be extended far from the structure so that the radiation condition can be imposed, resulting in an increased number of unknowns. Incorporating absorbing boundary conditions in order to reduce the region outside the antenna (see e.g. [Mittra and Ramahi 1990]), can solve the open boundary problem. However, conformal antenna related problems often include complex and electrically large geometries, which increases the number of elements (unknowns). As a consequence, the conventional FEM itself has seldom been used to study conformal antennas. One exception is [Özdemir and Volakis 1997], in which absorbing boundary conditions are used. However, FEM in combination with integral equations results in a much more useful method (FE-BI), to be discussed next. Another promising possibility is to combine FEM with FDTD as presented in [Rylander and Bondesson 2002, Rylander and Jin 2003].

It should also be mentioned that there are powerful commercial FEM codes that can be used to study conformal antennas. One example is given in [Vogel 2001] where the commercial finite element software, Ansoft HFSS, is used.

4.3.5.2 Hybrid FE-BI Method.
Since FEM does not incorporate the Sommerfeld radiation condition, the domain of discretization must be extended far from the scatterer/antenna, resulting in a large number of unknowns and long computation time. One possibility to get around this is to use an integral equation in combination with FEM, since the integral equation automatically incorporates the radiation condition. This hybrid method is called the finite element boundary integral (FE-BI) method.

The basic idea of this approach was introduced in the beginning of the 1970s [Silvester and Hsieh 1971, McDonald and Wexler 1972]. The method has then been further developed and solutions to general three-dimensional problems can be found in [Jin and Volakis 1991a–c, Jin et al. 1991, Jin 1993, Volakis et al. 1998].

To use the FE-BI method a fictitious surface S enclosing the geometry is defined (see Figure 4.10). Within this surface, the finite element method is used to find the fields by discretizing the volume in small (volume) elements suitable for the geometry. The electric

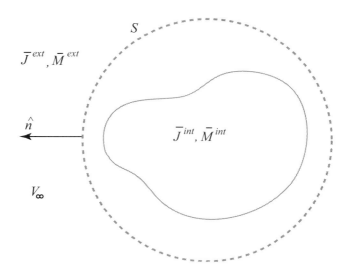

Figure 4.10. Geometry of the scattering/radiation problem.

and magnetic fields within S are then expanded using edge-based, vector basis functions.*
In the exterior region, the fields are given by external sources (if this is a scattering problem) and the equivalent sources on S. On S, an IE is used with the free space Green's function. The fields in the interior and exterior regions are finally coupled by enforcing continuity of the tangential fields on S.

Using this approach, a system of equations is obtained for the fields. The FE-BI equation for the electric field can be written in matrix form as

$$[A]\begin{bmatrix} \{E^{ap}\} \\ \{E^{int}\} \end{bmatrix} + \begin{bmatrix} \{G\} & \{0\} \\ \{0\} & \{0\} \end{bmatrix}\begin{bmatrix} \{E^{ap}\} \\ \{E^{int}\} \end{bmatrix} = \begin{bmatrix} \{f^{ext}\} \\ \{f^{int}\} \end{bmatrix} \qquad (4.19)$$

In Equation (4.19), the matrix $[A]$ is due to the finite element formulation and $\{G\}$ is the boundary integral submatrix; E^{ap} and E^{int} denote the unknowns. E^{ap} is associated with the aperture field and E^{int} is the interior (volume) electric field used, for example, when a cavity-backed antenna is considered. The right-hand side contains the sources: f^{int} is the internal sources and f^{ext} is due to external sources ($\neq 0$ only if scattering is considered).

This system must be solved numerically and there are various techniques to do this. However, according to the reference literature (see, for example, [Jin et al. 1991]), a combination of the biconjugate gradient method (BiCG) and the fast Fourier transform (FFT) is preferred. This gives a numerical scheme that takes advantage of the fact that the FE matrix is sparse whereas the boundary integral submatrix is fully populated, thus minimizing the computational requirements.

When applying this method to curved surfaces, some differences appear compared to planar cases. First, the surface S is assumed to perfectly fit the curved surface. This requires that the external free space Green's function must be replaced with the proper Green's function. With this Green's function, the boundary conditions become implicitly included and the numerical computation is tractable. Unfortunately, the proper Green's function based on the modal solution discussed in Section 4.3.2 is not suitable for electrically large surfaces. In this case, an asymptotic Green's function is preferred, to be discussed next in Section 4.4. Furthermore, the shape of the elements used to discretize the interior region must fit the geometry. When applied to a circular cylinder, shell-shaped elements have been used, as reported in [Kempel et al. 1993, 1994a,b, 1995]. This was later extended to handle elliptic cylinders using elliptic shell elements [Wu et al. 2002]. The FE-BI method has also been applied to a doubly curved surface with cavity-backed patch antennas on a prolate spheroid [Macon et al. 2001, 2002]. In this case, tetrahedral elements could be used to model the doubly curved surface. A patch antenna mounted on a cylinder with a wing has been studied in [Liu and Jin 2003a,b].

4.4 ELECTRICALLY LARGE SURFACES

4.4.1 Introduction

As mentioned in the previous section, many of the discussed methods become intractable for electrically large surfaces due to numerical problems. One way of avoiding these problems is to use a high-frequency approach. This means that asymptotic techniques are

*Traditional node-based finite elements have been proven unsatisfactory for three-dimensional electromagnetics applications; instead, edge-based elements are used [Kempel and Volakis 1994a].

used to find an approximate solution. The most important limitation of high-frequency solutions is that the surface must be smoothly curved and electrically large. This is not a serious limitation in many cases, however, since the minimum radius of curvature of the surface can be quite small. A commonly used requirement is $kR \geq 2 - 5$ (k is the wave number and R is the radius of the cylinder) for accurate results [McNamara et al. 1990]. Furthermore, edge diffraction can be included to account for discontinuities along the surface edges [Kouyoumjian and Pathak 1974].

There are a number of different asymptotic techniques and solutions available; some of them were listed in Section 4.1. The reason for the numerous asymptotic formulations is that an asymptotic technique is often specialized for a certain problem and cannot be generalized easily. Furthermore, an (uniform) asymptotic solution is not unique, but the difference between two solutions is usually small.* However, a general formulation is very desirable for efficient analyses of various realistic conformal antennas. The work by Robert Kouyoumjian and his group at the ElectroScience Laboratory at Ohio State University has been successful and their solution is valid for different types of convex surfaces. The solution is known as the uniform theory of diffraction (UTD), which is a ray-based theory. We have chosen to concentrate the discussion about high-frequency methods on this UTD formulation since we think it is a very powerful concept used worldwide. Furthermore, this concept (in addition to, e.g., the solution by [Lee 1978]) is well suited to study arbitrary surfaces since UTD is explicit in its way of describing the different effects that govern the behavior of wave propagation along curved surfaces.

Even though the UTD concept is a general approach, there are minor differences; for example, when a coated or noncoated surface is considered as well as if the far field radiation problem or the surface field problem is considered. There are also differences if the surface is convex or concave. Due to space limitations, this book will only consider smooth convex surfaces. For concave surfaces, the interested reader is referred to the available literature. Conformal antennas have been convex in most of their applications and will probably be so in the future as well. However, concave surfaces can find use on spacecraft and some RF lenses (see Chapter 9). An asymptotic (ray-based) theory can be used also for concave structures, but there are some additional features to consider such as caustics and whispering gallery modes [Inami et al. 1982; Felsen and Green 1976; Ishihara and Felsen 1978, 1988; Tomasic et al. 1989a,b, 1993a,b].

In Section 4.4.2, high-frequency methods for perfectly electric conducting (PEC) surfaces are discussed. This section presents the basic principles behind the ray-based theory, and some important features are discussed. In Section 4.4.3, the theory is applied to coated PEC surfaces. It is noted that the theory is not complete in the latter case.

Appendices 4A.1-5 are included at the end of the chapter for those who want to get a deeper understanding of the physics behind the ray theory and its connection to surface waves and the underlying modal solution.

4.4.2 High-Frequency Methods for PEC Surfaces

The literature contains numerous papers and reports describing asymptotic solutions for PEC surfaces, many of them specialized to a certain problem or geometry. One of the ear-

---

*For example, two different asymptotic expansions can be used to solve an integral with a stationary point near a pole. This is the case of a wedge, in which the uniform theory of diffraction (UTD) of Kouyoumjian and Pathak [Kouyoumjian and Pathak 1974] is one possible solution and the uniform asymptotic theory (UAT) of Lee and Deschamps [Lee and Deschamps 1976] is another possibility.

liest references is about the diffraction of radio waves around the earth, studied by V. A. Fock and presented in 1945 [Fock 1945, 1965]. Other references are [Bird 1984, 1985, 1988; Boersma and Lee 1978; Borovikov and Kinber 1994; Bouche et al. 1993, 1997; Chang et al. 1976; Golden et al. 1974; Lee and Mittra 1976; Lee and Safavi-Naini 1978; Lee 1978; Pridemore-Brown, 1972, 1973; Shapira et al. 1974; Stewart and Golden 1971]. See also [Hansen 1981a,b, Hansen 1998, Kouyoumjian 1965] for a more extensive list of references.

Here, we will concentrate on the uniform theory of diffraction (UTD), not mentioned in the list above. The history of UTD begins with Luneburg, Kline, and Keller who included concepts such as phase, polarization, and diffraction in the optic ray theory used by the ancient Greeks [Kouyoumjian 1965, Josefsson 1981, Keller 1985, McNamara et al. 1990]. The concept of ray theory means that the field is assumed to propagate along rays: along straight lines in free space and along so-called geodesics (see Chapter 5) on surfaces. The space is divided into regions, with different ray field descriptions. Figure 4.11 shows the different regions often used when discussing geometrical optics (GO), the geometrical theory of diffraction (GTD), and UTD. The difference between the three versions is basically that they are valid in different regions. The GO solution [Luneburg 1944, Kline 1951] does not include diffraction and is, therefore, only valid in the deep lit region. The diffraction was included in the GTD formulation presented by Joseph Keller in the mid 1950s. This formulation, also valid in the deep shadow region, was published in 1962 [Keller 1962]. Unfortunately, GTD fails in the shadow boundary region (or penumbra region). To overcome this, a uniform theory of diffraction was developed, namely UTD. UTD retains all the advantages of GTD and overcomes the failure of GTD in the shadow boundary transition regions. This generalization was mainly done by Kouyoumjian's group at Ohio State University [Kouyoumjian and Pathak 1974; Pathak and Wang 1978, 1981; Pathak et al. 1980, 1981; Pathak 1992]. The reader is referred to these papers for explicit equations of the surface and radiated fields due to sources on smoothly shaped convex PEC surfaces. However, we will discuss some of

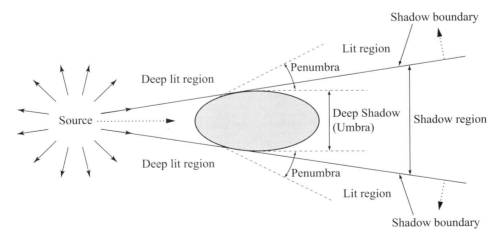

Figure 4.11. Definition of the different regions in space when a source is illuminating a smooth cylindrical surface. If the source is moving toward the surface (dotted arrow) the lit region will be concentrated to the "left" since the shadow boundary will move as well (dotted arrows). After [McNamara et al. 1990.]

the properties behind the ray-based theory. A more detailed discussion is found in Appendices 4A.1–4A.4.

The UTD formulation is a powerful tool for the analysis of smooth PEC surfaces. The ray-based theory also gives an attractive physical description of wave propagation along curved surfaces. To describe the underlying physics, we will consider diffraction by a smooth convex conducting surface shown in Figure 4.12. The ray connecting the source point (P_S) and the observation point (P) satisfies the generalized Fermat principle. This means that the surface diffracted ray path is such that the path $P_S Q' Q P$ is an extremum, although not necessarily a minimum. Thus, $P_S Q'$ and QP are straight lines in free space and they will be tangent to the surface at Q' and Q, respectively. The ray path along the surface between at Q' and Q follows a geodesic of the surface. Thus, geodesics turn out to be key parameters in the analysis and will be discussed in Chapter 5.

Due to the surface diffraction, the polarization of the incoming ray field may be altered when the ray is attached to the surface at Q'. This is determined by the torsion factor and rays propagating along an arbitrary surface have, in general, nonzero torsion. When the ray leaves the surface at Q, the polarization will, once again, be constant, that is, it is a ray with zero torsion.

The amplitude and phase variations of the field along the straight lines in free space follows from the GO representation, whereas the field variation along the surface is more

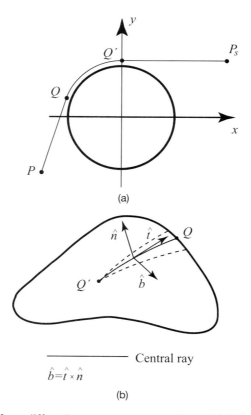

(a)

(b)

Figure 4.12. Ray surface diffraction on a convex surface. (*a*) Overview. (*b*) A surface ray strip connecting Q' and Q together with the surface ray coordinates at a point along the geodesic.

complicated. First of all, a ray propagating on the surface will lose energy continuously due to rays diffracted from the surface. The radius of curvature along the geodesic primarily determines the amount of energy that is lost. In addition, the strength of the surface ray field depends on the width of the surface ray strip, which changes along the path [see Figure 4.12 (b)]. This effect is described by the divergence factor, which can be obtained from differential geometry.

To obtain a (UTD) formulation valid for arbitrarily shaped surfaces, two canonical problems, the PEC circular cylinder and the PEC sphere, were studied. The two solutions have then been combined (heuristically) to be valid for arbitrarily shaped convex surfaces. This is justified by diffraction, as well as reflection and transmission, being a local phenomenon at wavelengths that are small in comparison to the size of the radiating object. Thus, the solutions are heuristically generalized with the aid of the local properties of high-frequency wave propagation. In other words, the total solution for an arbitrarily shaped convex surface may be viewed as a summation of a number of "circular cylinder and sphere solutions" (see also Appendix 4A.5).

The general solution presented above is very useful. However, some complications exist in the UTD formulation, especially in the paraxial region of the cylinder (along the axial direction). For this particular area, minor modifications are needed for UTD to be valid. There is also an alternative, more accurate, solution available for the paraxial region obtained by including higher-order terms in the asymptotic evaluation of the integrals [Boersma and Lee 1978]. Another disadvantage with UTD is that small details cannot be included. For example, due to the approximations used in the derivation, the distance between the source and field points (Q' and Q above) must be larger than 0.5 λ. To overcome this, UTD is often combined with the method of moments (MoM). Some recent references in which this hybrid method is used are [Abou-Jaoude and Walton 1998; Burnside and Pathak 1980; Coffey 1985; Ertürk 2000; Ertürk et al. 2004; Greenwood and Jin 1998; Persson 2001; Persson and Josefsson 1998, 1999a,b, 2000a,b, 2001a,b; Persson and Rojas 2001; Persson et al. 2001; Theron et al. 2000; Thors 2003; Thors and Josefsson 2000, 2003; Thors and Rojas 2003; Thors et al. 2003a–b].

4.4.3 High-Frequency Methods for Dielectric Coated Surfaces

The reason for discussing coated surfaces separately is that some of the approximations used for noncoated surfaces turn out to give nonsatisfactory results when the coating is applied. Furthermore, the problem is more complicated, resulting in more advanced mathematical functions and integrals, and new physical phenomena appear, such as surface and leaky waves that are excited on curved coated surfaces (as they are on planar surfaces). This is discussed in more detail in Appendix 4A.4. As a consequence, there is no complete high-frequency solution for dielectric coated surfaces yet available. Sometimes, impedance boundary surfaces valid for thin coatings are used as an approximation to overcome these difficulties.

The major difference between a noncoated and a coated cylinder is that it is not possible to obtain a single high-frequency solution valid in both the paraxial and nonparaxial regions. For the nonparaxial region, the same approach as used for noncoated surfaces can be used. The first attempt without considering impedance boundary surfaces was presented by Munk [Munk 1996]. In this work, a heuristically UTD-based Green's function was derived with the microstrip patch antenna application in mind. The approach consists of a two-step procedure in which the leading term of the potentials A_z and F_z are first developed. The fields are then obtained by taking the second derivative of A_z and F_z and dropping higher-order terms. This solution was later implemented by Demirdag and Rojas

[Demirdag and Rojas 1997, Rojas and Demirdag 1997], who studied the radiation pattern and the surface field (nonparaxial region) due to a source on the dielectric layer. The results were only reasonable when the separation between source and field points was large.

To obtain more satisfactory results, another approach has been considered. In this approach, the exact fields are first obtained directly from the dyadic Green's function. These expressions are then evaluated asymptotically, resulting in a formulation valid in the non-paraxial region only. This approach avoids the need for taking derivatives of the asymptotic potentials. Furthermore, higher-order terms are included in the asymptotic evaluation of the integrals. Thus, better results were obtained, including small separations between source and field points (distances $\geq 0.3 \lambda$). Unfortunately, the efficiency and accuracy of the method is strongly dependent on the numerical evaluation of some integrals. The solution to the microstrip patch antenna problem is presented in [Ertürk and Rojas 2000, Ertürk et al. 2004], and the coated aperture antenna is solved in [Persson and Rojas 2003, Thors and Rojas 2003]. Improvements have later been made to maximize the efficiency of the numerical evaluation of the integrals [Persson et al. 2003b]. This asymptotic solution is combined with the MoM for an accurate analysis. Due to the complexity, only single-layer problems have been investigated so far. If multilayered structures are considered, very few possibilities exist. In [Sipus et al. 1998], an approach is discussed in which a modal solution based on the superposition of one-dimensional spectral solutions and the moment method are combined together with asymptotic evaluations of special functions. With this approach, multilayered circular cylinders and spheres with a maximum radius of at least 20λ–25λ can be analyzed.

With the asymptotic approach, only dielectric layers with moderate thickness can be treated. This is due to some of the approximations used for the Bessel/Hankel ratios in the Green's function. The limiting thickness of the dielectric layer is approximately $0.15 \lambda_d$ (λ_d is the wavelength in the dielectric layer). This is probably not a serious limitation from a practical point of view and can probably be solved if better approximations are used.

The paraxial region of the circular cylinder requires special treatment; in fact, coated cylinders are more problematic than noncoated. The reason is that the approximations used are no longer valid when the field point moves toward the axial direction of the cylinder (the paraxial region). To overcome this, a completely different method has been used. The essence of this formulation is based on the fact that the circumferentially propagating series representation of the Green's function is periodic in one of its variables, and can be approximated by a Fourier series representation. This part is still under development and so far only the microstrip antenna case has been solved completely for the circular cylinder [Ertürk and Rojas 2002]. Lately, a modified version of the paraxial region solution has been presented [Erdöl and Ertürk 2005]. It is a closed-form asymptotic representation, making the analysis more attractive.

Generalizing the model to other singly or doubly curved surfaces is still under development. So far, no numerically verified asymptotic method has been presented for more general cases. Munk extended his formulas for a circular cylinder and a sphere in the same fashion as was done for PEC surfaces to handle general convex surfaces [Munk 1996], but these expressions have not been verified numerically. This also applies to the work presented by Hussar et al. [Hussar and Smith-Rowland 2002, Campbell et al. 2002, Hussar and Smith-Rowland 2003]. The work by Hussar is similar to the work by Munk.

Thus, despite the efforts presented above, there is still no satisfactory asymptotic solution for coated (convex) surfaces. Maybe a modified or completely different approach is needed!

Alternative approaches based on impedance boundary conditions (IBC) have been studied in the past, especially in connection with the theory of radio wave propagation around the earth; see, for example, [Wait 1956a, 1962, 1974; Spies and Wait 1967]. Later researchers [Ersoy and Pathak 1988, Wait et al. 1990] considered the radiation pattern due to a magnetic line source, or a magnetic line dipole source, located on a uniform impedance surface patch that partly covered an electrically large PEC convex cylinder. More recently, a general approach has been discussed in [Tokgöz 2002, Tokgöz et al. 2005]. However, using formulations based on impedance boundary conditions (IBC) also involve limitations such as the thickness of the layer.

4.5 TWO EXAMPLES

4.5.1 Introduction

In this section, we will discuss two examples in order to provide some more information about how to proceed when analyzing conformal antennas. The approach is based on the integral equation formulation, solved by the method of moments (MoM). As will become clear, the general analysis is not any different from the planar case. However, it is much more difficult to find the field representation outside a curved surface. In some cases, a high-frequency solution must be used to find the field, that is, a hybrid UTD-MoM approach is used. We will discuss this briefly in Section 4.5.2.

The two cases are the waveguide-fed aperture antenna array and the microstrip-patch antenna array. These two canonical examples will also be discussed extensively in the following chapters, where several numerical examples illustrate the characteristics of conformal antennas.

4.5.2 The Aperture Antenna

An aperture antenna can, generally speaking, be any of the following types: slits, slots, waveguide openings, horns, reflectors, and lenses. However, in this context we consider an aperture antenna in the form of a waveguide aperture flush mounted on or integrated into a curved surface (see Figure 4.13). The cross section can be of any kind, but from a practical point of view rectangular and circular cross sections are common.

Aperture antennas can be flush mounted on a surface and are therefore very suitable for conformal antenna applications. Furthermore, the opening can be covered with a dielectric material for protection without disturbing the overall shape of the surface. This is very important when considering, for example, aircraft, in which the aerodynamic profile is important.

Figure 4.13 shows the waveguide-fed aperture problem (for aperture m out of M apertures) with the computational domain divided into two regions: the exterior and interior regions. Here, a noncoated PEC surface is assumed, but the formulation is equivalent if a coating is present. The main idea of the solution presented here is to start with an exact formulation in terms of an integral equation for the boundary between the exterior and the interior regions. By using the field equivalence theorem [Harrington 1961, page 106] an equivalent (exterior) problem is obtained by covering the apertures with a perfectly conducting surface and introducing unknown equivalent magnetic current moments on the surface. These infinitesimal magnetic current moments then radiate in the presence of a perfectly conducting surface.

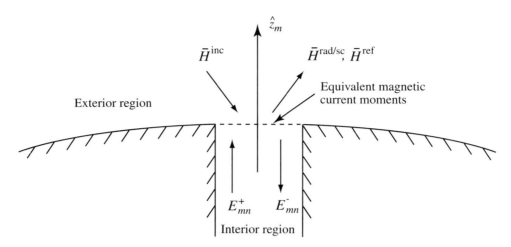

Figure. 4.13. The exterior and interior regions at aperture m for a noncoated PEC surface with the nth waveguide mode.

The IE is obtained by applying the boundary condition for the total tangential magnetic field, that is, by equating the tangential magnetic fields in the exterior and interior regions at the boundary. We get

$$\left(\underbrace{\overline{H}_{\text{tan}}^{\text{inc}} + \overline{H}_{\text{tan}}^{\text{ref}} + \overline{H}_{\text{tan}}^{\text{rad/sc}}}_{\overline{H}_{\text{tan}}^{\text{ext}}}\right)\Bigg|_{\overline{r}\in S_{\text{ap}}^{+}} = \overline{H}_{\text{tan}}^{\text{int}}\big|_{\overline{r}\in S_{\text{ap}}^{-}} \tag{4.20}$$

where S_{ap} is the set of all apertures, $\overline{H}_{\text{tan}}^{\text{inc}}$ is the incident plane wave in the presence of the surface without apertures, and $\overline{H}_{\text{tan}}^{\text{ref}}$ is the reflected plane wave in the absence of the apertures (both $\neq 0$ only if the scattering problem is considered). Furthermore, $\overline{H}_{\text{tan}}^{\text{ref}}$ equals zero if scattering from an infinitely long cylinder is considered. $\overline{H}_{\text{tan}}^{\text{rad/sc}}$ is the radiated/scattered field caused by the equivalent magnetic current moments. Finally, $\overline{H}_{\text{tan}}^{\text{int}}$ is the magnetic field in the interior region given by

$$\overline{H}_{\text{tan}}^{\text{int}}\big|_{\overline{r}\in S_{\text{ap}}^{-}} \approx \sum_{m=1}^{M}\sum_{n=1}^{N}[Y_{g}^{n}E_{mn}^{+}(\hat{z}_{m}\times\overline{e}_{m}^{n}) - Y_{g}^{n}E_{mn}^{-}(\hat{z}_{m}\times\overline{e}_{m}^{n})] \tag{4.21}$$

Here, we use the waveguide modes in the waveguide as basis functions with unknown amplitudes E_{mn}^{+} and E_{mn}^{-}. Y_{g}^{n} is the modal admittance and $\overline{e}_{m}^{n}$ is the modal function of the nth waveguide mode defined in waveguide m. The plus sign indicates a wave propagating toward the aperture and the minus sign a wave propagating away from the aperture. For the scattering problem, $E_{mn}^{+} = 0\ \forall\ m, n$ if the waveguides are matched. It should be noted that (throughout this book) the transmitting apertures are fed by the dominant waveguide mode only. However, at the different waveguide openings (see Figure 4.13) a reflection occurs, causing infinitely many reflected evanescent modes to be generated. Also, at the receiving apertures infinitely many modes are generated from the external field. A single/dominant mode approximation is often used, but in some examples a finite number of higher-order modes will be included for a more accurate analysis. The approximate equal sign in Equation (4.21) indicates this.

An important observation is that the IE is referred to the curved aperture plane. However, the basis functions ($\overline{e}_m^n$) (waveguide modes) are valid for a planar surface in a cross section of the waveguide. See Figure 4.14, where the planar surface in the waveguide is shaded. Despite this fact, the fields at the convex surface are often assumed to be the same as in the waveguide and the gap seen in Figure 4.14 is disregarded.

In many applications, this approximation may be tolerable, especially if the radius of curvature is large in the "wide" plane of the aperture. But if the radius of curvature becomes smaller, as illustrated in Figure 4.14, this approximation may lead to larger errors. One way of reducing the error is to add a phase to the aperture field at the planar surface to include the small distance to the curved surface. This is done in the simulations presented in the following chapters.

To proceed with the solution to the problem, the discretized integral equation (4.20) is transferred to a matrix equation by using the MoM procedure. Thus, using Galerkin's method the equation is multiplied with test or weighting functions $\overline{w}_p^q \, (= \hat{z}_p \times \overline{e}_p^q$, i.e., same as the basis functions) and integrated by using the inner product defined as $\langle \overline{f}, \overline{g} \rangle = \iint (\overline{f} \cdot \overline{g}) dS$. We get

$$
\langle \overline{H}_{\text{tan}}^{\text{inc}} + \overline{H}_{\text{tan}}^{\text{ref}}, \hat{z}_p \times \overline{e}_p^q \rangle |_{\overline{r} \in S_{\text{ap}}^+}
$$

$$
+ \sum_{m=1}^{M} \sum_{n=1}^{N} \left[E_{mn}^+ \underbrace{\langle \overline{H}_{\text{tan}}^{\text{rad/sc}}(\overline{e}_m^n), \hat{z}_p \times \overline{e}_p^q \rangle}_{Y_{pm}^{qn}} + E_{mn}^- \underbrace{\langle \overline{H}_{\text{tan}}^{\text{rad/sc}}(\overline{e}_m^n), \hat{z}_p \times \overline{e}_p^q \rangle}_{Y_{pm}^{qn}} \right] \Bigg|_{\overline{r} \in S_{\text{ap}}^+} = \qquad (4.22)
$$

$$
\sum_{m=1}^{M} \sum_{n=1}^{N} \left[Y_g^n E_{mn}^+ \underbrace{\langle \hat{z}_m \times \overline{e}_m^n, \hat{z}_p \times \overline{e}_p^q \rangle}_{\delta_{pm}^{qn}} - Y_g^n E_{mn}^- \underbrace{\langle \hat{z}_m \times \overline{e}_m^n, \hat{z}_p \times \overline{e}_p^q \rangle}_{\delta_{pm}^{qn}} \right] \forall pq
$$

where Y_{pm}^{qn} is the mutual admittance between the modes $n \to q$ in the apertures $m \to p$ [see Equation (4.2)]. The unknown modal amplitudes (E_{mn}^+, E_{mn}^-) can now be found if $\overline{H}_{\text{tan}}^{\text{rad/sc}}(\overline{e}_m^n)$ can be found. To obtain the radiated/scattered field, a suitable method has to be chosen from the methods presented earlier in this chapter. For a canonical surface like a circular cylinder, a modal representation can be used. If this is not possible, we can use an asymptotic technique such as UTD to calculate if $\overline{H}_{\text{tan}}^{\text{rad/sc}}(\overline{e}_m^n)$. If so, this method is described as a hybrid UTD–MoM method.

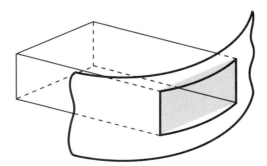

Figure 4.14. The geometry at the aperture.

When the modal amplitudes are found, the far-field radiation pattern can be obtained. Furthermore, other interesting parameters such as the scattering matrix are also found easily according Equation (4.3) since the mutual admittances are directly identified in Equation (4.22).

4.5.3 The Microstrip-Patch Antenna

When conformal and low-profile antennas are required, the microstrip-patch antenna is a very attractive solution. This type of antenna has the advantages of low cost and weight, design flexibility, and ease of installation. An interesting observation is that it seems that the conformal application was one of the main reasons for developing microstrip antennas. Gupta and Benalla [1988] state in the preface to their book: "The fact that printed 'microstrip' structures radiate has been known since the mid-1950s, but the application of this phenomenon to design useful antennas started only in the early 1970s when conformal antennas were required for missiles."

The common microstrip-patch antenna has a narrow bandwidth of only a few percent. The bandwidth can, however, be increased considerably by increasing the height of the substrate, by adding a parasitic element ("stacked patches"), and by introducing a balanced probe feed. A complication is that microstrip antennas require a full wave analysis

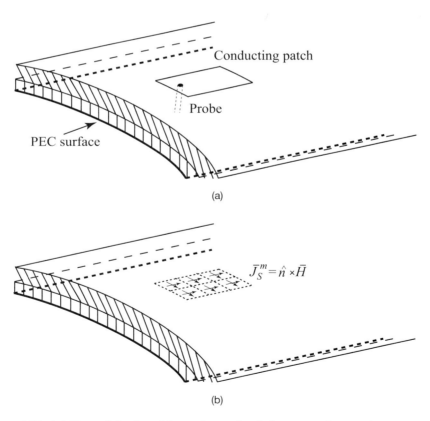

Figure 4.15. (*a*) The original problem of a probe-fed rectangular-patch antenna on a coated multilayer surface. (*b*) The equivalent problem.

for accurate results. This is not a major problem in the planar case, but becomes more difficult for curved surfaces, as will be discussed below.

As for aperture antennas, a microstrip patch can take different shapes. We will here only consider rectangular patches, but the principles are the same for other shapes.

The typical case is shown in Figure 4.15 (a), where a probe-fed microstrip-patch antenna m is seen on a multilayer substrate. Solving this problem is similar to the approach presented for the aperture antenna. Thus, using the field equivalence theorem [Harrington 1961, page 106] an equivalent (exterior) problem is obtained by replacing the infinitely thin microstrip patch with unknown equivalent electric surface current densities $\overline{J}_S$ on the surface. The equivalent problem is seen in Figure 4.15 (b).

For a microstrip-patch antenna array, the IE is given by the boundary condition at the curved metallic microstrip patches, that is, the total tangential electric field must vanish. We get

$$\left(\underbrace{\overline{E}_{\tan}^{\text{inc}} + \overline{E}_{\tan}^{\text{ref}} + \overline{E}_{\tan}^{\text{rad/sc}}}_{\overline{E}_{\tan}^{\text{ext}}}\right)\Bigg|_{\overline{r}\in S_{\text{ap}}^+} = -\overline{E}_{\tan}^{\text{int}}|_{\overline{r}\in S_{\text{ap}}^-} \qquad (4.23)$$

where S_{ap} is the set of all microstrip patches, $\overline{E}_{\tan}^{\text{inc}}$ is the incident plane wave in the presence of the surface without patches, and $\overline{E}_{\tan}^{\text{ref}}$ is the reflected plane wave in the absence of the patch antennas (both are $\neq 0$ only if the scattering problem is considered). Furthermore, $\overline{E}_{\tan}^{\text{ref}}$ equals zero if scattering from an infinitely long cylinder is considered. $\overline{E}_{\tan}^{\text{rad/sc}}$ is the radiated/scattered field caused by the unknown equivalent surface current distribution $\overline{J}_S^m$. The field is given by

$$\overline{E}_{\tan}^{\text{rad/sc}}|_{\overline{r}\in S_{\text{ap}}^+} = \sum_{m=1}^{M} \int_{S_m} (\overline{\overline{G}}_{\tan} \cdot \overline{J}_S^m)dS \qquad (4.24)$$

where $\overline{\overline{G}}_{\tan}$ represents the tangential components of the dyadic Green's function. Finally, $\overline{E}_{\tan}^{\text{int}}$ is the electric field generated by the known probe current density $\overline{J}_{\text{int}}^m$, according to

$$\overline{E}_{\tan}^{\text{int}}|_{\overline{r}\in S_{\text{ap}}^-} = \sum_{m=1}^{M} \int_{S_{\text{source}_m}} (\overline{\overline{G}}_{\tan} \cdot \overline{J}_{\text{int}}^m)dS \qquad (4.25)$$

Note that the probe-fed model is an idealized model, but reasonably valid for thin substrates [Pozar 1987]. $\overline{J}_{\text{int}}^m$ is often approximated by a unit current source on the probe, thus not including the feeding network in the model.

By using the MoM procedure, the unknown surface current densities are expanded in a set of N basis functions ($\overline{J}_S^m = \Sigma_{n=1}^N a_{mn}\overline{J}_m^n$), where $\overline{J}_m^n$ is the nth basis function at patch m and a_{mn} is the unknown amplitude. The basis functions $\overline{J}_m^n$ may represent currents in either of the two orthogonal tangential directions on the patch surface. In general, there are two types of basis functions: entire basis, and subsectional basis (see, for example [Pozar 1982]). Both types have been used for studies of conformal microstrip-patch antennas; see, for example, [Ertürk 2000] for subsectional basis functions and [Rafaelli et al. 2005] for entire domain basis functions.

The discretized integral equation (4.23) is finally transferred to a matrix equation by first multiplying the equation with test or weighting functions $\overline{J}_p^q$ (same as the basis functions, i.e. Galerkin's method) and integrating by using an inner product (defined as in the previous section). We get

$$\sum_{m=1}^{M} \sum_{n=1}^{N} a_{mn} \underbrace{\left\langle \int_{S_m} (\bar{\bar{G}}_{\tan} \cdot \bar{J}_m^n) dS, \bar{J}_p^q \right\rangle}_{Z_{pm}^{qn}} = \sum_{m=1}^{M} \underbrace{\left\langle -\int_{S_{\text{source}_m}} (\bar{\bar{G}}_{\tan} \cdot \bar{J}_m^{\text{int}}) dS, \bar{J}_p^q \right\rangle}_{V_p^q} \qquad (4.26)$$

$$+ \langle -\bar{E}_{\tan}^{\text{inc}} - \bar{E}_{\tan}^{\text{ref}}, J_p^q \rangle|_{\bar{r} \in S_{\text{ap}}^+} \;\forall pq$$

This formulation is exact if the Green's function ($\bar{\bar{G}}_{\tan}$) can be found. Since we now have to consider a coated surface, this is quite difficult. The possible solutions are mainly limited to circular and spherical surfaces as discussed earlier.

Finally, the characteristics of the microstrip-patch antenna arrays, such as input impedance, mutual coupling between patches, and radiation patterns, can be obtained in a similar fashion as we described before for the aperture case, once the unknown patch currents are found from Equation (4.26).

4.6   A COMPARISON OF ANALYSIS METHODS

Of all the methods discussed, which one is the best and should, therefore, be used? The answer to this question is somewhat subjective since the answer is (probably strongly) related to the experience of the analyst as well as the computer code(s) that he/she has access to. Nevertheless, in this section we will try to give some guidelines for the analysis of conformal antennas. We will initially disregard individual experience and code availability. Instead, the discussion will concentrate on the methods themselves and their ability to solve certain problems. At the end, experience will have to be included as a factor as well.

The discussions in the previous sections have led to one major conclusion: For the analysis of conformal antennas there are two alternatives to choose between. On one hand we have methods based on asymptotic techniques (UTD/MoM and FE-BI-UTD*), and on the other we have purely numerical methods (FE-BI* and FDTD). Keeping this in mind, we turn to the problem at hand. A number of typical examples are listed in Table 4.1 with comments regarding what to consider before choosing a method. The geometries that are listed represent typical canonical problems. However, in the real world, modifications of a canonical problem may be needed, or maybe a combination of two canonical problems will give the solution. Note that in all examples edges are not included. Adding edges complicates the solution, but can be handled both with asymptotic techniques and with numerical methods.

Finally, the experience and the availability of computer codes are of vital importance when choosing a method. As an example, if an asymptotic technique solves the problem, the experience in MoM and FEM will decide whether an UTD/MoM or FE-BI-UTD approach is chosen. This is, of course, a very subjective area. Table 4.2 is included to give some additional guidelines that may help in the decision.

APPENDIX 4A—INTERPRETATION OF THE RAY THEORY

In the presentation of the high-frequency methods in Section 4.4, the field propagating along a curved surface is interpreted as rays propagating along geodesics. This appendix is included to give the interested reader a deeper understanding of the principles behind

---

*The term FE-BI-UTD is introduced here to indicate that an asymptotic Green's function is used. The term FE-BI indicates a purely numerical version in which the free space Green's function is used.

Table 4.1. Guidelines for which method to use when taking the properties of the methods into account

Geometry	Comments
Singly curved convex PEC surfaces:	
Circular cross section	If the surface is electrically small, almost any kind of method can be used. The main difference between the methods presented in this chapter is the time required. UTD has a limitation, but radii as small as $1\ \lambda$ have been analyzed with satisfying results (see Section 6.3.2).
	If the surface is electrically large, an asymptotic-based method is preferable. Another possibility is FDTD in cylindrical coordinates, especially if there is symmetry in the antenna array.
Analytically specified cross section	An extension to general cross sections can almost only be analyzed with an asymptotic method. For electrically small surfaces, numerical codes such as FEM or MoM may be useful. However, when increasing the frequency, an asymptotic Green's function is needed. Thus, a hybrid UTD-MoM or FE-BI-UTD method can/must be used.
Numerically specified cross section	The key in using an asymptotic solution is to find the geodesics and the related ray-based parameters. This may be difficult and time-consuming for this type of surface. Thus, a purely numerical method such as FE-BI is probably more suitable. And, as always, if the surface is electrically large, numerical problems will appear.
Cone	An asymptotic solution is possible, which is probably the preferred method. For some special cases, a modal solution or a numerical solution may be used.
Singly curved coated convex PEC surfaces:	
Circular cross section	For electrically small surfaces, the modal solution is possible, as well as MoM, FEM, or FDTD. Unfortunately, the problem becomes more difficult if the surface is electrically large. If the surface is too large for a numerical approach, the key is to see if the Green's function can be computed easily. There is an asymptotic version that can be used quite efficiently for a single layer. If a multilayer coating is present, there are a very limited number of possible solutions; one example is the G1DMULT software [Sipus et al. 1998].
Analytically specified cross section	Difficult. Numerical methods are (probably) the only choice.
Numerically specified cross section	Very difficult problem. A purely numerical method is the only choice, but may require a large memory and long computing time.
Cone	Difficult. Numerical methods are probably the only choice.
Doubly curved convex PEC surfaces:	
Sphere	Canonical problem, solvable with almost any method if electrically small. Otherwise, an asymptotic-based method is preferable. If there is symmetry, some simplifications can be used in the solution.
Surface of revolution	There are two major choices: an asymptotic-based technique or a numerical approach. The key when using an asymptotic method is to find the geodesics. However, they can be found quite easily for surfaces of revolution. *(continued)*

Table 4.1. *Continued*

Geometry	Comments
Doubly curved convex PEC surfaces: (cont.) Numerically specified	The only possibility is a numerical method, preferably FE-BI or FEM/FDTD.
Doubly curved coated convex PEC surfaces: Any geometry	Very difficult. The only possibility is a numerical method, preferably FE-BI or FEM/FDTD.

the ray theory and its connection to the modal solution and to different types of waves. At the end, the generalization of the canonical solutions for the cylinder and sphere are discussed in more detail.

A full derivation of an asymptotic method is, of course, beyond the scope of this book. The details of the mathematical operations presented here can be found in different textbooks such as [Collin and Zucker 1969, Felsen and Marcuvitz 1973, Ishimaru 1991, Pathak 2001, Wait 1959]. Furthermore, a general summary of the surface field problem can be found in [Persson 2001] and the radiation problem is discussed in [Thors 2003].

The idea with asymptotic techniques is to approximate functions and integrals in such a way that they become increasingly accurate as some parameter approaches a limiting value. An asymptotic solution to an electromagnetic problem becomes increasingly accurate as the free space wave number k (or angular frequency ω) approaches infinity. An asymptotic representation is typically semiconvergent, that is, it tends to converge with the first few terms but then it diverges with the addition of further terms.

Table 4.2. Guidelines for which method to use when taking experience and available code(s) into account

Category	Comments
No experience, no code(s) available	Assuming that a UTD formulation exists for the problem at hand, the UTD/MoM method is probably the best choice for a beginner. If no UTD formulation is available, the best choice is an open question and is determined by the knowledge of the different methods (see Table 4.1) and EM and programming skills.
No experience, code(s) available	Use, and if necessary extend the existing code(s) with respect to the problem at hand. If not realistic, see previous comment.
Experience on a medium level, no code(s) available	The choice will probably depend on EM and programming skills in combination with Table 4.1.
Experience on a medium level, code(s) available	Use, and if necessary extend the existing code(s) with respect to the problem at hand. If not possible, see previous comment.
Experience on a high level, no code(s) available	Almost anything will work! Use the knowledge about the different methods (see Table 4.1) and decide which method to use.
Experience on a high level, code(s) available	Modify the code(s) to fit the problem at hand. If not realistic, see previous comment.

The modal field solution for a noncoated PEC circular cylinder (presented in Section 4.3.2) is given by a combined azimuthal Fourier series and an axial Fourier integral representation. This solution represents a formulation in terms of outward radially propagating waves. A similar expression can be found for a coated PEC circular cylinder as well. Under certain restrictions, such a representation can be evaluated asymptotically, resulting in a more useful formulation with faster convergence. To discuss the relations between the modal solution and the ray theory, a general expression of a modal solution will be used, written in the following form [S is any of the field components; see also Equation (4.5)]

$$S = \int_{-\infty}^{\infty} \left(\sum_{n=-\infty}^{\infty} T(n, k_z) e^{-jn\varphi} \right) e^{-jk_z z} dk_z \tag{A.1}$$

This expression, as well as the general theory below, is valid both for a coated and noncoated circular cylinder. If there are any differences, they will be pointed out clearly.

4A.1 Watson Transformation

By applying the Watson transformation [Watson 1918] Equation (A.1) can also be written in terms of circumferentially propagating waves:

$$S = \int_{-\infty}^{\infty} \left(\frac{1}{2j} \left[\int_{-\infty-j\varepsilon}^{\infty-j\varepsilon} \frac{T(v, k_z) e^{-jv(\varphi-\pi)}}{\sin v\pi} \, dv + \int_{-\infty+j\varepsilon}^{\infty+j\varepsilon} \frac{T(v, k_z) e^{-jv(\varphi-\pi)}}{\sin v\pi} \, dv \right] \right) e^{-jk_z z} dk_z \tag{A.2}$$

where ε is a small distance. Depending on how the integral in the v plane is evaluated, a solution valid in different regions (see Figure 4.11) will be obtained. If the field point is in the illuminated (lit) region, the integrand can be evaluated by using the method of stationary phase. In that case, the solution is often referred to as the geometrical optics (GO) solution. If a uniform solution valid in the boundary region (and the shadow and illuminated regions as well) is to be obtained, the integral must be evaluated along the original contour. This is an approach originally used by V. A. Fock [Fock 1945, 1965]. Using Fock's approach, S can be expressed in terms of the so-called Fock functions, known from the UTD solution for a noncoated surface. With the aid of some mathematical operations Equation (A.2) is rewritten as

$$S = \int_{-\infty}^{\infty} \left(\int_{-\infty-j\varepsilon}^{\infty-j\varepsilon} \left[\sum_{m=0}^{\infty} [T(v, k_z) e^{-jv(\varphi+2\pi m)} + T(-v, k_z) e^{-jv((2\pi-\varphi)+2\pi m)}] \right] dv \right) e^{-jk_z z} dk_z \tag{A.3}$$

Here, the ray interpretation of the solution is evident. In Equation (A.3), each exponential term, for a given m, can be identified as a wave that has crept around the cylinder m times along geodesics in the $+\hat{\varphi}$ and $-\hat{\varphi}$ direction, respectively. However, this expression is often simplified since in most problems the term $m = 0$ is assumed to be enough when kR is large (R is the radius of the cylinder and k is the wave number). Furthermore, only waves in one direction, $+\hat{\varphi}$ in general, are often considered.

4A.2 Fock Substitution

Following the approach made by Fock, a substitution is introduced which transforms the v plane to a τ plane by $v = k_r r_s + (k_r r_s/2)^{1/3} \tau$, called the Fock substitution.* An interesting

---

*Originally, this substitution was used by Fock to be able to approximate the Hankel functions in the noncoated solution by the so-called third-order approximation, that is, they are approximated as proportional to the Airy functions. As a consequence, the integral in the τ plane is called the Fock function. Note that there are different Fock functions for different problems (radiation, surface field, etc.) [McNamara et al. 1990].

effect of the Fock substitution is that the surface field can be interpreted as propagating at a fixed radius r_s. Thus, for each of these various substitutions, the ray paths lie on a cylindrical surface whose radius of curvature is the chosen "suitable" radius r_s.

For a noncoated PEC circular cylinder, the suitable choice is the radius of the PEC circular cylinder; see Figure 4A.1 (*a*). However, for coated surfaces the number of choices is infinite in general and any choice between the radius R of the PEC surface and the outer radius R_1 is possible; see Figure 4A.1 (*b*). Depending on the choice, a different ray interpretation follows. In practice, the choice is often determined by the problem. If microstrip-patch antennas are considered, the most reasonable choice for r_s is the outer radius R_1. On the other hand, if the source is located on the PEC surface, as in the case of dielectric-coated-aperture antennas, the preferred choice is the inner radius R.

4A.3 SDP Integration

The above steps have made it easier to analyze electrically large cylinders. Additionally, the computation of the field can be performed more efficiently if the original contour in

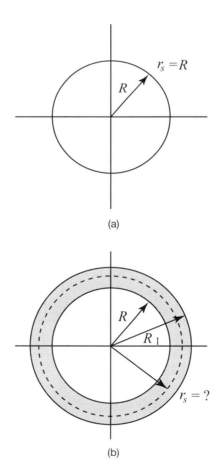

Figure 4A.1. The interpretation of r_s for (*a*) the noncated PEC circular cylinder, (*b*) the coated PEC circular cylinder.

the k_z plane is deformed into its steepest descent path (SDP) on which the integrand decays most rapidly. To obtain the UTD solution, valid for a noncoated cylinder, only the saddle point contribution is included and the k_z integral can be solved in closed form:

$$\int_{-\infty}^{\infty} f(\tau, k_z)dk_z \rightarrow \int_{C_{SDP}} f(\psi)e^{ksq(\psi)}d\psi \sim \sqrt{\frac{-2\pi}{ksq''(\psi_S)}}f(\psi_S)e^{ksq(\psi)}, \ k \rightarrow \infty \quad (A.4)$$

where $f(\tau, k_z)$ is given in Equation (A.3) after using the Fock substitution and s is the geodesic length, that is, the length of the ray path along the surface connecting the source and field points.

For the coated cylinder (nonparaxial region only!) the saddle point contribution shown in Equation (A.4) is, in general, not enough. The result in Equation (A.4) must, therefore, be modified with higher-order terms by including the contributions from points away from the saddle point. For mutual coupling calculations, when the surface field is considered, a three-point approach is often satisfactory. However, the number of points required may depend on the geometrical parameters [Ertürk and Rojas 2000, Persson and Rojas 2003]. The radiation problem still seems to give accurate results if the saddle point only is used [Thors and Rojas 2003].

4A.4 Surface Waves

In the previous sections, the wave propagation along the surface was interpreted as rays propagating along geodesics. Another way of interpreting the wave propagation is to use the concept of surface waves related to the poles of the Green's function. In this section, we will give a short overview of the surface wave interpretation for curved surfaces. Note that it is not possible to mix the interpretations, that is, by saying that the rays consist of different types of waves or vice versa. The two interpretations are solutions to the same problem, but interpreted differently.

To be able to interpret the wave propagation in terms of waves, we will go back to the circular cylinder solution and, in particular, the v plane integration in Equation (A.2). (This equation is valid both for coated and noncoated surfaces if the appropriate Green's function is used.) As discussed in Appendix 4A.1, there are two ways of solving the v plane integrals, resulting in solutions valid in different regions. There is also a third option in which a contour deformation is made in the v plane by closing the line integrals with semicircles in the upper and lower half planes. The integrals are then evaluated with residue series expansions, that is, the poles of $T(v, k_z)$ (containing the Green's function) must be found. This will give a solution valid mainly in the (deep) shadow region. The number of terms/poles needed is difficult to know a priori since it depends on the geometry, thickness of the dielectric layer, and so on. Furthermore, if the field point is moving toward the boundary region between the lit and shadow regions, more terms must be included in the residue expansion. Therefore, this is not a tractable solution, in general.

However, the interesting part of this solution is that the poles can be identified as waves propagating in opposite azimuthal directions along the surface. From now on, the coated cylinder will be considered first and the noncoated cylinder will be seen as a special case. Investigations of the Green's function pole locations for a coated cylinder have been performed before. The two-dimensional problem has been investigated by, for example, Paknys and Wang [Paknys and Wang 1986, 1987; Wang 1982, 1985]. The full three-dimensional problem has been treated by [Pearson 1987, Felsen and Naishadham 1991, Naishadham and Felsen 1993]; see also [Richard et al. 1997, 1999].

The classification of the different waves is related to the locations of the poles. Thus, it is necessary to solve for the complex pole locations $v_p(k_z)$, $p = 1, 2, \ldots$ (the azimuthal propagation constants). As an example, Figure 4A.2 shows the positions of some poles for a coated cylinder when k_z is real.

To study the characteristics of the poles, the complex roots are broken down into their real and imaginary parts: $v_p = v_p' + j v_p''$. The following criteria are then used for a physical classification of the different waves that are supported by the geometry [Felsen and Naishadham 1991, Naishadham and Felsen 1993]:

If $v_p' < k_r R_1$, then they correspond to (essentially) *leaky waves*
If $v_p' \approx k_r R_1$, then they correspond to (essentially) *creeping waves*
If $v_p' > k_r R_1$, then they correspond to (essentially) *trapped waves*

Here, k_r is the transverse wave number defined in Section 4.3.2 and R_1 is the outer radius of the cylinder; see Figure 4A.1 (*b*). As indicated, it is difficult to give a 100% clear distinction between waves of different types and their properties for curved surfaces. This is in contrast to the corresponding planar case in which a complete description of possible wave types and their properties can be defined; see, for example [Collin and Zucker 1969].

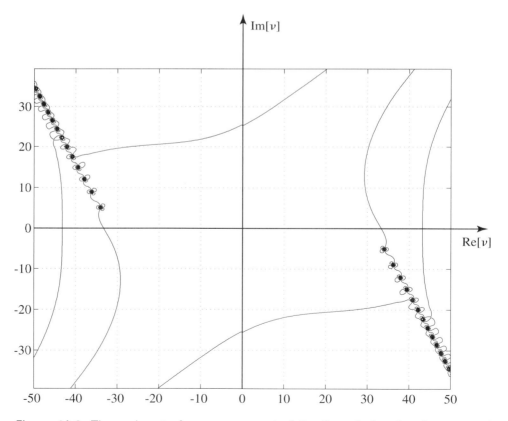

Figure 4A.2. The real part of one component of the Green's function for a coated PEC cylinder with some pole locations shown in the complex *v* plane. Here, k_z is real.

The term "creeping wave" may be familiar since it is often used for a wave on conducting surfaces where an extra attenuation due to the curvature is present. A creeping wave is characterized by a rapidly increased attenuation when the radius is decreased. It is a fast wave traveling almost at the speed of light, and this wave is the only wave possible for a noncoated surface. By a complete study of the properties of the poles, it can be shown that a thin coating will mainly support creeping waves. Sometimes, a creeping wave is also referred to as a Watson-type creeping wave [Paknys and Wang 1986]. This type of wave is unique for the curved surface and cannot be found in the planar case.

If the layer becomes thicker, a transition region takes place at a thickness of about $d/\lambda_0 \approx 0.23$. At this thickness, the waves begin to become trapped. Trapped waves are slow waves whose phase velocity is less than the speed of light and they possess low attenuation. These waves are often referred to as Elliott-type creeping waves and can only exist at coated surfaces [Elliott 1955].

If the radius is decreased (keeping the frequency constant) and the layer is thicker than the transition thickness ($d/\lambda_0 \approx 0.23$), the initially (well) trapped wave becomes transitional and then leaky. A leaky wave is a wave with high velocity and the attenuation increases with decreasing outer radius. The leaky waves are analogous to those in the planar case, but with additional leakage due to curvature.

Thus, the resemblance between the waves in the planar and cylindrical cases becomes clear. When $R_1 \rightarrow \infty$ in the cylindrical case, the effect of the curvature will decrease, that is, the different pole locations in the cylindrical case will recover the pole locations in the planar case. As a consequence, the pole locations in the planar case can be used as a starting point to find the poles in the cylindrical case, which are more difficult to find.

4A.5 Generalization

A general formulation for arbitrarily shaped convex surfaces cannot be based only on the circular cylinder solution described above. One reason for this is that the principal radii of curvature are different and finite for a general surface. Furthermore, the torsion is generally not a constant as it is for the cylinder. These arguments lead to the solution of another canonical problem, namely, the sphere. The sphere has a constant radius of curvature since both principal radii of curvature are equal and finite, that is, the rays possess no torsion. Hence, the circular cylinder and the sphere do not possess all of the general properties associated with an arbitrary convex surface. Nevertheless, they provide useful information and an asymptotic solution for an arbitrarily convex and smooth surface can be developed on the basis of these two solutions with some additional assumptions (some of them are completely heuristic in nature). These two canonical solutions are then combined with the aid of the local properties of high-frequency wave propagation. This is possible because diffraction, as well as reflection and transmission, is a local phenomenon at wavelengths that are small in comparison to the size of the radiating object.

To perform this generalization both the canonical circular cylinder and sphere solutions are expressed in terms of the surface ray coordinates: the binormal ($\hat{b}$), tangential ($\hat{t}$), and normal ($\hat{n}$) unit vectors at the surface, defined in Figure 4.12 (b). Thereby, a straightforward comparison between the two solutions can be obtained. It turns out that the major differences between the circular cylinder and sphere solutions are the torsion associated with the surface rays and the different Fock functions, which appear in the two solutions. This means that the sphere solution can be obtained from the cylindrical solution if the torsion is set to zero and the cylinder Fock function is replaced with the sphere Fock func-

Figure 4A.3. The development of UTD for a general smooth convex surface? [Gordon 1985]

tion [Pathak and Wang 1978]. Therefore, the idea is to use the circular cylinder solution as the basic building block for the general solution.

To finally accomplish the generalization, different circular cylinders approximate the neighborhood of each point along the geodesic. Each radius is set equal to the local radius of curvature at each point. This means that the general solution can be viewed as a summation of "circular cylinder solutions," with some additional features included due to the geometrical properties of the surface, c.f. Figure 7.4, page 272. Thus, by using a combination of differential geometry, reciprocity, local properties of wave propagation at high frequencies, and heuristically defined functions, an asymptotic solution valid for a general convex and smooth surface is obtained, see Figure 4A.3.

The procedure described above has been used for a perfectly electrical conducting (PEC), smooth, convex surface. It is based on the fact that the canonical asymptotic solutions can be expressed in surface ray coordinates. This requires that only the saddle point contribution be taken into account in the SDP integration [Equation (A.4)]. However, if more points are needed along the SDP path, it seems impossible to rewrite the asymptotic solution in a ray parameter form. Thus, a generalization will be much more difficult in that case. This is the typical situation for a dielectric coated surface since, in general, more points beside the saddle point are needed in the SDP integration, as discussed in Appendix 4A.3.

REFERENCES

Abou-Jaoude R. and Walton E. K. (1998), "Numerical Modeling of On-Glass Conformal Automobile Antennas," *IEEE Transactions on Antennas and Propagation,* Vol. 46, No. 6, pp. 845–852, June.

Adams A. T. (1974), "An Introduction to the Method of Moments," Syracuse University, Report RADCTR-73–217, Vol. 1, August.

Amitay N., Galindo V., and Wu C. P. (1972), *Theory and Analysis of Phased Array Antennas,* New York, Wiley-Interscience.

Bailin L. L. and Silver S. (1956), "Exterior Electromagnetic Boundary Value Problems for Spheres and Cones," *Transactions of IRE,* Vol. AP-4, pp. 5–16, January.

Bailin L. L. and Silver S. (1957), corrections to "Exterior Electromagnetic Boundary Value Problems for Spheres and Cones," *Transactions of IRE,* Vol. AP-5, p. 313, July

Balanis C. A. (1997), *Antenna Theory, Analysis and Design,* 2nd ed., New York, Wiley.

Balzano Q. (1974), "Analysis of Periodic Arrays of Waveguide Apertures on Conducting Cylinders Covered by a Dielectric," *IEEE Transactions on Antennas and Propagation,* Vol. AP-22, No. 1, pp. 25–34, January.

Belinka M. G. and Wainstein L. A. (1957), "Radiation Characteristics of Spherical Surface Antennas," in *Diffraction of Electromagnetic Waves on Certain Bodies of Revolution,* Moscow, Soviet Radio Press.

Bennett C. L. and Mieras H. (1981), "Time-Domain Scattering from Open Thin Conducting Surfaces," *Radio Science,* Vol. 16, pp. 1231–1239, Nov.–Dec.

Bertuch T., Vecchi G., Matekovits L., and Orefice M. (1999), "Full-wave Analysis of Conformal Antennas of Arbitrary Shapes Printed on Circular Cylinders," in *Proceedings of 1st European Workshop on Conformal Antennas,* Karlsruhe, Germany, 29 October, pp. 8–11.

Bird T. S. (1984), "Comparison of Asymptotic Solutions for the Surface Field Excited by a Magnetic Dipole on a Cylinde*r,"* *IEEE Transactions on Antennas and Propagation,* Vol. AP-32, No. 11, pp. 1237–1244, June.

Bird T. S. (1985), "Accurate Asymptotic Solution for the Surface Field Due to Apertures in a Conducting Cylinder," *IEEE Transactions on Antennas and Propagation,* Vol. AP-33, No. 10, pp. 1108–1117, October.

Bird T. S. (1988), "Admittance of Rectangular Waveguide Radiating from a Conducting Cylinder," *IEEE Transactions on Antennas and Propagation,* Vol. 36, No. 9, pp. 1217–1220, September.

Boersma J. and Lee S. W. (1978), "Surface Field Due to a Magnetic Dipole on a Cylinder: Asymptotic Expansion of Exact Solution," Electromagnetics Laboratory, Technical Report No. 78-17, University of Illinois, December.

Borgiotti G. V. (1968), "Modal Analysis of Periodic Planar Phased Arrays of Apertures," *Proceedings of IEEE,* Vol. 56, No. 11, pp. 1881–1892, November.

Borgiotti G. V. and Balzano Q. (1970), "Mutual Coupling Analysis of a Conformal Array of Elements on a Cylindrical Surface," *IEEE Transactions on Antennas and Propagation,* Vol. AP-18, No. 1, pp. 55–63, January.

Borgiotti G. V. and Balzano Q. (1972), "Analysis and Element Pattern Design of Periodic Arrays of Circular Apertures on Conducting Cylinders," *IEEE Transactions on Antennas and Propagation,* Vol. AP-20, No. 5, pp. 547–555, September.

Borovikov V. A. and Kinber B. Y. (1994), *Geometrical Theory of Diffraction,* IEE Electromagnetic Waves Series 37, London.

Bouche D. P., Molinet F. A., and Mittra R. (1993), "Asymptotic and Hybrid Techniques for Electromagnetic Scattering," *Proceedings of IEEE,* Vol. 81, No. 12, pp. 1658–1684, December.

Bouche D. P., Molinet F. A., and Mittra R. (1997), *Asymptotic Methods in Electromagnetics,* Berlin–Heidelberg, Springer-Verlag.

Burnside W. D. and Pathak P. H. (1980), "A Summary of Hybrid Solutions Involving Moment Methods and GTD," in *Applications of the Method of Moments to Electromagnetic Fields,* St. Cloud, FL, SCEEE Press.

Byun J., Lee B., and Harackiewicz F. J. (1999), "FDTD Analysis of Mutual Coupling between Microstrip Patch Antennas on Curved Surfaces," in *IEEE AP-S International Symposium,* Vol. 2, August, pp. 886–889.

Calander N. and Josefsson L. (2001), "A Look at Polarization Properties of Cylindrical Array Antennas," in *Proceedings of Second European Workshop on Conformal Antennas,* The Hague, The Netherlands, April 24–25, 2001.

Cangellaris A. C. and Wright D. B. (1991), "Analysis of the Numerical Error Caused by the Stair-stepped Approximation of a Conducting Boundary in FDTD Simulations of Electromagnetic Phenomena," *IEEE Transactions on Antennas and Propagation,* Vol. AP-39, No. 10, pp. 1518–1525, October.

Campbell B., Hussar P. E., and Smith-Rowland E. M. (2002), "High-Frequency Radiation Pattern Analysis for Antennas Conformal to Convex Platform Surfaces," in *Proceedings of IEEE AP-S International Symposium,* Vol. 1, San Antonio, Texas, June 16–21, p. 29.

Carter P. S. (1943), "Antenna Arrays Around Cylinders," *Proceedings of IRE,* Vol. 31, pp. 671–693, December.

Carver K. R. and Mink J. W. (1981), "Microstrip Antenna Technology," *IEEE Transactions on Antennas and Propagation,* Vol. AP-29, No. 1, pp. 2–24, January.

Chang Z. W., Felsen L. B., and Hessel A. (1976), "Surface Ray Methods for Mutual Coupling in Conformal Arrays on Cylindrical and Conical Surfaces," Final Report, Contract N00123–76–C-0236, Polytechnic Institute of New York, July.

Coffey E. L. (1985), "Efficient In-Place Antenna Modeling with MoM and GTD," in *Proceedings of IEEE AP-S International Symposium,* Vol. 3, pp. 783–786, June.

Collin R. E. and Zucker F. J. (1969), *Antenna Theory, Part 2,* McGraw-Hill.

Courant R. (1943), "Variational Methods for the Solution of Problems of Equilibrium and Vibrations," *Bulletin American Mathematical Society,* Vol. 49, pp.1–23.

Craddock I. J., Paul D. L., Railton C. J., Ball G., and Watts J. (2001), "Cylindrical-Cartesian FDTD Model of a 17-Element Conformal Antenna Array," *Electronics Letters,* Vol. 37, No. 24, pp. 1429–1431, 22 November.

Demirdag C. and Rojas R. G. (1997), "Mutual Coupling Calculations on a Dielectric Coated PEC Cylinder using UTD-based Green's Function," *IEEE Antennas and Propagation Society International Symposium Digest,* Vol. 3, pp. 1525–1528, July.

Derneryd A. G. (1976), "Linearly Polarized Microstrip Antennas," *IEEE Transactions on Antennas and Propagation,* Vol. AP-24, pp. 846–851, November.

Desai C. S. and Abel J. F. (1972), *Introduction to the Finite Element Method: A Numerical Approach for Engineering Analysis,* New York, Van Nostrand Reinhold.

Dey S. and Mittra R. (1994), "A Locally Conformal Finite-Difference Time-Domain (FDTD) Algorithm for Modeling Three-dimensional Perfectly Conducting Objects," *IEEE Microwave Guided Wave Letters,* Vol. 7, pp. 273–275, March.

Elliott R. S. (1955), "Azimuthal Surface Waves on Circular Cylinders," *Journal of Applied Physics,* Vol. 26, No. 4, pp. 368–376.

Erdöl T. and Ertürk V. B. (2005), "An Asymptotic Closed-Form Paraxial Formulation for Surface Fields on Electrically Large Coated Circular Cylinders," in *Proceedings of the 4th European Workshop on Conformal Antennas,* Stockholm, Sweden, May 23–24, pp. 19–22.

Ersoy L. and Pathak P. H. (1988), "An Asymptotic High-Frequency Analysis of the Radiation by a Source on a Perfectly Conducting Convex Cylinder with an Impedance Surface Patch," *IEEE Transactions on Antennas and Propag.,* Vol. AP-36, pp. 1407–1417, October.

Ertürk V. B. (2000), *Efficient Hybrid MoM/Greens's Function Technique to Analyze Conformal Microstrip Antennas and Arrays.* PhD Thesis, Ohio State University, Dept. of Electrical Engineering.

Ertürk V. B. and Rojas R. G. (2000), "Efficient Computation of Surface Fields Excited on a Dielectric Coated Circular Cylinder," *IEEE Transactions on Antennas and Propagation,* Vol. AP-48, pp. 1507–1516, October.

Ertürk V. B. and Rojas R. G. (2002), "Paraxial Space-Domain Formulation for Surface Fields on a Large Dielectric Coated Circular Cylinder," *IEEE Transactions on Antennas and Propagation,* Vol. AP-50, No. 11, pp. 1577–1587, November.

Ertürk V. B., Rojas R. G., and Lee K. W. (2004), "Analysis of Finite Arrays of Axially Directed

Printed Dipoles on Electrically Large Circular Cylinders," *IEEE Transactions on Antennas and Propagation,* Vol. 50, No. 10, pp. 2586–2595, October.

Felsen L. B. and Marcuvitz N. (1973), *Radiation and Scattering of Waves.* Prentice-Hall.

Felsen L. B. and Green A. (1976), "Excitation of Large Concave Surfaces," in *Proceedings of AP-S International Symposium,* Univ. of Massachusetts, Amherst, October, pp. 374–377.

Felsen L. B. and Naishadham K. (1991), "Ray Formulation of Waves Guided by Circular Cylindrically Stratified Dielectrics," *Radio Science,* Vol. 26, No. 1, pp. 203–209, January–February.

Fock V. A. (1945), "Diffraction of Radio Waves Around the Earth's Surface," *Journal of Physics U.S.S.R,* Vol. 9, pp. 256–266.

Fock V. A. (1965), "Diffraction of Radio Waves Around the Earth's Surface," in *Electromagnetic Diffraction and Propagation Problems,* pp. 191–212, Pergamon Press.

Fusco M. A. (1990), "FDTD Algorithm in Curvilinear Coordinates," *IEEE Transactions on Antennas and Propagation,* Vol. AP-38, No. 1, pp. 76–89, January.

Fusco M. A., Smith M. V., and Gordon L. W. (1991), "A Three-dimensional FDTD Algorithm in Curvilinear Coordinates," *IEEE Transactions on Antennas and Propagation,* Vol. AP-39, No. 10, pp. 1463–1471, October.

Gerini G. and Zappelli L. (2004), "Phased Arrays of Rectangular Apertures on Conformal Cylindrical Surfaces: A Multimode Equivalent Network Approach," *IEEE Transactions on Antennas and Propagation,* Vol. AP-52, No. 7, pp. 1843–1850, July.

Golden K. E., Stewart G. E., and Pridmore-Brown D. C. (1974), "Approximation Techniques for the Mutual Admittance of Slot Antennas on Metallic Cones," *IEEE Transactions on Antennas and Propagation,* Vol. AP-22, No. 1, pp. 43–48, January.

Gordon W. E. (1985), "A Hundred Years of Radio Propagation," *IEEE Transactions on Antennas and Propagation,* Vol. AP-33, No. 2, p. 129, February.

Gottwald G. and Wiesbeck W. (1994), "Near-field Coupling Calculation in Cylindrical Structures for Conformal Patch Antenna Applications," in *International Symposium on Antennas, JINA,* Nov. 8–10, 1994, Nice, France, pp. 123–126.

Greenwood A. D. and Jin J.-M. (1998), "Hybrid MoM/SBR Method to Compute Scattering From a Slot Array Antenna in a Complex Geometry," *Applied Computer Electromagnetic Society Journal,* Vol. 13, No. 1, pp. 43–51, March.

Gupta K. C. and Benalla A. (eds.) (1988), *Microstrip Antenna Design,* Artech House.

Habashy T. M., Ali. S. M., and Kong J. A. (1990), "Input Impedance and Radiation Pattern of Cylindrical-Rectangular and Wraparound Microstrip Antennas," *IEEE Transactions on Antennas and Propagation,* Vol. AP-38, No. 5, pp. 722–731, May.

Hall R. C., Thng C. H., and Chang D. C. (1995), "Mixed Potential Green's Functions for Cylindrical Microstrip Structures," in *Proceedings of IEEE Antennas and Propagation Society International Symposium 1995,* Vol. 4, 18–23 June, pp. 1776–1779.

Hall R. C. and Wu D. I. (1996), "Modeling and Design of Circularly Polarized Cylindrical Wraparound Microstrip Antennas," in *Proceedings of IEEE Antennas and Propagation Society International Symposium 1996,* Vol. 1, 21–26 July, pp. 672–675.

Hall R. C. (1997), "Modeling Conformal Antennas for Wireless Personal Communications," in *Proceedings of IEEE Aerospace Conference,* Vol. 1, 1–8 Feb., pp. 71–82.

Hansen R. C. (ed.) (1981a), *Conformal Antenna Array Design Handbook,* Dept. of the Navy, Air Systems Command, September, AD A110091.

Hansen R. C. (ed.) (1981b), *Geometrical Theory of Diffraction,* New York, IEEE Press.

Hansen R. C. (1998), Phased Array Antennas, Wiley.

Harms P. H., Lee J.-F., and Mittra R. (1992), "A Study of the Nonortogonal FDTD Method Versus the Conventional FDTD Technique for Computing Resonant Frequencies of Cylindrical Cavities," *IEEE Transactions on Microwave Theory and Techniques,* Vol. 40, No. 4, pp. 741–746, April.

Harrington R. F. and Lepage W. R. (1952), "Directional Antenna Arrays of Elements Circularly Disposed about a Cylindrical Reflector," *Proceedings of IRE,* Vol. 40, pp. 83–86, January.

Harrington R. F. (1961), *Time Harmonic Electromagnetic Fields,* Prentice-Hall, 1961.

Harrington R. F. (1968), *Field Computation by Moment Methods,* New York, Macmillan.

Herper J. C., Hessel A., and Tomasic B. (1985), "Element Pattern of an Axial Dipole in a Cylindrical Phased Array, Part I: Theory," *IEEE Transactions on Antennas and Propagation,* Vol. AP-33, No. 3, pp. 259–272, March.

Holland R. (1983), "Finite Difference solution of Maxwell's Equations in Generalized Nonorthogonal Coordinates," *IEEE Transactions on Nuclear Science,* Vol. NS-30, No. 6, pp. 4589–4591, December.

Hussar P. E. and Smith-Rowland E. M. (2002), "An Asymptotic Solution for Boundary-Layer Fields Near a Convex Impedance Surface," *Journal of Electromagnetic Waves Applications,* Vol. 16, No. 2, pp. 185–208.

Hussar P. E. and Smith-Rowland E. M. (2003), "A Dyadic Green's Function Representation of Fields Near a Convex Impedance Surface," in *Proceedings of IEEE AP-S International Symposium,* Columbus, Ohio, June 22–27.

Inami K., Sawaya K., and Mushiake Y. (1982), "Mutual Coupling Between Rectangular Slot Antennas on a Conducting Concave Spherical Surface," *IEEE Transactions on Antennas and Propagation,* Vol. AP-30, No. 5, pp. 927–933, September.

Ishihara T. and Felsen L. B. (1978), "High-Frequency Fields Excited by a Line Source Located on a Perfectly Conducting Concave Cylindrical Surface," *IEEE Transactions on Antennas and Propagation,* Vol. AP-26, No. 6, pp. 757–767, November.

Ishihara T. and Felsen L. B. (1988), "High-Frequency Propagation at Long Ranges Near a Concave Boundary," *Radio Science,* Vol. 23, No. 6, pp. 997–1012, November–December.

Ishimaru A. (1991), *Electromagnetic Wave Propagation, Radiation and Scattering.* New Jersey: Prentice-Hall.

Jakobsen K. R. (1984), "The Radiation from Microstrip Antennas Mounted on Two-dimensional Objects," *IEEE Transactions on Antennas and Propagation,* Vol. AP-32, No. 11, pp. 1255–1259, November.

Jakobus U. and van Tonder J. J. (2003), "Conformal Antenna Simulation by Means of an Advanced Method of Moments Code Using Adaptive Frequency Interpolation Techniques," in *Proceedings of 3rd European Workshop on Conformal Antennas,* Bonn, Germany, October 22–23, pp. 57–60.

Josefsson L. (ed.) (1981), "Teoretiska Metoder för Antennanalys, del 2," Kurslitteratur: CTH/elkretsteknik, FOA 3, L M Ericsson. In Swedish

Jin. J.-M. and Volakis J. L. (1991a), "A Finite Element-Boundary Integral Formulation for Scattering by Three-Dimensional Cavity-Backed Apertures," *IEEE Transactions on Antennas and Propagation,* Vol. AP-39, No. 1, pp. 97–104, January.

Jin. J.-M. and Volakis J. L. (1991b), "Electromagnetic Scattering by and Transmission Through a Three-Dimensional Slot in a Thick Conducting Plane," *IEEE Transactions on Antennas and Propagation,* Vol. AP-39, No. 4, pp. 543–550, April.

Jin. J.-M. and Volakis J. L. (1991c), "A Hybrid Finite Element Method for Scattering and Radiation by Microstrip Patch Antennas and Arrays Residing in a Cavity," *IEEE Transactions on Antennas and Propagation,* Vol. AP-39, No. 11, pp. 1598–1604, November.

Jin. J.-M., Volakis J. L. and Collins J. D. (1991), "A Finite-Element Boundary-Integral Method for Scattering and Radiation by Two- and Three-Dimensional Structures," *IEEE Antennas and Propagation Magazine,* Vol. 33, No. 3, pp. 22–32, June.

Jin J.-M. (1993), *The Finite Element Method in Electromagnetics,* Wiley.

Jurgens T. G., Taflove A., Umashankar K., and Moore T. (1992), "Finite-Difference Time-Domain Modeling of Curved Surfaces," *IEEE Transactions on Antennas and Propagation,* Vol. AP-40, No. 4, pp. 357–366, April.

Jurgens T. G. and Taflove A. (1993), "Three-dimensional Contour FDTD Modeling of Scattering from Single and Multiple Bodies," *IEEE Transactions on Antennas and Propagation,* Vol. AP-41, No. 12, pp. 1703–1708, December.

Kantorovich L. V. and Krylov V. I. (1964), *Approximate Methods of Higher Analysis,* Wiley. (Translated by C. D. Benster from the fourth Russian edition.)

Karr P. R. (1951), "Radiation Properties of Spherical Antennas as a Function of Location of the Driving Force," *Journal of Research of National Bureau of Standards,* Vol. 46, pp. 422–436.

Kashiwa T., Onishi T., and Fukai I. (1993), "Analysis of Microstrip Antennas on a Curved Surface Using the Conformal Grids FD-TD Method," in *Proceedings of IEEE AP-S International Symposium,* Vol. 1, 28 June–2 July, pp. 34–37.

Kashiwa T., Onishi T., and Fukai I. (1994), "Analysis of Microstrip Antennas on a Curved Surface Using the Conformal Grids FD-TD Method," *IEEE Transactions on Antennas and Propagation,* Vol. AP-42, No. 3, pp. 423–427, March.

Keller J. B. (1962), "Geometrical Theory of Diffraction," *Journal of of Optical Society of America,* Vol. 52, No. 2, pp. 116–130, February.

Keller J. B. (1985), "One Hundred Years of Diffraction Theory," *IEEE Transactions on Antennas and Propagation,* Vol. AP-33, No. 2, pp. 123–126, February.

Kempel L. C. and Volakis J. L. (1993), "A Finite Element-Boundary Integral Method for Cavities in a Circular Cylinder," in *Proceedings of IEEE AP-S International Symposium,* pp. 292–295.

Kempel L. C. and Volakis J. L. (1994a), "Scattering by Cavity-Backed Antennas on a Circular Cylinder using the FE-BI Method," in *Proceedings of IEEE AP-S International Symposium,* pp. 1382–1385.

Kempel L. C. and Volakis J. L. (1994b), "Radiation by Patch Antennas on a Circular Cylinder using the FE-BI Method," in *Proceedings of IEEE AP-S International Symposium,* pp. 182–185.

Kempel L. C., Volakis J. L., and Silva R. J. (1995), "Radiation by Cavity-backed Antennas on a Circular Cylinder," *IEE Proceedings on Microwave Antennas Propagation,* Vol. 142, No. 3, pp. 233–239, June.

Kline M. (1951), "An Asymptotic Solution of Maxwell's Equations," *Communications Pure Applied Mathematics,* Vol. 4, pp. 225–262.

Knudsen H. L. (1959), "Antennas on Circular Cylinders," *Transactions of IRE,* Vol. AP-7, pp. 361–370, December.

Kouyoumjian R. G. (1965), "Asymptotic High-Frequency Methods," *Proceedings of the IEEE,* Vol. 53, pp. 864–876, August.

Kouyoumjian R. G. and Pathak P. H. (1974), "A Uniform Geometrical Theory of Diffraction for an Edge in a Perfectly Conducting Surface," *Proceedings of the IEEE,* Vol. 62, No. 11, pp. 1448–1461, November.

Krowne C. M. (1983), "Cylindrical-Rectangular Microstrip Antenna," *IEEE Transactions on Antennas and Propagation,* Vol. AP-31, No. 1, pp. 194–199, January.

Kunz K. and Luebbers R. (1993), *The Finite Difference Time Domain Method for Electromagnetics,* CRC Press.

Lee J. F., Palandech R. and Mittra R. (1992), "Modeling Three-dimensional Discontinuities in Waveguides Using Nonorthogonal FDTD Algorithm," *IEEE Transactions on Microwave Theory Techniques,* Vol. MTT-40, No. 2, pp. 346–352, February.

Lee S. W. and Deschamps G. A. (1976), "A Uniform Asymptotic Theory of Electromagnetic Diffraction by a Curved Wedge," *IEEE Transactions on Antennas and Propagation,* Vol. AP-24, pp. 25–34, January.

Lee S. W. and Mittra R. (1976), "Study of Mutual Coupling Between Two Slots on a Cylinder," Final Report July 16–November 15, University of Illinois at Urbana-Champaign, AD-A034095.

Lee S-.W. and Safavi-Naini S. (1978), "Approximate Asymptotic Solution of Surface Field Due to a Magnetic Dipole on a Cylinder," *IEEE Transactions on Antennas and Propagation,* Vol. AP-26, No. 4, pp. 593–598, July.

Lee S.-W. (1978), "Mutual Admittance of Slots on a Cone: Solution by Ray Technique," *IEEE Transactions on Antennas and Propagation,* Vol. AP-26, No. 6, pp. 768–773, November.

Levin M. L. (1951), "The Theory of a Slot Cut in a Waveguide," *Journal of Technical Physics,* Vol. 21, pp. 772–786.

Liu J. and Jin J. M. (2003a), "Analysis of Conformal Antennas on a Complex Platform," *Microwave and Optical Technology Letters,* Vol. 36, No. 2, pp. 139–142, January.

Liu J. and Jin J. M. (2003b), "Analysis of Conformal Antennas on a Complex Platform," in *Proceedings of IEEE AP-S International Symposium,* pp. 421–424.

Lo Y. T., Solomon D., and Richards W. F. (1979), "Theory and Experiment on Microstrip Antennas," *IEEE Transactions on Antennas and Propagation,* Vol. AP-27, No. 2, pp. 137–145, March.

Luk K. M., Lee K. F., and Dahele J. S. (1987), "Input Impedance and Q factors of Cylindrical-Rectangular Microstrip Patch Antennas," in *Proceedings of Fifth International Conference on Antennas and Propagation ICAP 87,* Part 1, York, England, pp. 95–98.

Luneberg R. M. (1944), Mathematical Theory of Optics, Brown University Press.

Macon C. A., Kemple L. C., and Schneider S. W. (2001), "Modeling Conformal Antennas on Prolate Spheroids Using the Finite Element-Boundary Integral Method," in *Proceedings of IEEE AP-S International Symposium,* pp. 358–361.

Macon C. A., Kemple L. C., and Schneider S. W. (2002), "Modeling Cavity-Backed Apertures Conformal to Prolate Spheroids Using the Finite Element-Boundary Integral Technique," in *Proceedings of IEEE AP-S International Symposium,* pp. 550–553.

Marcano D., Duran F., and Torres W. (1998), "Analysis of a Conformal Microstrip Antenna for Microwave Hyperthermia Using FDTD," in *Proceedings of the 1998 Second IEEE International Caracas Conference on Devices, Circuits and Systems,* 2–4 March, pp. 302–306.

Martin H. C. and Carey G. F. (1973), *Introduction to Finite Element Analysis: Theory and Application,* New York, McGraw-Hill.

Martin T. and Pettersson L. (2003a), "Cylindrical FDTD Using Phase Shift Boundaries for Simulation of Large Cylindrical Antenna Arrays," in *Proceedings of the Nordic Antenna Symposium 2003 (Antenn03),* Kalmar, Sweden, 13–15 May.

Martin T. and Pettersson L. (2003b), "Cylindrical FDTD with Phase Shift Boundaries for Simulation of Cylindrical Antennas," in *Proceedings of 3rd European Workshop on Conformal Antennas,* Bonn, Germany, October 22–23, pp. 49–52.

McDonald B. H. and Wexler A. (1972), "Finite-Element Solution of Unbounded Field Problems," *IEEE Transactions Microwave Theory Technology,* Vol. MTT-20, pp. 841–847.

McNamara D. A., Pistorius C. W. I., and Malherbe J. A. G. (1990), *Introduction to the Uniform Geometrical Theory of Diffraction,* Artech House.

Mei K. K., Cangellaris A., and Angelakos D. J. (1984), "Conformal Time Domain Finite-Difference Method," *Radio Science,* Vol. 19, pp. 1145–1147.

Miller E. K. and Burke G. J. (1992), "Low-Frequency Computational Electromagnetics for Antenna Analysis," *Proceedings of IEEE,* Vol. 80, No. 1, pp. 24–43, January.

Mittra R. and Ramahi O. (1990), "Absorbing Boundary Conditions for the Direct Solution of Partial Differential Equations Arising in Electromagnetic Scattering Problems," in M. A. Morgan (ed.), *PIER 2: Finite Element and Finite Difference Methods in Electromagnetic Scattering,* New York, Elsevier.

Munger A. D., Provencher J. H. and Gladman B. R. (1971), "Mutual Coupling on a Cylindrical Array of Waveguide Elements," *IEEE Transactions on Antennas and Propagation,* Vol. AP-19, No. 1, pp.131–134, January.

Munk P. (1996), *A Uniform Geometrical Theory of Diffraction for the Radiation and Mutual Coupling Associated With Antennas on a Material Coated Convex Conducting Surface.* PhD Thesis, Ohio State University, Dept. of Electrical Engineering.

Munson R. E. (1974), "Conformal Microstrip Antennas and Microstrip Phased Arrays," *IEEE Transactions on Antennas and Propagation,* Vol. AP-22, No. 1, pp. 74–78.

Mushiake Y. and Webster R. E. (1957), "Radiation Characteristics with Power Gain for Slots on a Sphere," *IRE Transactions on Antennas and Propagation,* Vol. AP-5, pp. 47–55, January.

Nakayama I., Kawano T., and Nakano H. (1999), "A Conformal Spiral Array Antenna Radiating an Omnidirectional Circularly-polarized Wave," in *Proceedings of IEEE AP-S International Symposium,* Orlando FL, USA, Vol. 2, 11–16 July, pp. 894–897.

Naishadham K. and Felsen L. B. (1993), "Dispersion of Waves Guided Along a Cylindrical Substrate-Superstrate Layered Medium," *IEEE Transactions on Antennas and Propagation,* Vol. AP-41, No. 3, pp. 304–313, March.

Norrie D. H. and de Vries G. (1973), *The Finite Element Method: Fundamentals and Applications,* New York, Academic Press.

Özdemir T. and Volakis J. L. (1997), "Triangular Prisms for Edge-Based Vector Finite Element Analysis of Conformal Antennas," *IEEE Transactions on Antennas and Propagation,* Vol. AP-45, No. 5, pp.788–797, May.

Paknys R. and Wang N. (1986), "Creeping Wave Propagation Constants and Modal Impedance for a Dielectric Coated Cylinder," *IEEE Transactions on Antennas and Propagation,* Vol. AP-34, No. 5, pp. 674–680, May.

Paknys R. and Wang N. (1987), "Excitation of Creeping Waves on a Circular Cylinder with a Thick Dielectric Coating," *IEEE Transactions on Antennas and Propagation,* Vol. AP-36, No.12, pp. 1487–1489, December.

Pathak P. H. and Wang N. N. (1978), "An Analysis of the Mutual Coupling Between Antennas on a Smooth Convex Surface," Final Rep. 784583-7, Ohio State University, ElectroScience Lab., Dept. of Electrical Engineering, Ohio, USA, October.

Pathak P. H., Burnside W. D. and Marhefka R. J. (1980), "A Uniform GTD Analysis of the Diffraction of Electromagnetic Waves by a Smooth Convex Surface," *IEEE Transactions on Antennas and Propagation,* Vol. AP-28, No. 5, pp. 631–642, September.

Pathak P. H. and Wang N. (1981), "Ray Analysis of Mutual Coupling Between Antennas on a Convex Surface," *IEEE Transactions on Antennas and Propagation,* Vol. AP-29, No. 6, pp. 911–922, November.

Pathak P. H., Wang N., Burnside W. D., and Kouyoumjian R. G. (1981), "A Uniform GTD Solution for the Radiation from Sources on a Convex Surface," *IEEE Transactions on Antennas and Propagation,* Vol. AP-29, No. 4, pp. 609–622, July.

Pathak P. H. (1992), "High-Frequency Techniques for Antenna Analysis," *Proceedings of IEEE,* Vol. 80, No. 1, pp. 44–65, January.

Pathak P. H. (2001), "Advanced Electromagnetics—2," Unpublished EE818 Class Notes, ElectroScience Lab., Ohio State University, Columbus, Ohio, USA.

Paul D. L. and Craddock I. J. (2001), "Simulation of Finite Conformal Facetted Microstrip Patch Antenna Array by Locally Distorted FDTD Technique", in *Proceedings of Second European Workshop on Conformal Antennas,* The Hague, The Netherlands, April.

Paul D. L., Craddock I. J., and Railton C. J. (2003), "Simulation of Circular Conformal Facetted Stacked Patch Antenna Array with Tight Radius of Curvature by Hybrid Cartesian/Cylindrical FDTD Approach," in *Proceedings of Third European Workshop on Conformal Antennas,* Bonn, Germany, October 22–23, pp. 53–56.

Pearson L. W. (1987), "A Ray Representation of Surface Diffraction by a Multilayer Cylinder," *IEEE Transactions on Antennas and Propagation,* Vol. AP-35, No. 6, pp. 698–707, June.

Persson P. (2001), *Analysis and Design of Conformal Array Antennas,* Ph.D. Thesis, Royal Institute of Technology, Stockholm, Sweden.

Persson P. and Josefsson L. (1998), "Mutual Coupling Effects on Conformal Arrays," in *Proceedings of EMB 98,* Linköping, Sweden, November, pp. 124–131.

Persson P. and Josefsson L. (1999a), "Calculating the Mutual Coupling between Apertures on Convex Cylinders Using a Hybrid UTD-MoM Method," in *Proceedings of IEEE AP-S International Symposium,* Orlando FL, USA, July, pp. 890–893.

Persson P. and Josefsson L. (1999b), "Calculating the Mutual Coupling between Apertures on Convex Surfaces Using a Hybrid UTD-MoM Method," in *Proceedings of First European Workshop on Conformal Antennas,* Karlsruhe, Germany, pp. 60–63, October.

Persson P. and Josefsson L. (2000a), "Calculating the Mutual Coupling between Apertures on Doubly Curved Convex Surfaces Using a Hybrid UTD-MoM Method," paper No. 0404, in *Proceedings of AP2000,* Davos, Switzerland, April 200.

Persson P. and Josefsson L. (2000b), "Investigation of the Mutual Coupling between Apertures on Doubly Curved Convex Surfaces Using a Hybrid UTD–MoM Method," in *Proceedings of Antenn 00,* Lund, Sweden, pp. 297–302, September.

Persson P. and Josefsson L. (2001a), "Calculating The Mutual Coupling between Apertures on a Convex Circular Cylinder Using a Hybrid UTD–MoM Method," *IEEE Transactions on Antennas and Propagation,* Vol. 49, No. 4, pp. 672–677, April.

Persson P. and Josefsson L. (2001b), "Mutual Coupling Effects on Doubly Curved Surfaces," in *Proceedings of EMB 01,* Uppsala, Sweden, pp. 35–42, November.

Persson P., Josefsson L., and Lanne M. (2001), "Investigation of the Mutual Coupling between Apertures on a General Paraboloid of Revolution: Theory and Measurements," in *Proceedings of Second European Workshop on Conformal Antennas,* The Hague, The Netherlands, April.

Persson P. and Rojas R. G. (2001), "Development of a High Frequency Technique for Mutual Coupling Calculations between Apertures on a PEC Circular Cylinder Covered with a Dielectric Layer," in *Proceedings of EMB 01,* Uppsala, Sweden, pp. 51–58, November.

Persson P. and Rojas R. G. (2003), "High-frequency Approximation for Mutual Coupling Calculations between Apertures on a Perfect Electric Conductor Circular Cylinder Covered with a Dielectric Layer: Nonparaxial Region," *Radio Science,* Vol. 38, No. 4, 1079, doi:10.1029/2002RS002745.

Persson P., Josefsson L., and Lanne M. (2003a), "Investigation of the Mutual Coupling between Apertures on Doubly Curved Convex Surfaces: Theory and Measurements," *IEEE Transactions on Antennas and Propagation,* Vol. AP-51, No. 4, pp. 682–692, April.

Persson P., Thors B., and Rojas R. G. (2003b), "An Improved Numerical Approach for Surface Field Calculations on Large Dielectric Coated Circular Cylinders," Technical Report TRITA-TET 03-4, Royal Institute of Technology, Division of Electromagnetic Theory, Stockholm, Sweden, June.

Peterson A. F. and Mittra R. (1989), "Mutual Admittance Between Slots in Cylinders of Arbitrary Shape," *IEEE Transactions on Antennas and Propagation,* Vol. AP-37, No. 7, pp. 858–864, July.

Pistolkors A. A. (1947), "Radiation from Longitudinal and Transverse Slots in a Circular Cylinder," *Journal of Technical Physics (U.S.S.R.),* Vol. 17, pp. 365–385.

Poggio A. J. and Miller E. K. (1973), "Integral Equation Solution of Three-Dimensional Scattering Problems" in *Computer Techniques for Electromagnetics,* R. Mittra (ed.), New York, Pergamon Press.

Pozar D. M. (1982), "Input Impedance and Mutual Coupling of Rectangular Microstrip Antennas," *IEEE Transactions on Antennas and Propagation,* Vol. AP-30, No. 11, pp. 1191–1196, November.

Pozar D. M. (1987), "Radiation and Scattering from a Microstrip Patch on a Uniaxial Substrate," *IEEE Transactions on Antennas and Propagation,* Vol. AP-35, No. 6, pp. 613–621, June.

Pridemore-Brown D. C. and Stewart G. (1972), "Radiation from Slot Antennas on Cones," *IEEE Transactions on Antennas and Propagation,* Vol. AP-20, No. 1, pp. 36–39, January.

Pridemore-Brown D. C. (1972), "Diffraction Coefficients for a Slot-Excited Conical Antenna," *IEEE Transactions on Antennas and Propagation,* Vol. AP-20, No. 1, pp. 40–49, January.

Pridemore-Brown D. C. (1973), "The Transition Field on the Surface of a Slot-Excited Conical Antenna," *IEEE Transactions on Antennas and Propagation,* Vol. AP-21, No. 6, pp. 889–890, November.

Raffaelli S., Sipus Z., and Kildal P.-S. (2001), "Analysis of Arbitrarily Oriented Patches on Multilayer Circular Cylindrical Structures," in *Proceedings of Second European Workshop On Conformal Antennas,* The Hague, The Netherlands, April.

Raffaelli S., Sipus Z., and Kildal P.-S. (2005), "Analysis and Measurements of Conformal Patch Array Antennas on Multilayer Circular Cylinder," *IEEE Transactions on Antennas and Propagation,* Vol. 53, No. 3, pp. 1105–1113, March.

Railton C. J. (1993), "An Algorithm for the Treatment of Curved Metallic Laminas in the Finite-Difference Time-Domain Method," *IEEE Transactions on Microwave Theory and Techniques,* Vol. 41, No. 8, pp. 1429–1438, August.

Rao S. M. and Wilton D. R. (1991), "Transient Scattering by Conducting Surface of Arbitrary Shape," *IEEE Transactions on Antennas and Propagation,* Vol. 39, No. 1, pp. 56–61, January.

Richard L., Nosich A. I., and Daniel J. P. (1997), "Surface-Impedance Model Analysis of a Coated Cylinder with Application to Wave Propagation and Conformal Antennas," in *Proceedings of Tenth International Conf. on Antennas and Propagation,* 14–17 April, pp. 1123–1125.

Richard L., Nosich A. I., and Daniel J. P. (1999), "Revisiting the Waves on a Coated Cylinder by Using Surface-Impedance Model," *IEEE Transactions Antennas Propagation Letters,* Vol. 47, No. 8, pp. 1374–1375, August.

Richards W. F., Lo Y. T., and Harrison D. D. (1981), "An Improved Theory for Microstrip Antennas and Applications," *IEEE Transactions on Antennas and Propagation,* Vol. AP-29, No. 1, pp. 39–46, January.

Rojas R. G. and Demirdag C. (1997), UTD-based Mutual Coupling and Radiation Pattern Calculations for a Source Excited Dielectric Coated PEC Circular Cylinder, Tech. Rep. 731716-1, Ohio State University, ElectroScience Lab., Dept. of Electrical Engineering, September.

Rousseau P. R. and Pathak P. H. (1995a), "Time-Domain Uniform Geometrical Theory of Diffraction for a Curved Wedge," *IEEE Transactions on Antennas and Propagation,* Vol. AP-43, No. 12, pp. 1375–1382, December.

Rousseau P. R. and Pathak P. H. (1995b), "TD-UTD Slope Diffraction for a Perfectly Conducting Curved Wedge," in *Proceedings of IEEE AP-S International Symposium,* pp. 856–859.

Rousseau P. R. and Pathak P. H. (1996), "TD-UTD for Scattering from a Smooth Convex Surface," in *Proceedings of IEEE AP-S International Symposium,* pp. 2084–2087.

Rylander T. and Bondesson A. (2002), "Application of Stable FEM-FDTD Hybrid to Scattering Problems," *IEEE Transactions on Antennas and Propagation,* Vol. AP-50, No. 2, pp. 141–144, February.

Rylander T. and Jin J.-M. (2003), "Conformal Perfectly Matched Layers for the Time Domain Finite Element Method," in *Proceedings of IEEE AP-S International Symposium,* pp. 698–701.

Rynne B. P. (1991), "Time Domain Scattering from Arbitrary Surface Using the Electric Field Integral Equation," *Journal of Electromagnetic Waves Applications,* Vol. 5, No. 1, pp. 93–112.

Safavi-Naini S. and Lee S. W. (1976), "Calculations of Mutual Admittance between Two Slots on a Cylinder"; Attachment A in Lee et al. (1976).

Sengupta D. L. (1984), "Transmission Line Model Analysis of Rectangular Patch Antennas," *Electromagnetics,* Vol. 4, pp. 355–376.

Shapira J., Felsen L. B., and Hessel A. (1974), "Ray Analysis of Conformal Antenna Arrays," *IEEE Transactions on Antennas and Propagation,* Vol. AP-22, No. 1, pp. 49–63, January.

Silva F. C., Fonseca S. B. A., Soares J. M., and Giarola A. J. (1991), "Analysis of Microstrip Antennas on Circular-Cylindrical Substrates with a Dielectric Overlay," *IEEE Transactions on Antennas and Propagation,* Vol. AP-39, No. 9, pp. 1398–1403, September.

da Silva C. M. and Lacava J. C. (1995), "Mutual Impedance of Conformal Cylindrical Microstrip Antenna Arrays with a Protection Layer," in *Proceedings of IEEE MTT-S International Microwave and Optoelectronics Conference,* Vol. 1, Rio de Janeiro, Brazil, 24–27 July, pp. 314–319.

Silver S. and Saunders W. K. (1950), "The External Field Produced by a Slot on an Infinite Circular Cylinder," *Journal of Applied Physics,* Vol. 21, pp. 153–158, February.

Silvester P. P. and Hsieh M. S. (1971), "Finite-Element Solution of 2–dimensional Exterior Field Problems," *IEE Proceedings,* Vol. 118, pp. 1743–1747, December.

Sipus Z., Kildal P.-S., Leijon R., and Johansson M. (1998), "An Algorithm for Calculating Green's Functions of Planar, Circular Cylindrical, and Spherical Multilayer Substrates," *Applied Computational Electromagnetics Society Journal,* Vol. 13, pp. 243–254, November.

Sipus Z., Rupcic S., Lanne M., and Josefsson L. (2001a), "Moment Method Analysis of Circular-Cylindrical Array of Waveguide Elements Covered with a Radome," in *Proceedings of IEEE AP-S International Symposium,* Vol. 2, Boston, pp. 350–353, July.

Sipus Z., Rupcic S., Lanne M., Josefsson L., and Persson P. (2001b), "Analysis of Circular Cylindrical Array of Waveguide Elements Using Moment Method," in *Proceedings of 16th International Conference on Applied Electromagnetics and Communications—ICECOM'01,* Dubrovnik, Croatia, 1–3 October, pp. 50–53.

Sipus Z., Bartolic J., and Burum N. (2003), "Theoretical and Experimental Study of Spherical Rectangular Microstrip Patch Arrays," in *Proceedings of Third European Workshop on Conformal Antennas,* Bonn, Germany, October 22–23, pp. 69–72.

Sohtell E. V. (1986), "Microstrip Patch Antennas on Cylindrical Structures," Journées Internationales de Nice sur les Antennas, *JINA86 Conference Proceedings,* Nice, France, pp. 216–220.

Sohtell E. V. (1987), *Microwave Antennas on Cylindrical Structures,* Ph.D. Thesis, Chalmers University of Technology, Gothenburg, Sweden.

Spies K. P. and Wait J. R. (1967), "On the Calculation of Antenna Patterns for an Inhomogeneous Spherical Earth," *Radio Science,* Vol. 2, No. 11, pp. 1361–1378, November.

Stewart G. E. and Golden K. E. (1971), "Mutual Admittance for Axial Rectangular Slots in a Large Conducting Cylinder," *IEEE Transactions on Antennas and Propagation,* Vol. AP-19, No. 1, pp. 120–122, January.

Sun E.-Y. and Rusch W. V. T. (1994a), "Time-Domain Physical-Optics," *IEEE Transactions on Antennas and Propagation,* Vol. 42, No. 1, pp. 9–15, January.

Sun E.-Y. and Rusch W. V. T. (1994b), "EFIE Time-Marching Scattering from Bodies of Revolution and Its Applications," *IEEE Transactions on Antennas and Propagation,* Vol. AP-42, No. 3, pp. 412–417, March.

Svezhentsev A. and Vandenbosch G. A. E. (2001), "Effective Approach to the Problem of a Cylindrical Patch over a Metal Cylindrical Surface," in *Proceedings of Second European Workshop on Conformal Antennas,* The Hague, The Netherlands, April.

Svezhentsev A. (2003), "Analysis of Conformal Array Antenna by Moment Method in the Spatial Domain," in *Proceedings of Third European Workshop on Conformal Antennas,* Bonn, Germany, October 22–23, pp. 61–64.

Taflove A. (1995), *Computational Electrodynamics, The Finite-Difference Time-Domain Method,* Artech House.

Taflove A. (ed.) (1998), *Advances in Computational Electrodynamics,* Artech House.

Teixeira F. L. and Chew W. C. (1997), "Systematic Derivation of Anistropic PML Absorbing Media in Cylindrical and Spherical Coordinates," *IEEE Microwave and Guided Wave Letters,* Vol. 7, No. 11, pp. 371–373, November.

Theron I. P., Jackson D. B., and Jakobus U. (2000), "Extensions to the Hybrid Method of Moments/Uniform GTD Formulation for Sources Located Close to a Smooth Convex Surface," *IEEE Transactions on Antennas and Propagation,* Vol. AP-48, No. 6, pp. 940–945, July.

Thiel M. and Dreher A. (1999), "Perturbed Dyadic Green's Function for Quasi-cylindrical Multilayer Microstrip Structures," in *Proceedings of First European Workshop on Conformal Antennas,* Karlsruhe, pp. 32–35, Germany, October.

Thiel M. and Dreher A. (2001a), "Microstrip Antennas on Cylindrical Sector Structures," in *Proceedings of Second European Workshop on Conformal Antennas,* The Hague, The Netherlands, April.

Thiel M. and Dreher A. (2001b), "Microstrip Antennas on Multilayer Cylindrical and Quasi-Cylindrical Structures," in *Proceedings of IEEE AP-S International Symposium,* Vol. 3, 8–13 July, pp. 264–267.

Thiel M. and Dreher A. (2001c), "Eigensolution Expansion of Dyadic Green's Function for the Analysis of Microstrip Antennas on Cylindrical Sector Multilayer Structures," in *Proceedings of IEEE AP-S International Symposium,* Vol. 3, 8–13 July, pp. 272–275.

Thiel M. (2002), "Design Considerations for Microstrip Antennas on Cylindrical Sector Structures," in *Proceedings of IEEE AP-S International Symposium,* Vol. 1, 16–21 June, pp. 88–91.

Thiel M. and Dreher A. (2002), "Dyadic Green's Function of Multilayer Cylindrical Closed and Sector-Structures for Waveguide, Microstrip-Antenna, and Network Analysis," *IEEE Transactions on Microwave Theory and Techniques,* Vol. 50, No. 11, pp. 2576–2579, November.

Thomas W., Hall R. C., and Wu D. I. (1997), "Effects of Curvature on the Fabrication of Wraparound Antennas," in *Proceedings of IEEE Antennas and Propagation Society International Symposium 1997,* Vol. 3, 13–18 July, pp. 1512–1515.

Thors B. (2003), *Radiation and Scattering from Conformal Array Antennas,* Ph.D. Thesis, Royal Institute of Technology, Stockholm, Sweden.

Thors B. and Josefsson L. (2000), "Scattering from A Cylindrical Conformal Array Antenna with Waveguide Apertures," Technical Report: TRITA-TET 00-14, Royal Institute of Technology, Div. of Electromagnetic Theory, Sweden.

Thors B. and Josefsson L. (2003), "Radiation and Scattering Trade-off Design for Conformal Arrays," *IEEE Transactions on Antennas and Propagation,* Vol. AP-51, No. 5, pp. 1069–1076, May.

Thors B. and Rojas R. G. (2003), "Uniform Asymptotic Solution for the Radiation from a Magnetic Source on a Large Dielectric Coated Circular Cylinder: Non-paraxial Region," *Radio Science,* Vol. 38, No. 5.

Thors B., Josefsson L., and Rojas R. G. (2003a), "The RCS of a Cylindrical Array Antenna Coated with a Dielectric Layer," *IEEE Transactions on Antennas and Propagation,* Vol. AP-52, No. 7, pp. 1851–1858, July.

Thors B., Josefsson L., and Norgren M. (2003b), "Radiation Characteristics and Matching Properties for a Cylindrical Conformal Finite Array Antenna Coated with a Dielectric Layer," Technical Report: TRITA-TET 03-05, Royal Institute of Technology, Div. of Electromagnetic Theory, Sweden.

Tokgöz C. (2002), *Asymptotic High Frequency Analysis of the Surface Fields of a Source Excited Circular Cylinder with an Impedance Boundary Condition,* Ph.D. Thesis, ElectroScience Lab., Ohio State University, Columbus, Ohio, USA.

Tokgöz C., Pathak P. H., and Marhefka R. J. (2005), "An Asymptotic Solution for the Surface Magnetic Field within the Paraxial Region of a Circular Cylinder with an Impedance Boundary Condition," *IEEE Transactions on Antennas and Propagation,* Vol. 53, No. 4, pp. 1435–1443, April.

Tomasic B. and Hessel A. (1989a), "Periodic Structure Ray Method for Analysis of Coupling Coefficients in Large Concave Arrays—Part I: Theory," *IEEE Transactions on Antennas and Propagation,* Vol. AP-37, No. 11, pp. 1377–1385, November.

Tomasic B. and Hessel A. (1989b), "Periodic Structure Ray Method for Analysis of Coupling Coefficients in Large Concave Arrays—Part II: Application," *IEEE Transactions on Antennas and Propagation,* Vol. AP-37, No. 11, pp. 1386–1397, November.

Tomasic B., Hessel A., and Ahn H. (1993a), "Asymptotic Analysis of Mutual Coupling in Concave Circular Cylindrical Arrays—Part I: Far-Zone," *IEEE Transactions on Antennas and Propagation,* Vol. AP-41, No. 2, pp. 121–136, February.

Tomasic B., Hessel A., and Ahn H. (1993b), "Asymptotic Analysis of Mutual Coupling in Concave Circular Cylindrical Arrays—Part II: Near-Zone," *IEEE Transactions on Antennas and Propagation,* Vol. AP-41, No. 2, pp. 137–145, February.

Vecchi G., Bertuch T., and Orefice M. (1997a), "Spectral-domain Analysis of Printed Antennas of General Shape on Cylindrical Substrates," in *Proceedings of IEEE AP-S International Symposium,* Vol. 3, 13–18 July, pp. 1496–1499.

Vecchi G., Bertuch T., and Orefice M. (1997b), "Analysis of Cylindrical Printed Antennas with Subsectional Basis Functions in the Spectral Domain," in *Proceedings of ICEAA 7,* Torino, Italy, pp. 301–304, September.

Veruttipong T. W. (1990), "Time Domain Version of the Uniform GTD," *IEEE Transactions on Antennas and Propagation,* Vol. AP-38, No. 11, pp. 1757–1764, November.

Vogel M. H. (2001), "Analysis of Arbitrarily-Shaped Conformal Antennas with the Finite-Element Method," in *Proceedings of Second European European Workshop on Conformal Antennas,* The Hague, The Netherlands, April.

Volakis J. L., Chatterjee A., and Kempel L. C. (1998), *Finite Element Method for Electromagnetics: Antennas, Microwave Circuits, and Scattering Applications,* New York, IEEE Press.

Vorobyev Y. V. (1965), *Method of Moments in Applied Mathematics,* New York, Gordon and Breach. (Translated from the Russian by B. Seckler.)

Wait J. R. (1956a), "Radiation from a Vertical Antenna over a Curved Stratified Ground," *Journal of Research, National Bureau Standards,* Vol. 56, No. 4, pp. 237–244, April.

Wait J. R. (1956b), "Radiation Pattern of an Antenna Mounted on a Surface of Larger Radius of Curvature," *Proceedings of Institute of Radio Engineers,* Vol. 44, p. 694, May.

Wait J. R. (1959), *Electromagnetic Radiation from Cylindrical Structures,* London, Pergamon Press.

Wait J. R. (1962), *Electromagnetic Waves in Stratified Media,* New York, Pergamon Press.

Wait J. R. (1974), "Recent Analytical Investigations of Electromagnetic Ground Wave Propagation over Inhomogeneous Earth Models," *Proceedings of IEEE,* Vol. 62, No. 8, pp. 1061–1072, August.

Wait J. R., Ersoy L., and Pathak P. H. (1990), Comments on "An asymptotic High-frequency Analysis of the Radiation by a Source on a Perfectly Conducting Convex Cylinder with an Impedance Surface Patch" [and reply], *IEEE Transactions on Antennas Propagation,* Vol. AP-38, No. 4, pp. 585–587, April.

Wang N. (1982), "Regge Poles, Natural Frequencies, and Surface Wave Resonance of a Circular Cylinder with a Constant Surface Impedance," *IEEE Transactions on Antennas and Propagation,* Vol. AP-30, No. 6, pp. 1244–1247, November.

Wang N. (1985), "Electromagnetic Scattering from a Dielectric-Coated Circular Cylinder," *IEEE Transactions on Antennas and Propagation,* Vol. AP-33, No. 9, pp. 960–963, September.

Wang J. J. H. (1991), *Generalized Moment Methods in Electromagnetics,* New York, Wiley.

Watson G. N. (1918), "The Diffraction of Radio Waves by the Earth," *Proceedings of Royal Society, London,* Vol. A95, pp. 83–99.

Wong K.-L., Cheng Y.-T., and Row J.-S. (1993), "Resonance in a Superstrate-loaded Cylindrical-

Rectangular Microstrip Structure," *IEEE Transactions Microwave Theory Tech.,* Vol MTT-41, No. 5, pp. 814–819, May.

Wu C.-W., Kempel L. C. and Rothwell E. J. (2002), "Mutual Coupling between Microstrip Antennas on an Elliptic Cylinder," in *Proceedings of IEEE AP-S International Symposium,* Vol. 1, pp. 554–557.

Wu K.-Y. and Kauffman J. F. (1983), "Radiation Pattern Computations for Cylindrical-Rectangular Microstrip Antennas," in *Proceedings of IEEE AP-S International Symposium,* Vol. 1, pp. 39–42.

Yee K. S. (1966), "Numerical Solution of Initial Boundary Value Problems Involving Maxwell's Equations in Isotropic Media," *IEEE Transactions on Antennas and Propagation,* Vol. AP-14, No. 3, pp. 302–307, May.

Yu W., Dey S., and Mittra R. (2000), "On the Modeling Periodic Structures using FDTD Method," *Microwave and Optical Technology Letters,* Vol. 22, No. 2, February.

Yu W. and Mittra R. (2000), "A Conformal FDTD Software Package for Modeling of Antennas and Microstrip Circuit Components," *IEEE Antennas and Propagation Magazine,* Vol. 42, No. 5, pp. 28–39, October.

Yu W. and Mittra R. (2001), "A Conformal Finite Difference Time Domain Technique for Modeling Curved Dielectric Surfaces," *IEEE Microwave and Guided Wave Letters,* Vol. 11, No. 1, pp. 25–27, January.

Yu W., Farahat N., and Mittra R. (2001), "Application of FDTD Method to Conformal Patch Antennas," in *Proceedings of IEEE AP-S International Symposium,* Vol. 2, pp. 362–365 June.

Zakharyev L. N., Lemanski A. A., and Shcheglov K. S. (1970), *Radiation from Apertures in Convex Bodies (Flush-Mounted Antennas),* Boulder, CO, The Golem Press. (Translated from the Russian by Petr Beckmann).

Zienkiewicz O. C. and Taylor R. L. (1989), *The Finite Element Method* (4th ed.). Vol. 1: *Basic Formulation and Linear Problems,* New York, McGraw-Hill, 1989.

5

GEODESICS ON CURVED SURFACES

In this chapter we will discuss geodesics on curved surfaces. This area is important for the analysis of conformal antennas using asymptotic techniques, as discussed in Chapter 4. Finding the proper rays connecting points on a curved surface has been studied for a long time within the area of differential geometry. For singly curved surfaces, the procedure can be reduced significantly since the singly curved surface is developable. Thus, the proper ray is a straight line on the developed surface. For arbitrarily curved surfaces, the full geodesic equation must be solved (numerically). However, for certain types of doubly curved surfaces the geodesic equation can be reduced, making the ray tracing manageable for many practically useful shapes of conformal antennas. This will be discussed thoroughly in this chapter, starting with the governing equations. Then we will give explicit equations for the geodesics on different curved surfaces and provide figures showing the properties of geodesics on curved surfaces, including ray splitting among other features.

5.1 INTRODUCTION

In Section 4.4, asymptotic ray-based techniques where described. They turned out to be very efficient tools when combined with the MoM or FE-BI methods for the analysis of conformal antennas. However, there is a fundamental question to be answered before these techniques can be applied: How does one find the proper ray(s) that connect(s) arbitrarily located points on a smooth surface? For a correct solution, the rays must obey certain conditions in order to be valid geodesics on the surface.

Conformal Array Antenna Theory and Design. By Lars Josefsson and Patrik Persson
© 2006 Institute of Electrical and Electronics Engineers, Inc.

The geodesics are not only useful for the study of fields on the surface on the structure, for example, when the mutual coupling between elements in an array is considered, but also for the analysis of radiation and scattering properties. Finding the geodesics is only one part of the problem, however; the other part is to find how an incoming wave or the radiated field is reflected/diffracted at the object/surface. The solution to this can be found by applying, for example, a shoot and bounce ray (SBR) method using the laws of reflection and diffraction.

The SBR method and other similar procedures have been treated in several antenna-related references; some examples are [Lee et al. 1988, Ling et al. 1989, Pérez et al. 1999, González et al. 2003, Zhou and Ling 2003]. But the ray tracing technique is also widely used within three-dimensional game programming and computer graphics. Here, ray tracing is a technique for obtaining three-dimensional graphics with advanced light interactions for creating pictures with mirrors, shadows, and so on. Obviously, there are a lot of ray tracing techniques developed in this area that can be used within the antenna community as well. Some examples of literature with an introduction to ray tracing for image synthesis are [Glassner 1989, Rademacher 1997]; a more solid mathematical background is given by Lengyel [Lengyel 2001].

In this chapter, we will not discuss the SBR method or any similar procedures. Instead, let us return to the problem of finding the geodesics between two points along a curved surface. This problem is also often referred to as a "ray tracing" problem. We will use this term frequently in the following text, but interpret it as finding the geodesic between two points on a surface. A complete treatment of this problem is beyond the scope of this book; the interested reader is referred to textbooks on differential geometry, but the basics will be described here. The history with geodesics begins with John Bernoulli's solution of the problem of the shortest distance between two points on a convex surface (1697–1698). Euler was the first to formulate the geodesic equation for rays on a surface in his article in 1732. However, it was the French mathematician J. Liouville who first used the term "geodesic line" in its present meaning [Struik 1988]. The geodesic concept has subsequently been used extensively in other disciplines, for example, the study of gravitation and light propagation in the universe [Rosquist 2001]. Ray theory and the geodesic concept were introduced for the analysis of antennas in the 1950s when GTD was introduced by Keller (see Section 4.4.2 for more details). In the beginning, only singly curved surfaces, for which ray tracing is simple, were treated. Geodesics on doubly curved surfaces were first used in the beginning of 1970s. One example is the study of radiation patterns due to sources on a prolate spheroid [Burnside 1972]. In this work, an approximate numerical ray tracing procedure was used. For surfaces of revolution, it is possible to do an accurate ray tracing based on analytical expressions, as we will demonstrate here.

The geodesic lines are given by a second-order differential equation, the geodesic equation, based on differential geometry satisfying the generalized Fermat's principle, as will be discussed below. Unfortunately, the solution to the differential equation is often difficult to find explicitly and, instead, some kind of numerical (ray tracing) procedure has to be used. However, the ray tracing procedure can be simplified for certain geometries for which the geodesic equation can be written and solved in closed form. This considerably reduces the numerical computations. The class of surfaces that can be analyzed in this way belongs to the geodesic coordinate system, that is, the parameter lines of the surface are orthogonal to each other. A particular advantage of analytically described geodesics is that all the ray parameters required in the GTD/UTD formalism can be written in closed form. This helps a lot in the analysis of conformal antennas using high-frequency approximations.

In this chapter, the discussion of geodesics is limited to convex surfaces, which are those mainly used for conformal antennas. This is not a severe limitation, however, since geodesics along concave or mixed convex–concave surfaces follow the same principles. First, some background in differential geometry necessary for the understanding of geodesics will be presented. The theory will then be applied to different convex surfaces, and closed-form expressions for the geodesics on both singly and rotational symmetric doubly curved surfaces will be given. In principle, these equations are ready to use in the GTD/UTD formulation when studying antennas on these surfaces. However, in some cases numerical computations of certain functions will be required. Special features of geodesics will also be discussed and figures showing these features will be provided. Finally, arbitrary surfaces will be discussed briefly.

5.1.1 Definition of a Surface and Related Parameters

Let us first define a general surface. For the analysis presented in this chapter, it is preferable if the surface can be expressed in an explicit form, that is, the coordinates of a point on the surface are given by two parameters u and v within a certain closed interval. Thus, we assume that the parametric equations of the surface are of the form:

$$x_i = X_i(u, v), \qquad \chi \leq u \leq \delta, \qquad \tau \leq v \leq \xi, \qquad i = 1 \ldots 3 \qquad (5.1)$$

where u and v take real values. The functions X_i are single, real valued functions and they are continuous. When keeping v constant in Equation (5.1), $\bar{x}$ depends on only one parameter u and thus determines a curve on the surface—a parametric curve, $v = $ constant. Another parametric curve is represented by $u = $ constant. When the constants vary, the surface is covered with a grid of parametric curves. The derivatives $\bar{x}_u$ and $\bar{x}_v$ are the tangents to the curve $v = $ constant and $u = $ constant, respectively. u and v are often called the curvilinear coordinates of a particular point (see Figure 5.1). For example, on a sphere latitude and longitude may be selected as the curvilinear coordinates of the spherical surface.

In Figure 5.2, an arc C is passing through a point P on a surface. Assume that the arc has endpoints A and B, and is given by equations of the form $u = u(t)$ and $v = v(t)$. Is this arc a geodesic or not?

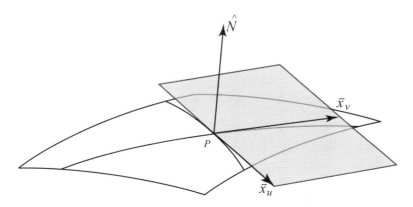

Figure 5.1. The u and v parameter lines with the tangent plane and the surface unit normal vector $\hat{N}$.

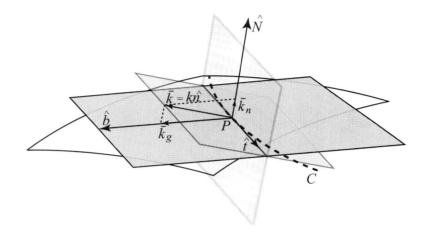

<u>Figure 5.2.</u> An arc C passing through a point P with the curvature vector $\overline{k}$ and unit tangential vector $\hat{t}$ shown.

Before this question can be answered, some additional parameters must be defined. The first fundamental form in differential geometry gives the length of the arc:

$$s(C) = \int_0^1 \sqrt{Eu'^2 + 2Fu'v' + Gv'^2}\, dt \qquad (5.2)$$

where $t = 0$ at A and $t = 1$ at B, and E, F, and G are the first fundamental coefficients given by

$$E = \overline{x}_u \cdot \overline{x}_u, \qquad F = \overline{x}_u \cdot \overline{x}_v, \qquad G = \overline{x}_v \cdot \overline{x}_v \qquad (5.3)$$

and the derivatives are taken with respect to the parameter t. Observe that a prime will denote differentiation with respect to t, unless otherwise stated.

The curvature vector at a point P, through which an arc C is passing, has to be defined as well. Decomposed into normal and tangential components to the surface, the curvature vector is given as

$$d\hat{t}/ds = \overline{k} = \overline{k}_n + \overline{k}_g = k\hat{n} \qquad (5.4)$$

where $\hat{n}$ is the unit normal vector at P of the arc C. $\overline{k}_n$, the normal curvature vector, can be expressed in terms of the unit surface normal vector $\hat{N}$ as

$$\overline{k}_n = \kappa_n \hat{N} \qquad (5.5)$$

where κ_n is the normal curvature and will not be given explicitly here. Similarly, $\overline{k}_g$ is the tangential curvature vector or geodesic curvature vector. By introducing the tangential or geodesic curvature κ_g, the vector can be written as

$$\overline{k}_g = \kappa_g \hat{b} \qquad (5.6)$$

It can be shown (see, e.g. [Struik 1988]) that the geodesic curvature, contrary to the normal curvature above, depends only on E, F, G, and their derivatives together with the equation $\psi(u, v) = 0$ defining the ray. This relation for κ_g is given as

$$\kappa_g = [\Gamma^2_{11}(u')^3 + (2\Gamma^2_{12} - \Gamma^1_{11})(u')^2 v' + (\Gamma^2_{22} - 2\Gamma^1_{12})u'(v')^2 - \Gamma^1_{22}(v')^3$$
$$+ u'v'' - u''v'] \sqrt{EG - F^2} \tag{5.7}$$

where the root is taken to be positive and the gamma functions are the Christoffel symbols [Struik 1988, pp. 107–108], depending only on the coefficients of the first fundamental form and their first derivatives.

5.1.2 The Geodesic Equation

As mentioned earlier, the details of the derivation of the geodesic equation are beyond the scope of this book. But, by using the defined parameters above, a schematic derivation of the geodesic equation will be shown, which will end with a definition of a geodesic.

Geodesics are often referred to or defined as lines of the shortest distance between points on a surface. This is also the oldest way of describing geodesics, as was mentioned in the beginning of this section. Therefore, the geodesic equation has often been found from a study of a variational problem. Another way is to consider the geodesic curvature in Equation (5.6) and define geodesics as curves of zero geodesic curvature. Both ways give the same answer and it turns out that they are connected to each other, as will be described below. First, the variation problem will be considered since it may give a more intuitive way of describing geodesics.

Let us once again consider the arc C shown in Figure 5.2, given by the equations $u = u(t)$ and $v = v(t)$. Assume now that an arc C' is obtained by deforming C slightly, but keeping its end points fixed at A and B. Thus, C' is given as

$$u = u(t) + \varepsilon\xi(t), \qquad v = v(t) + \varepsilon\zeta(t) \tag{5.8}$$

where ε is small and ξ and ζ are arbitrary functions in $0 \le t \le 1$ that satisfy $\xi = \zeta = 0$ at $t = 0$ and $t = 1$. Now, $s(C)$ is said to be stationary if the variation in $s(C)$, that is, $s(C) - s(C') = \delta s$, is at most of order ε^2 for small variations in C [Willmore 1984]. If this is the case, C is a geodesic. Geometrically, this can be explained by stretching an elastic band between two points and making sure that the band remains on the surface.

It can be shown by a more complete analysis that arcs that solve the variation problem are solutions to the Euler–Lagrange equations. Thus, we have

$$\frac{\partial f}{\partial u} - \frac{d}{dt}\left(\frac{\partial f}{\partial u'}\right) = 0 \tag{5.9}$$

$$\frac{\partial f}{\partial v} - \frac{d}{dt}\left(\frac{\partial f}{\partial v'}\right) = 0 \tag{5.10}$$

where $f = \sqrt{Eu'^2 + 2Fu'v' + Gv'^2}$. However, it follows also that the two equations are not independent; they can be reduced to one equation with a solution of the form $v = v(u)$. Thus, the new equation is a second-order differential equation:

$$\frac{d^2 v}{du^2} = \Gamma^1_{22}\left(\frac{dv}{du}\right)^3 + (2\Gamma^1_{12} - \Gamma^2_{22})\left(\frac{dv}{du}\right)^2 + (\Gamma^1_{11} - 2\Gamma^2_{12})\left(\frac{dv}{du}\right) - \Gamma^2_{11} \tag{5.11}$$

This is the *geodesic equation* in its most general form. The solution to Equation (5.11) with given A and B positions, $v = v(u)$, is the geodesic connecting A and B on the given surface. In fact, every solution $v = v(u)$ that fulfills the requirements above is called a geodesic, whether it is an arc of shortest distance or not. Thus, geodesics may be regarded as curves of stationary rather than strictly shortest distances on the surface. As an example, two points on the surface of a circular cylinder can be connected with infinitely many curves, and all of them are geodesics but only one of them is the shortest one. If the points are located at different positions along the axial direction, every geodesic is a helix. But there are many helices with different inclinations, so there are a lot of geodesics connecting the two points. In addition, there are curves encircling the cylinder in both the counter-clockwise and clockwise directions.

When the expression for the variation in $s(C)$, that is, δs above, is studied in more detail, it is found to be related to the geodesic curvature κ_g. Furthermore, the condition that gives the Euler–Lagrange equations turns out to be fulfilled if the geodesic curvature κ_g vanishes along the arc C between A and B. Thus, the same equation as given in Equation (5.11) can be obtained from the condition $\kappa_g = 0$!

To summarize, a definition of a geodesic that is only based on the length is not a useful one. A more general definition of a geodesic given in [Struik 1988] is "a curve of zero geodesic curvature." This definition can also be expressed as the osculating plane of a geodesic contains the surface normal, or along the geodesic the principal normal ($\hat{n}$) coincides with the surface normal ($\hat{N}$). Figure 5.3 illustrates a situation when a curve along the surface is not a geodesic.

Note that the normal property strictly fails in the case of a straight line on a (plane) surface, since the principle normal $\hat{n}$ is indeterminate. But, still, the straight line is a geodesic! This fact, according to some authors writing about geodesics, has been taken as a reason for not using the above general definition of a geodesic. However, here we will adopt this definition since it is useful and gives an intuitive way of defining geodesics on curved surfaces.

5.1.3 Solving the Geodesic Equation and the Existence of Geodesics

As can be understood from the above, it is not trivial, neither from an analytical nor numerical point of view, to solve the second-order geodesic differential equation for an arbi-

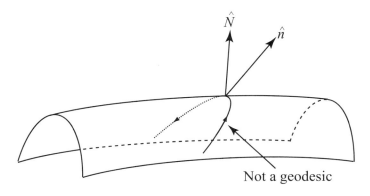

Figure 5.3. A curve along a surface, but not a geodesic surface ray.

trary surface. In addition, the question arises as to whether a solution exists or not, as well as whether the solution is unique.

Starting with the existence of geodesics, it is known from Equation (5.11) that when a direction dv/du is given at a point $P(u, v)$, d^2v/du^2 is determined. This tells us the way in which the ray of a given direction is continued. And from the existence theorem* for the solutions of ordinary differential equations of the second order, the following properties are known: Through every point of the surface passes a geodesic in every direction. A geodesic is uniquely determined by an initial point and tangent at that point.

Thus, it is to be expected that it is possible to find a direction at P such that the geodesic through P also passes through another point Q. However, from a mathematical point of view, there is another theorem that says that Q can be joined to P if it is sufficiently near P; that is, it might be possible to find two points on a surface that cannot be connected with a geodesic. However, we also have to define the surface. If the surface is complete it will appear that any two points can be connected by at least one geodesic [Willmore 1984], that is, it is possible to find at least one (unique) solution to the geodesic equation.

With this in mind, the remaining part of this section will be about solving the geodesic equation in general for different geometries. For an arbitrary surface, described either numerically or analytically, the geodesic equation must be solved using numerical methods. However, the procedure can be simplified for certain geometries. In these cases, the geodesic equation can be written as a first-order differential equation and, thus, the geodesics can be found more easily. Actually, the class of geometries that can be solved with the first-order equation is quite large. Thus, a faster analysis can be performed for many geometries of interest within the conformal antenna area. Here, we will not treat the purely numerical solution. The emphasis will be on surfaces that can be analyzed with first-order differential equations, for example, cylinders and paraboloids. These surfaces are very useful shapes for many types of conformal antennas since they can model, for example, base station antennas or the nose or fuselage of an aircraft. The geodesics in these cases can be found by quadratures.

If E, F, and G are functions of one parameter only, say u, the geodesic equation in (5.11) is reduced to

$$\frac{dv}{du} = -\frac{F}{G} \pm \frac{\alpha\sqrt{EG - F^2}}{G\sqrt{G - \alpha^2}} \Rightarrow v(u) = \int \left[-\frac{F}{G} \pm \frac{\alpha\sqrt{EG - F^2}}{G\sqrt{G - \alpha^2}} \right] du + \beta \quad (5.12)$$

where E, F, and G are the first fundamental coefficients given in Equation (5.3), and α is a constant of integration, known as the first geodesic constant. The $\pm$ sign in front of α depends on whether v is monotonically increasing or decreasing with u. The solution to Equation (5.12), which relates the geodesic coordinates u and v along the geodesic, can, in general, be written as

$$v(u) = \Im(u, \alpha) + \beta \quad (5.13)$$

where $\Im(u, \alpha)$ is a real-valued function of u and α, and β is the second geodesic constant. The physical significance of α and β is that they uniquely characterize a geodesic. With

---

*The existence theorem states that a differential equation $d^2v/du^2 = g(u, v, dv/du)$, where g is a single-valued and continuous function of its independent variables inside a certain given interval (with a Lipschitz condition satisfied in this case), has inside this interval a unique continuous solution for which dv/du takes a given value $(dv/du)_0$ at a point $P(u_0, v_0)$ [Struik 1988].

this equation known, and for many cases the solution is also known, the ray tracing procedure becomes much faster.

Equation (5.12) can be simplified even further if the surface can be expressed in the so-called geodesic coordinate system [Struik 1988]. A coordinate surface belongs to the geodesic coordinate system if it can be expressed in such a way that the parameter lines are orthogonal, that is, $F = 0$, see Equation (5.3). Examples of surfaces that belong to the geodesic coordinate system are any of the eleven surfaces defined by the Eisenhart coordinate system [Spiegel 1994]. For these surfaces, the first-order geodesic differential equation given in Equation (5.12) is simplified to

$$\frac{dv}{du} = \frac{\pm\alpha\sqrt{E}}{\sqrt{G}\sqrt{G-\alpha^2}} \Rightarrow v(u) = \int \frac{\pm\alpha\sqrt{E}}{\sqrt{G}\sqrt{G-\alpha^2}}\, du + \beta \qquad (5.14)$$

As will be shown in Sections 5.2 and 5.3, the integral in Equation (5.14) can be solved in closed form for the surfaces defined by the Eisenhart coordinate system. Actually, this approach can be extended to other surfaces as well; for example the ogive was considered in [Jha et al. 1989a]. In fact, any rotational symmetric surface with an analytically described cross section can be solved with Equation (5.14) [Rosquist 2001]. This will be discussed in Section 5.4.

The solution to Equation (5.14) provides the general solution for the specific geometry. To find the particular solution that connects A and B on a surface, the geodesic constants α and β have also to be found. For many of the applications related to conformal antennas, the points between which the geodesics traverse are known a priori, since the positions of the different antenna elements are known, as well as the field points or diffraction points. Thus, α and β can be found by solving the following system of equations:

$$\begin{aligned} v_{\text{start}} &= \Im(u_{\text{start}}, \alpha) + \beta \\ v_{\text{stop}} &= \Im(u_{\text{stop}}, \alpha) + \beta \end{aligned} \qquad (5.15)$$

When the geodesics are known, various well-known properties and standard results from differential geometry follow directly. We are thus ready to apply our ray-based theory for the analysis of conformal antennas. It turns out that both the ray parameters of the differential type as well as the integral type can be found very easily for the geometries where $\Im(u, \alpha)$ has been found. In fact, the only unknown in these expressions is the first geodesic constant α. Hence, the accuracy of the results obtained with a ray-based theory depends critically on the accuracy of the first geodesic constant α found from Equation (5.15). As a consequence, some authors have called this method the geodesic constant method (GCM) [Jha and Wiesbeck 1995].

This is as far as a general analytical approach is possible. In practice, numerical techniques are often necessary in order to determine the first and second geodesic constants. Note that from the theory above, we know that there can be more than one geodesic line connecting two given points on a surface, and each one of these possible geodesics is related to a unique set of geodesic constants α and β. However, the maximum number of geodesics between two points is different depending on the shape of surface. This will be discussed and shown in the following sections.

5.2   SINGLY CURVED SURFACES

In this section, the emphasis will be on solving the geodesic equation for singly curved surfaces. Singly curved surfaces may also be called developable surfaces since they can be unfolded and analyzed as a flat surface and the key parameters are obtained from a two-dimensional analysis (see Figure 5.4). As a consequence, the geodesics on any singly curved surface can always be found, at least numerically. Thus, the singly curved surfaces can be studied without solving Equation (5.15).

However, there is an advantage in actually solving the geodesic equation directly without unfolding the surface. The reason is that the first and second geodesic constants will be then known and with them also some very useful general ray parameter results from differential geometry. With this, a more general computer code can be developed for analyzing both singly and doubly curved surfaces.

The general expression for a singly curved surface in the geodesic coordinate system is

$$x = f(u), \qquad y = g(u), \qquad z = v \tag{5.16}$$

and the geodesic equation for this singly curved surface becomes

$$\frac{dv}{du} = \frac{\pm \alpha \sqrt{(f'(u))^2 + (g'(u))^2}}{\sqrt{1 - \alpha^2}} \tag{5.17}$$

with the solution

$$v(u) = \frac{\pm \alpha}{\sqrt{1 - \alpha^2}} \int \sqrt{(f'(u))^2 + (g'(u))^2} \, du + \beta = \Im(u, \alpha) + \beta \tag{5.18}$$

This equation defines all geodesics. The geodesic constants α and β for a particular geodesic are found from Equation (5.15). Table 5.1 shows a list of the explicit equations for

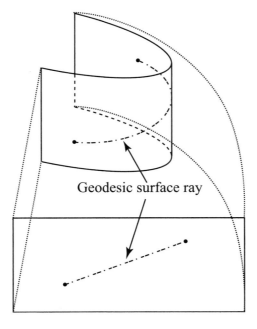

Figure 5.4. A developed cylindrical surface.

Table 5.1. Explicit equations for geodesics on different singly curved surfaces. See also [Jha et al. 1989b]

Geometry	Parametric equation	Geodesic coordinates	Solution to geodesic equation
Circular cylinder	$x = a \cos \varphi$ $y = a \sin \varphi$ $z = z$ $0 \leq \varphi \leq 2\pi$ $a > 0$	$u = \varphi$ $v = z$	$v = \dfrac{\alpha}{\sqrt{1 - \alpha^2}}[au] + \beta$
Elliptic cylinder	$x = a \cos \varphi$ $y = b \sin \varphi$ $z = z$ $0 \leq \varphi \leq 2\pi$ $a, b > 0$	$u = \varphi$ $v = z$	$v = \dfrac{\alpha}{\sqrt{1 - \alpha^2}}[bE(u, k)] + \beta$ $k = \dfrac{\sqrt{b^2 - a^2}}{b}$
Parabolic cylinder	$x = au$ $y = u^2$ $z = z$ $-\infty \leq u \leq \infty$ $a > 0$	$u = u$ $v = z$	$v = \dfrac{\alpha}{\sqrt{1 - \alpha^2}}\left[\dfrac{u\sqrt{a^2 + 4u^2}}{2} + \dfrac{a^2}{4} \right.$ $\left. \ln(2u + \sqrt{a^2 + 4u^2}) \right] + \beta$
Rectangular hyperbolic cylinder	$x = a \cosh \gamma$ $y = a \sinh \gamma$ $z = z$ $-\infty \leq y \leq \infty$ $a > 0$	$u = y$ $v = z$	$v = \dfrac{\alpha a}{\sqrt{1 - \alpha^2}}\left\{ \dfrac{1}{\sqrt{2}}[F(r, k) - 2E(r, k)] \right.$ $\left. + \dfrac{\sinh (2u)}{\sqrt{\cosh^2 (u)}} \right\} + \beta$ $r = \sin^{-1}\left[\dfrac{\cosh (2u) - 1}{\cosh (2u)} \right]$ $k = \dfrac{1}{\sqrt{2}}$
General hyperbolic cylinder	$x = a \cosh \gamma$ $y = b \sinh \gamma$ $z = z$ $-\infty \leq y \leq \infty$ $a, b > 0$	$u = y$ $v = z$	$v = \dfrac{\alpha}{\sqrt{1 - \alpha^2}}\{b[F(r, k) - E(r, k)] +$ $[\tanh (u)] \sqrt{a^2 \sinh^2 (u) + b^2 \cosh^2 (u)}\} + \beta$ $r = \sin^{-1}[\tanh (u)]$ $k = \dfrac{a}{b}$

geodesics on common singly curved surfaces in the Eisenhart coordinate system. Observe that u and v are chosen in such away that E, F, and G become functions of only one variable. In Table 5.1 $F(r, k)$ and $E(r, k)$ are the incomplete elliptic integrals of first and second kind [Bulirsch 1969, Erdelyi et al. 1953, Spiegel 1994].

In Figures 5.5 and 5.6 some examples of geodesics on singly curved surfaces are shown. Figure 5.5 shows the only possible geodesic connecting A and B on a (infinitely long) parabolic cylinder. Figure 5.6 shows the first- and second-order geodesics connecting A and B on an elliptic cylinder.

As mentioned earlier, there is at least one geodesic between two points A and B. For closed surfaces, for example, an elliptic cylinder, the geodesics between A and B can be

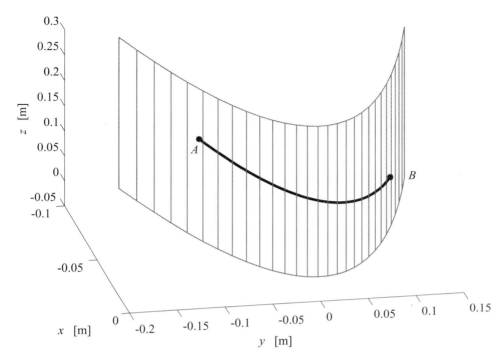

<u>Figure 5.5.</u> The geodesic connecting A and B on a parabolic cylinder. Here, $a = 0.5$, $\alpha = -0.414$, and $\beta = 0.13$.

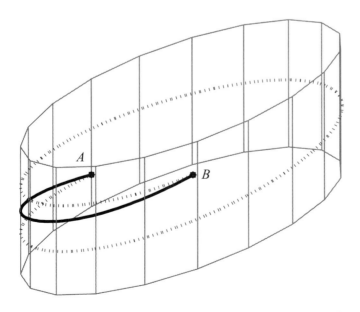

<u>Figure 5.6.</u> The first- and second-order geodesics connecting A and B on an elliptic cylinder with the following parameters: $a = 0.3$, $b = 0.8$, $\alpha_1 = 0.14$, $\beta_1 = 0$, $\alpha_2 = 0.0396$, and $\beta_2 = 0$.

drawn in the clockwise as well as the counterclockwise directions. Besides, in each direction there may be higher-order geodesics (multiple encircling surface ray paths) in addition to the primary geodesics. To obtain the geodesic constants for higher-order geodesics, the observation point coordinates for the mth-order geodesic in the counterclockwise direction are:

$$B_m(u_P, v_P) = (u_P + 2(m - 1)\pi, v_P) \qquad m \geq 1 \qquad (5.19)$$

Higher-order geodesics on a circular or elliptic cylinder are easily identified by a two-dimensional analysis. If the cylinder is developed into a plane, we find infinitely many images B_m of B. This is illustrated in Figure 5.7, in which a closed and the corresponding developed circular cylinder are shown.

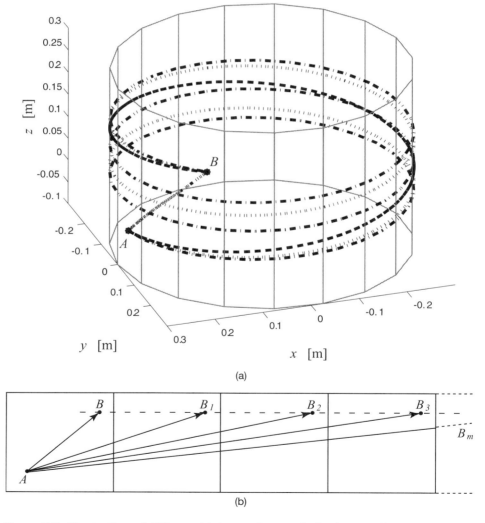

Figure 5.7. Illustration of different higher-order geodesics in one direction connecting A and B. (*a*) The closed circular cylinder. (*b*) The developed circular cylinder.

Our examples of Eisenhart surfaces above can be extended to other, more general surfaces. For many functions defining the surface, that is, $f(u)$ and $g(u)$, the integral in Equation (5.18) can be solved in closed form. If this is not possible, the integral can be computed numerically, or the ray tracing can be solved by developing the surface and solving it as a two-dimensional problem.

5.3 DOUBLY CURVED SURFACES

5.3.1 Introduction

In this section, doubly curved surfaces will be considered, in particular, rotationally symmetric surfaces. More general surfaces will be discussed in Section 5.4. Solving the geodesic equation for doubly curved surfaces is, of course, more difficult than for singly curved surfaces. Often, the equation has to be solved numerically, resulting in a time-consuming analysis. However, as will be shown in this section, the geodesic equation for rotationally symmetric doubly curved surfaces can in many cases be solved in closed form. This simplifies the analysis of doubly curved surfaces and makes it possible to use a ray-based theory for the study of complex, doubly curved conformal array antennas.

Once again, we start by describing the surface with a parametric equation in the geodesic coordinate system. For rotationally symmetric surfaces we get

$$x = f(u) \cos v, \qquad y = f(u) \sin v, \qquad z = g(u) \qquad (5.20)$$

and the solution to the geodesic equation for rotational symmetric doubly curved surfaces becomes

$$v(u) = \int \frac{\pm \alpha \sqrt{(f'(u))^2 + (g'(u))^2}}{f(u)\sqrt{(f(u))^2 - \alpha^2}} \, du + \beta = \Im(u, \alpha) + \beta \qquad (5.21)$$

As in the singly curved case, α and β are unique for a specific geodesic between the given points A and B. They can be obtained in the same way as before, that is, by solving

$$v_A = \Im(u_A, \alpha) + \beta$$
$$v_B = \Im(u_B, \alpha) + \beta \qquad (5.22)$$

The geodesics on any doubly curved rotationally symmetric surface can now be found by quadrature. However, analyzing doubly curved surfaces gets even more interesting when it turns out that the integral in Equation (5.21) can be solved in closed form for many rotationally symmetric, doubly curved surfaces, in particular, those belonging to the Eisenhart coordinate system.

5.3.2 The Cone

The cone can be viewed as a transition form between singly and doubly curved surfaces since it is a developable surface, but at the same time has properties related to doubly curved surfaces. For the closed, singly curved surfaces presented earlier in Section 5.2 the number of geodesics connecting two points are infinite. The cone, on the other hand, is an example of a developable surface in which the number of geodesics connecting two

Table 5.2. Explicit equation for geodesics on a circular cone. See also [Jha et al. 1989b]

Geometry	Parametric equation	Geodesic coordinates	Solution to geodesic equation
Circular cone	$x = r \sin \theta_0 \cos \varphi$ $y = r \sin \theta_0 \sin \varphi$ $z = -r \cos \theta_0$ $0 \le \varphi \le 2\pi$ $r \ge 0$	$u = r$ $v = \varphi$	$v = \dfrac{1}{\sin \theta_0} \sec^{-1}\left(\dfrac{u \sin \theta_0}{\alpha} \right) + \beta$
	θ_0 is the half-cone angle		

points can be more than one but is bounded. The same feature applies to doubly curved surfaces, as will be shown later.

In Table 5.2, the equations related to the circular cone are given, as well as the solution to the geodesic equation. Figure 5.8 shows a typical case of a geodesic on a cone.

Since the cone is a developable surface, a simple but illustrative way of viewing the geodesics is to unfold the surface. In the two-dimensional plane, the cone is unrolled into a set of plane sectors. In Figure 5.9, a cone is unrolled with A as the starting point and B (with its images) as the end point.

In Figure 5.9 (*a*) an example is shown in which it is possible to find two geodesics in the positive φ direction (A–B, A–B_1) and two in the negative φ direction (A–B_{-1}, A–B_{-2}). These are the only possible geodesics for this particular cone, since the lines A–B_i must all be on the right side of Q (which is not the case with A–B_2), and all lines A–B_{-i} must be on the left side of Q. Figure 5.9 (*b*) shows a case in which there is only one single geodesic connecting A and B. Clearly, the smaller half-cone angle of the cone, the greater the number of geodesics; but there cannot be infinitely many geodesics connecting the two given

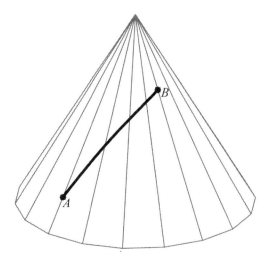

Figure 5.8. A solution to the geodesic equation for a cone with $\theta_0 = 30°$. Here $\alpha = -0.0869$ and $\beta = -221°$.

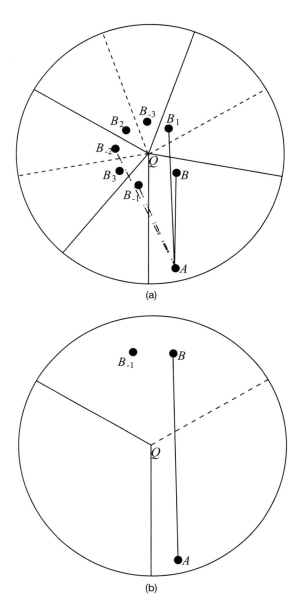

Figure 5.9. The unrolled cone for two different values of the half-cone angle θ_0. (a) θ_0 < 45°. (b) θ_0 > 45°.

points. Thus, the number of geodesics is limited for the cone, as is the case for general doubly curved surfaces, see Section 5.3.4.

5.3.3 Rotationally Symmetric Doubly Curved Surfaces

By extending the analysis to general rotationally symmetric doubly curved surfaces, we can study more surfaces of great practical interest. Table 5.3 shows the explicit equations

Table 5.3. Explicit equations for geodesics on rotationally symmetric doubly curved surfaces. See also [Jha et al. 1987, Jha et al. 1989c, Jha et al. 1991, Choudhury and Jha 1995]

Geometry	Parametric equation	Geodesic coordinates	Solution to geodesic equation
Sphere	$x = a \sin \theta \cos \varphi$ $y = a \sin \theta \sin \varphi$ $z = a \cos \theta$ $0 \leq \varphi \leq 2\pi$ $0 \leq \theta \leq \pi$ $a > 0$	$u = \theta$ $v = \varphi$	$v = -\tan^{-1}\left(\dfrac{a \cos u}{\sqrt{a^2 \sin^2 u - \alpha^2}} \right) + \beta$
Ellipsoid of revolution	$x = a \sin \theta \cos \varphi$ $y = a \sin \theta \sin \varphi$ $z = b \cos \theta$ $0 \leq \varphi \leq 2\pi$ $0 \leq \theta \leq \pi$ $a, b > 0$	$u = \theta$ $v = \varphi$	$v = \dfrac{-\alpha}{b}\left[\Pi(n, r, k) + \dfrac{b^2 - a^2}{a^2} F(r, k) \right] + \beta$ $n = \dfrac{a^2 - \alpha^2}{a^2}$ $r = \dfrac{a \cos u}{\sqrt{a^2 - \alpha^2}}$ $k = \dfrac{\sqrt{b^2 - a^2}\,\sqrt{a^2 - \alpha^2}}{ab}$
General paraboloid of revolution (GPOR)	$x = au \cos \varphi$ $y = au \sin \varphi$ $z = -u^2$ $0 \leq u \leq \infty$ $0 \leq \varphi \leq 2\pi$ $a > 0$	$u = u$ $v = \varphi$	$v = \dfrac{\alpha}{a^2} \ln \dfrac{a\sqrt{a^2 + 4u^2} + 2\sqrt{a^2 u^2 - \alpha^2}}{a\sqrt{a^2 + 4u^2} - 2\sqrt{a^2 u^2 - \alpha^2}} +$ $\sin^{-1}\left[\dfrac{a\sqrt{a^2 u^2 - \alpha^2}}{u\sqrt{a^4 + 4\alpha^2}} \right] + \beta$
General hyperboloid of revolution (GHOR)	$x = a \sinh \gamma \cos \varphi$ $y = a \sinh \gamma \sin \varphi$ $z = b \cosh \gamma$ $0 \leq \gamma \leq \infty$ $0 \leq \varphi \leq 2\pi$ $a, b > 0$	$u = \gamma$ $v = \varphi$	$v = \dfrac{\alpha}{b}\left\{ \Pi[n, r, k] - \dfrac{a^2 + c^2}{a^2} F[r, k] \right\} + \beta$ $n = -\dfrac{a^2 + \alpha^2}{a^2}$ $r = \dfrac{a \cosh u}{\sqrt{a^2 + \alpha^2}}$ $k = \dfrac{\sqrt{a^2 + b^2}\,\sqrt{a^2 + \alpha^2}}{ab}$

for geodesics on several rotationally symmetric doubly curved surfaces in the Eisenhart coordinate system. Observe that u and v are chosen in such away that E, F, and G are functions of only one variable.

Some of the equations presented above appear as functions of the incomplete elliptic integrals of the first $[F(r, k)]$ and third kind $[\Pi(n, r, k)]$ which require numerical computations; see, for example, [Bulirsch 1969, Erdelyi et al. 1953, Spiegel 1994].

The analysis presented in this section can also be extended to non-Eisenhart surfaces. The only requirement is that the surface can be defined in the geodesic coordinate system, that is, the parameterization is such that the parameter lines are orthogonal. One example of a non-Eisenhart surface is the ogive (Figure 5.10), which is of great interest in aerospace engineering since it can describe many of the shapes encountered in the area. The

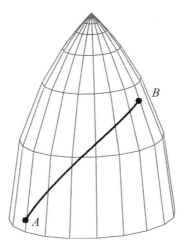

Figure 5.10. A solution to the geodesic equation for an ogive with $R = 0.5$ and $a = 3\pi/4$. Here $\alpha = 0.1329$ and $\beta = -6.7907°$.

ogive is not a coordinate surface of an Eisenhart coordinate system and cannot be parameterized as such. However, the ogive can be identified as the coordinate surface of the bispherical coordinate system [Jha et al. 1989a]. In Table 5.4, the parameterization of the ogive is presented. This parameterization can be shown to be an example of a geodesic coordinate system. In Figure 5.10, an example of an ogive is shown, with $R = 0.5$ and $a = 3\pi/4$; a geodesic connecting A and B is also shown.

5.3.4 Properties of Geodesics on Doubly Curved Surfaces

Since the properties of geodesics on doubly curved surfaces are less trivial than those on singly curved surfaces, a number of special cases will be discussed. To do this, the exam-

Table 5.4. The parameterization of the ogive; R and a are the shaping parameters of the ogive. See also [Jha et al. 1989a]

Geometry	Parametric equation	Geodesic coordinates	Solution to geodesic equation
Ogive	$x = \dfrac{R \sin a}{\cosh \gamma - \cos a} \cos \varphi$	$u = \gamma$ $v = \varphi$	$v(u) = \displaystyle\int \dfrac{\pm\alpha \sqrt{(f'(u))^2 + (g'(u))^2}}{f(u)\sqrt{(f(u))^2 - \alpha^2}}\, du + \beta$
	$y = \dfrac{R \sin a}{\cosh \gamma - \cos a} \sin \varphi$		where
	$z = \dfrac{R \sinh \gamma}{\cosh \gamma - \cos a}$		$f(u) = \dfrac{R \sin a}{\cosh u - \cos a}$
	$0 \le \varphi \le 2\pi$ $-\infty \le \gamma \le \infty$ $a > \pi/2$		$f'(u) = \dfrac{(-1)R \sinh u \sin a}{(\cosh u - \cos a)^2}$
			$g'(u) = \dfrac{R(1 - \cosh u \cos a)}{(\cosh u - \cos a)^2}$

ple of a general paraboloid of revolution (GPOR) is taken since it can also model parts of a sphere or an ellipsoid. Thus, the results obtained here can to some extent also be generalized to other doubly curved surfaces such as those shown in Table 5.3.

Geodesics between A and B on the GPOR can be drawn both in the clockwise (negative φ direction) and the counterclockwise (positive φ direction) directions. Besides, in each direction there can be higher-order geodesics (multiple encircling surface ray paths) in addition to the primary geodesics. The analysis presented here is valid for all geodesics on the surface, but we focus on different orders of geodesics in the positive φ direction. The general expression for geodesics on a GPOR was given in Table 5.3 and the equations to be solved for a specific geodesic between A and B are given by

$$\begin{aligned} \varphi_S &= \Im(u_S, \alpha) + \beta \\ \varphi_P &= \Im(u_P, \alpha) + \beta \end{aligned} \qquad (5.23)$$

More generally, the observation point coordinates for the mth-order geodesic (in the positive φ direction) are:

$$P_m(u_P, \varphi_P) = (u_P, \varphi_P + 2(m-1)\pi) \qquad m \geq 1 \qquad (5.24)$$

In the first example, the source point A is located at $u_A = 4$, $\varphi_A = 20°$, the observation point is located at $u_B = 6.5$, $\varphi_B = 140°$, and the shaping parameter is $a = 0.4$ (a determines the sharpness of the surface). Figure 5.11 (a)–(d) shows the first-, second-, third-, and fourth-order geodesics respectively. Except for these four geodesics, no higher-order geodesics in the positive φ direction exist for this particular value of a. Note that all these geodesics are monotonically increasing (or decreasing) functions of u.

Let us now consider the same example as above but with the shaping parameter a increased from 0.4 to 0.6. In this case, a third- and fourth-order geodesic no longer exists. If a is increased even further, the second-order geodesic vanishes as well. All of the geodesics discussed above are monotonically decreasing/increasing with u. However, in general, one could have A and B connected with a combination of both decreasing and increasing geodesics (as a function of the parameter u). In such cases, the geodesics may be treated in two parts: A to C as the decreasing/increasing geodesic, and C to B as the increasing/decreasing geodesic. For the case when $a = 1.6$, a monotonically decreasing or increasing second-order geodesic does not exist, but a second-order geodesic in the positive φ direction with an increasing and decreasing part is possible to find. Such a second-order geodesic is shown in Figure 5.12 (b). A similar analysis can be done for geodesics in the negative φ direction.

In the examples above, the number of higher-order geodesics for the GPOR depends on the shaping parameter a. This is explained by the shape of the surface; compare this to the cone considered earlier. When a is small, the surface looks like a circular cylinder, except for the upper part of the GPOR. Then it is possible to find several higher-order geodesics. When a is increased, the surface gradually becomes more sphere-shaped and the higher order (monotonically increasing or decreasing) geodesics vanish. In Section 5.3.5, a parameter study is done illustrating this for the case when $u_A = u_B$. The same result is obtained for a hyperboloid of revolution and for an ellipsoid. However, for the ellipsoid there can exist additional higher-order geodesics since the surface is closed. Thus, a geodesic starting at the upper part of the ellipsoid can turn back at the lower pole before it reaches the end point. In fact, along the principal planes of the ellipsoid there exist infinitely many geodesics since the surface is reduce to elliptic cylinders at these positions. The sphere is also a special case since infinitely many geodesics exist for any combination of given points on the surface.

5.3.5 Geodesic Splitting

There is one special combination of source and observation points that has not been discussed yet—the case when both A and B are at the same u value. For this case, a phenomenon called geodesic splitting has been observed. Before this phenomenon is described, it

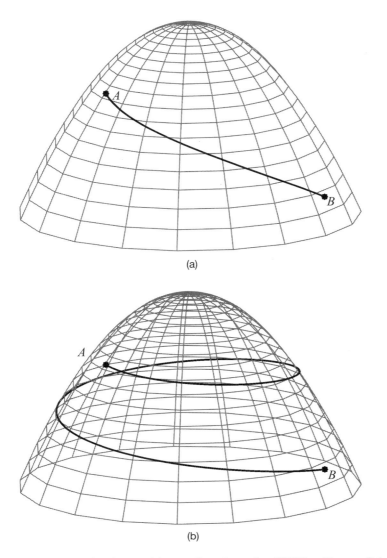

(a)

(b)

Figure 5.11. Geodesics in the positive φ direction of a GPOR with $a = 0.4$. (*a*) The first-order geodesic; $\alpha = 0.33981$ and $\beta = -529.5008°$. (*b*) The second-order geodesic; $\alpha = 1.1303$ and $\beta = -696.0058°$. (*continued*)

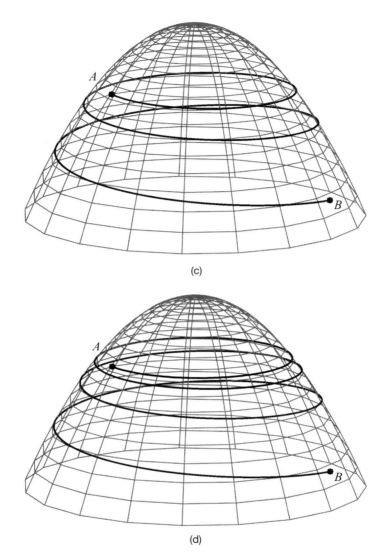

(c)

(d)

Figure 5.11. (*cont.*) Geodesics in the positive φ direction of a GPOR with $a = 0.4$. (*c*) The third-order geodesic; $\alpha = 1.4983$ and $\beta = -373.6902°$. (*d*) The fourth-order geodesic; $\alpha = 1.5997$ and $\beta = -3.563°$.

is worth pointing out that the geodesics that result when $u_A = u_B$ for a general paraboloid of revolution cannot be along the φ parameter line according to the definition of a geodesic. Thus, the geodesic between A and B in this case must be an increasing and decreasing geodesic. Such a case is illustrated in Figure 5.13 for the source point A at $u_A = 2$, $\varphi_A = 0°$, the observation point B at $u_B = 2$, $\varphi_B = 100°$, and $a = 1.6$.

Now, what about this geodesic splitting phenomenon? It appears for certain combinations of source and observation coordinates (A and B) for a particular shaping parameter a. In such a case, the geodesic equation results in two distinct values for α, thereby resulting in two distinct geodesics for a given direction and order. This phenomenon is called

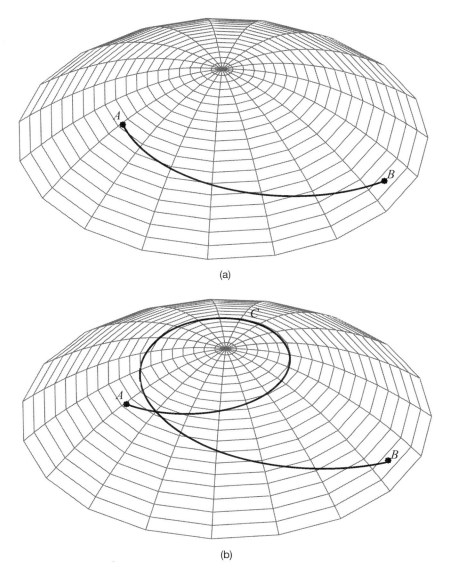

Figure 5.12. A GPOR in which $a = 1.6$ ($u_A = 4$, $\varphi_A = 20°$; $u_B = 6.5$, $\varphi_B = 140°$). (a) The first-order geodesic in the positive φ direction; $\alpha = 4.4851$ and $\beta = -165.218°$. (b) The nontrivial second-order geodesic in the positive φ direction; $\alpha = 3.4$ and $\beta = -217.4338°$.

geodesic splitting, reported within the antenna community by Jha et al. in 1989 [Jha et al. 1989d, Jha et al. 1992]. This was unexpected, since it has been known that between any two arbitrarily located points on a cone or a cylinder there exist primary and higher-order (of multiple encirclements) geodesics in both the counterclockwise and clockwise directions, but the number of geodesics of a given order and direction never exceeds one. In contrast, for general paraboloids of revolution, the geodesic of a given order and direction can be split into two!

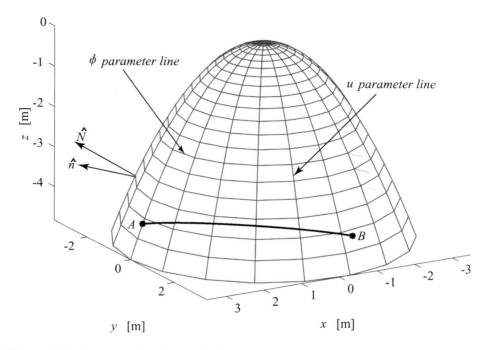

Figure 5.13. An example of a geodesic when $u_A = u_B$ but no geodesic splitting occurs. Furthermore, for the φ parameter lines the normal direction and the surface normal direction are indicated. This shows that the φ parameter lines are not geodesics.

Let us now increase the angular distance (φ_B) between A and B in the example shown in Figure 5.13. When the angular distance goes toward 180°, only one solution exists, but when the angular distance equals and exceeds 180° two solutions exist for the first order geodesic in the positive φ direction. Thus, a geodesic splitting has occurred. Figure 5.14 (a)–(d) shows all possible first-order geodesics in the positive φ direction for φ_B distances (170°, 180°, 200°, 220°). In the final subfigure, Figure 5.14 (e), $\varphi_B = 225.6°$. For larger φ_B values, no solutions exist to the geodesic equation except geodesics in the negative φ direction. In this case, no higher-order geodesics in the positive φ direction can be found.

This phenomenon gives rise to some questions: when do we have to worry about this and, when it occurs, which one should we choose when doing the analysis? Let us start with the first question. Figure 5.15 shows a parameter study of possible solutions to the geodesic equation when only geodesics in the positive φ direction are considered. The most interesting parameters to vary are the coordinates for the source and field points, and this is done for two different shaping parameters.

In the first example, $a = 0.5$, a pointed surface is considered. The second case is a less pointed surface with $a = 1.6$. In these examples, the source point is always located at $\varphi_A = 0°$, that is, $\Delta\varphi$ in the figures can be viewed as φ_B. From this analysis, we can see that the geodesic splitting occurs when $\Delta\varphi \geq 180°$ if u is large enough. If u is too small, no solution exists to the geodesic equation. These pictures also show the fact that higher-order encircling geodesics are possible if both $\Delta\varphi$ and u are large enough, but the number of encircling geodesics will depend on the shaping parameter. This was also discussed and shown in Section 5.3.4 when $u_A \neq u_B$. However, no parameter study was presented since

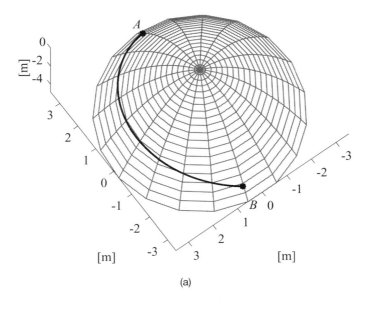

(a)

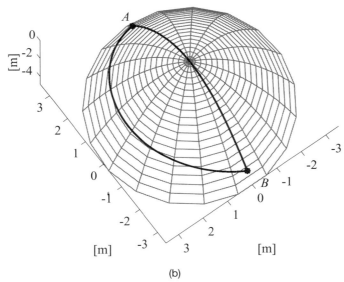

(b)

Figure 5.14. A sequence showing the splitting phenomenon and the appearance of the geodesics in the positive φ direction. As can be seen, the two geodesics move toward each other when the angular distance increases. Here, A is located at $u_A = 2$, $\varphi_A = 0°$, and $a = 1.6$. The observation point B is located at $u_B = 2$ and the angular distance is (a) $\varphi_B = 170°$, (b) $\varphi_B = 180°$, (c) $\varphi_B = 200°$, (d) $\varphi_B = 220°$, and (e) $\varphi_B = 225.6°$. Note that there are geodesics in the negative φ direction (not shown) in addition to the geodesics in the positive φ direction. (*continued*)

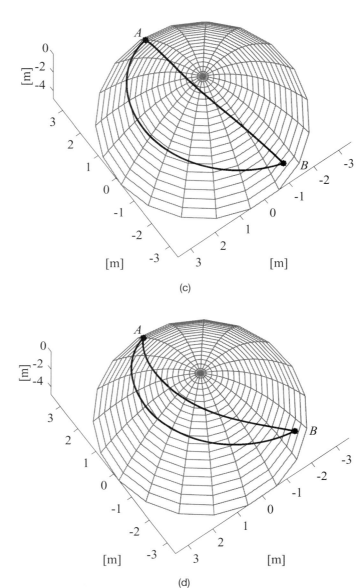

(c)

(d)

<u>Figure 5.14.</u> (*cont.*)

it is difficult to show a clear picture of all possible solutions versus coordinates and shaping parameter when $u_A \neq u_B$.

Since higher-order geodesics in the positive φ direction are possible it is also possible, to get geodesic splitting for higher-order geodesics. An example of this is shown in Figure 5.16. Obviously, the number of geodesics and the area for which splitting occurs depend on the shaping parameter of the surface. It is possible to derive a relation for this but the main principles are the same independent of the shaping parameter. This is expected since a pointed or more flat GPOR surface are in principle the same.

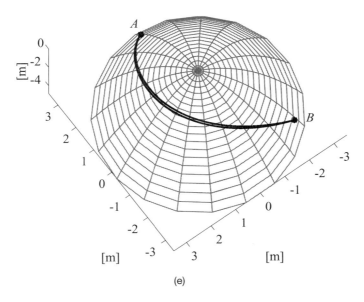

(e)

Figure 5.14. (*cont.*)

The second question referred to above (which is the most important geodesic?) is not easily answered. The reason is that there is no a priori method of identifying which geodesic(s) are the dominant contributors to the total surface ray field. Thus, the splitting phenomenon leads to an unavoidable doubling of the computations. However, in most cases, the arc length gives a hint about which geodesic is the dominant one since the surface ray field is decaying as a function of the arc length. But when A and B are moving apart in the angular direction, the geodesics are moving toward each other [see Figure 5.14 (*d*)–(*e*)], and their lengths become almost equal. From a practical point of view, this situation is not critical since the geodesics traveling in the opposite direction around the surface are much shorter. But for a more general surface, there can be situations in which two geodesics have about equal lengths. In such cases, the arc length is no longer a useful parameter for finding the dominant surface ray path. Instead, the radius of curvature can be used since the loss of energy is also due to diffraction from the surface ray. If the surface radius of curvature is small, more energy will be lost than for a surface ray with a large radius of curvature.

It has been stated that the first geodesic constant α must be computed accurately up to eight decimal places for a given arbitrary set of source and observation points [Jha et al. 1989d]. In the EM analysis of a surface using ray theory, the expressions are in a one-parameter-dependent form and their accuracy depends on the accuracy to which the first geodesic constant α has been determined; see, for example [Jha and Wiesbeck 1995]. To give some examples of the effect of the accuracy of α, some of the previous cases will be considered again. Let us start with the case when $u_A = u_B$. In the first example, $a = 0.5$, $u_B = 0.5$, and $|\varphi_A - \varphi_B| = 120°$. By changing the value of α from the correct result, the angular distance will change, resulting in a shorter or longer geodesic. In the analysis of conformal antennas, some surface-ray parameters will then change and in the worst case give unsatisfactory results. To keep the angular distance within one-tenth of a degree, α has to be computed accurately up to four decimal places for this example. If eight decimals are

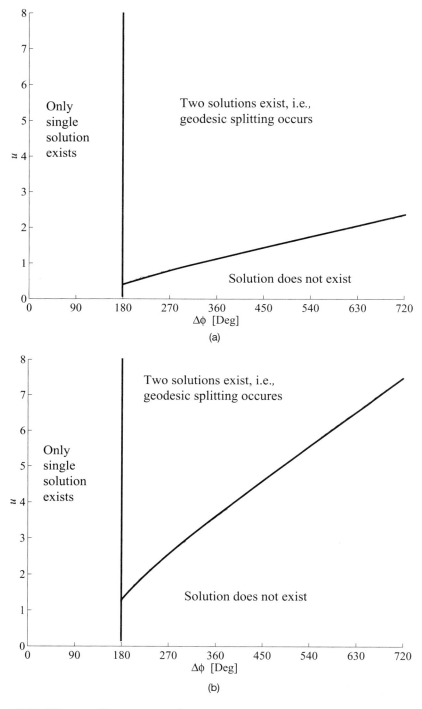

Figure 5.15. The set of parameters for which no solution exists, one solution exists, and two solutions exist to the geodesic equation when (*a*) *a* = 0.5 and (*b*) *a* = 1.6. Note that the second-order geodesics begin at $\Delta\varphi = 360°$ in both cases. Furthermore, it is also possible to find third-order geodesics in some cases ($\Delta\varphi \geq 720°$).

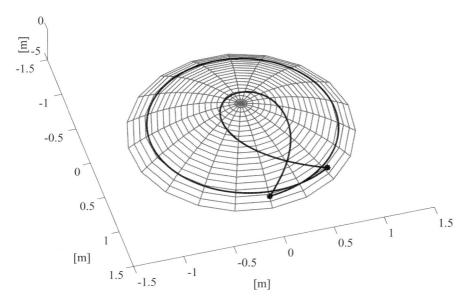

Figure 5.16. An example of geodesic splitting for second-order geodesics encircling the surface in the positive φ direction. In this case, $a = 0.5$, $u_A = 2$, $\varphi_A = 0°$, and $u_B = 2$, $\varphi_B = 400°$, respectively. Note, there is also a first-order geodesic in the positive φ direction that is not included in this picture.

correct, the angular difference will change no more than about 0.000005° in this case. Changing the angular distance while keeping the other parameters and using α with eight correct decimals leads to the following conclusion: If the angular distance is decreased, the error increases, and if the angular distance is increased, the error decreases. However, no matter what the angular distance, the change in the angular distance due to the inaccuracy of α is still less than 10^{-3} degrees. We get about the same result if a is changed instead of the angular distance. If the geodesic is a monotonically increasing/decreasing function of u, almost similar results are obtained. Thus, there should not be any problem in the practical analysis of conformal antennas.

The conclusion is that for the practical analysis of conformal antennas, the required accuracy in the geodesic constant α is not a big problem. An accuracy of about one thousandth of a percentage in α will keep the angular distance within one tenth of a degree of the correct value. However, if the error in α is as much as 1%, the angular distance can be extended or shortened by several degrees. One example is given in Figure 5.17, where $a = 1.0$, $u_A = 0.5$, $u_B = 2.5$, and $|\varphi_B - \varphi_A| = 130°$. In this example, the correct geodesic is shown together with two geodesics when the correct value of α is increased and decreased, respectively, by 1%. The result is an angular difference of about 3°.

5.4   ARBITRARILY SHAPED SURFACES

In this section, we will present some viewpoints on arbitrarily shaped surfaces. These surfaces include hybrid surfaces, analytically specified surfaces, and numerically specified surfaces. These surfaces can simulate a real vehicle or aircraft better than the generic sur-

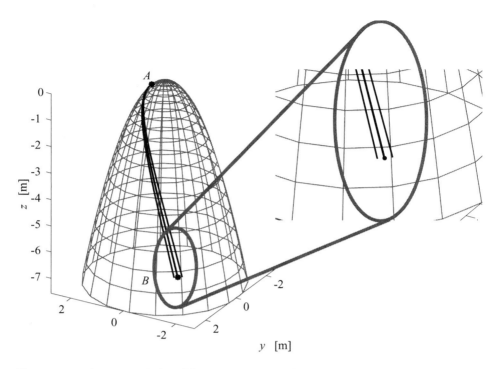

Figure 5.17. An example in which the angular difference has changed by 3° when α was changed by ±1%.

faces considered earlier in this chapter. However, they are in general much more difficult to analyze and it may be good enough to approximate the real surface with one of the simpler surfaces considered previously. Examples of arbitrarily shaped surfaces that we will consider in this section are a paraboloid with an elliptic cross section and a hybrid surface, which consists of a circular cylinder with a paraboloid or a cone at the top. Numerically specified surfaces are left for the reader to investigate.

5.4.1 Hybrid Surfaces

Let us first consider the hybrid surface consisting of a circular cylinder and a cone [Figure 5.18 (*a*)]. Assume that the source point A is on the cylinder, the observation point B is on the cone, and T is the junction between the two surfaces. In order to find the ray connecting A and B, we apply the extended Fermat's principle [Keller 1962]. This means that the geodesic should be continuous across the junction and the geodesic from A to T is the geodesic on the cylinder, whereas the geodesic from T to B is the geodesic on the cone. This becomes clear if the surfaces are developed; the geodesics over the developed surfaces map onto straight lines. As a consequence, the coordinates for the junction point T can be obtained in closed form.

For nondevelopable surfaces, it is more difficult to find the junction point T. In these cases, the junction point is defined as the point at which the geodesic crossing the junction line preserves the angle with respect to the junction line. Since this property is local, it can be used for nondevelopable surfaces. The property follows from Hertz's principle of par-

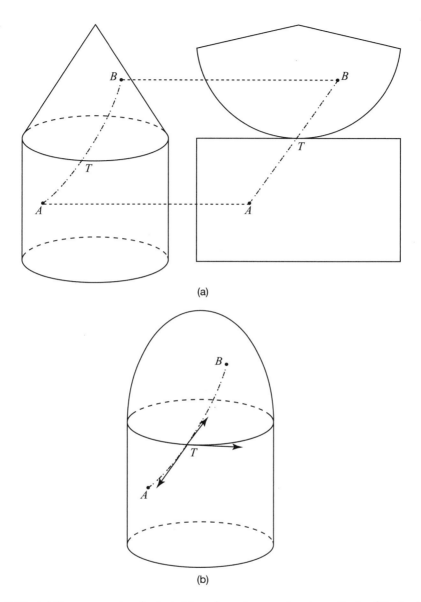

<u>Figure 5.18.</u> (*a*) One example of a hybrid surface, here a circular cylinder fitted with a circular cone. (*b*) A hybrid surface in which a circular cylinder and a general paraboloid of revolution (GPOR) are combined.

ticle dynamics [Lyusternik 1964]. To find the junction point T between two adjacent matched surfaces, the following relation must be fulfilled [see also Figure 5.18 (*b*)].

$$\hat{t}_B \cdot \hat{\varphi}_T = \hat{t}_A \cdot \hat{\varphi}_T \qquad (5.27)$$

As can be seen, the ray tracing for combinations of known surfaces follows the analysis presented earlier. Thus, it should not be very difficult to include this kind of surface in a

Table 5.5. The paraboloid with an elliptic cross section

Geometry	Parametric equation	Geodesic coordinates	Solution to geodesic equation
General paraboloid	$x = au \cos \varphi$ $y = bu \sin \varphi$ $z = -u^2$ $0 \leq u \leq \infty$ $0 \leq \varphi \leq 2\pi$ $a, b > 0$	$u = u$ $v = \varphi$	Very difficult to find the geodesics

code developed for nonhybrid surfaces. Note that if UTD is used, some functions must be redefined for consistency of the solution [Jha et al. 1994].

5.4.2 Analytically Described Surfaces

By relaxing the limitations on the types of surface shapes to be studied, the analysis becomes much more difficult. It is fascinating how easy it is to create a much more difficult problem just by changing one single parameter. One example is changing the circular cross section of the paraboloid studied in Section 5.3 to an elliptic cross section. By doing this, the analysis becomes much more difficult and it is no longer possible to find a closed expression for the geodesics. Such simple surfaces with a nonrotational symmetric cross section are of great interest since they can model, for example, the nose of some aircraft more accurately than a corresponding surface with a circular cross section.

We will consider the paraboloid with an elliptic cross section here (see Table 5.5) to illustrate how to extend the ray tracing described above. Many other interesting surfaces could be approached in the same fashion.

The first fundamental coefficients E, F, and G are found to be

$$E = a^2 \cos^2 \varphi + b^2 \sin^2 \varphi + 4u^2$$
$$F = u \sin \varphi \cos \varphi (b^2 - a^2) \qquad (5.28)$$
$$G = a^2 u^2 \sin^2 \varphi + b^2 u^2 \cos^2 \varphi$$

From this it is clear that this surface with the given parameterization does not belong to the geodesic coordinate system since $F \neq 0$ when $a \neq b$. Furthermore, neither E, F, or G are functions of only one parameter and this means that the geodesic equation cannot be reduced to a first-order differential equation as has been described earlier. Thus, we leave this case (and many similar ones) at this point, hoping that future investigations some day will produce useful solutions.

REFERENCES

Bulirsch R. (1969), "Numerical Calculation of Elliptic Integrals and Elliptic Functions," *Numerische Mathematik,* Vol. 13, pp. 305–315.

Burnside W. D. (1972), *Analysis of On-Aircraft Antenna Patterns,* Ph.D. Thesis, Ohio State University, Dept. of Electrical Engineering, 1972.

Choudhury R. and Jha R. M. (1995), "Ellipsoidal Surface Characterization for Validating The UTD Formulation," in *Proceedings of 1995 IEEE AP-S International Symposium,* June 18–23, pp. 1922–1925.

Erdelyi A., Magnus W., Oberhettinger F., and Tricomi F. G. (1953), *Higher Transcendental Functions,* New York: McGraw-Hill.

Glassner A. (ed.) (1989), *An Introduction to Ray Tracing,* New York, Academic Press.

González I., Delgado C., Saez de Adana F., Gutiérrez O., and Cátedra M. F. (2003), "3D Ray-tracing Acceleration Technique for the Analysis of Propagation and Radiation in Environments Modeled by Surfaces with Arbitrary Form," in *Proceedings of 2003 IEEE AP-S/URSI Symposium,* Columbus, OH, June 22–27.

Jha R. M., Sudhakar V., and Balakrishnan N. (1987), "Ray Analysis of Mutual Coupling between Antennas on a General Paraboloid of Revolution (GPOR)," *Electronic Letters,* Vol. 23, No. 11, pp. 583–584.

Jha R. M., Bokhari S. A., Sudhakar V., and Mahapatra P. R. (1989a), "Closed Form Surface Ray Tracing on Ogival Surfaces," in *Proceedings of 1989 IEEE AP-S International Symposium,* June 26–30, pp. 1294–1297.

Jha R. M., Bokhari S. A., Sudhakar V., and Mahapatra P. R. (1989b), "Closed Form Expressions for Integral Ray Geometric Parameters for Wave Propagation on General Quadric Cylinders," in *Proceedings of 1989 IEEE AP-S International Symposium,* June 26–30, pp. 203–206.

Jha R. M., Bokhari S. A., Sudhakar V., and Mahapatra P. R. (1989c), "Closed Form Expressions for Integral Ray Geometric Parameters for Wave Propagation on General Quadric Surfaces of Revolution," in *Proceedings of 1989 IEEE AP-S International Symposium,* June 26–30, pp. 207–210.

Jha R. M., Bokhari S. A., Sudhakar V., and Mahapatra P. R. (1989d), "Geodesic Splitting on General Paraboloid of Revolution and its Implications to the Surface Ray Analysis," in *Proceedings of 1989 IEEE AP-S International Symposium,* June 26–30, pp. 223–226.

Jha R. M., Edwards D. J., and Bhakthavathsalam R. (1991), "Application of Novel Method for Surface-Ray Tracing over Hyperboloidal Scatterers," *Electronics Letters,* Vol. 27, No. 17, pp. 1501–1502.

Jha R. M., Edwards D. J., and Sudhakar V. (1992), "Novel Geodesic Splitting on Paraboloid of Revolution with Applications to Ray-Theoretic Analysis," *Electronics Letters,* Vol. 28, No. 8, pp. 701–702.

Jha R. M., Mahapatra P. R., and Wiesbeck W. (1994), "Surface-Ray Tracing on Hybrid Surfaces of Revolution for UTD Mutual Coupling Analysis," *IEEE Transactions on Antennas and Propagation,* Vol. AP-42, No. 8, pp. 1167–1175, August.

Jha R. M. and Wiesbeck W. (1995), "The Geodesic Constant Method: A Novel Approach to Analytical Surface-Ray Tracing on Convex Conducting Bodies," *IEEE Antennas and Propagation Magazine,* Vol. 37, No. 2, pp. 28–38.

Keller J. B. (1962), "Geometrical Theory of Diffraction," *Journal of Optical Society of America,* Vol. 52, pp. 116–130, February.

Lee S. W., Ling H., and Chou R. (1988), "Ray Tube Integration in Shooting and Bouncing Ray Method," *Microwave and Optical Technology Letters,* Vol. 1, pp. 285–289, October.

Ling H., Chou R., and Lee S. W. (1989), "Shooting and Bouncing Rays: Calculating the RCS of an Arbitrarily Shaped Cavity," *IEEE Transactions on Antennas and Propagation,* Vol. AP-37, No. 2, pp. 194–205.

Lengyel E. (2001), *Mathematics for 3D Game Programming and Computer Graphics,* Charles River Media.

Lyusternik L. A. (1964), *The Shortest Lines: Variational Problems,* London, Pergamon.

Pérez J., Saez de Adana F., Gutierrez O., Gonzalez I., Cátedra M. F., Montiel I., and Guzmán J. (1999), "FASANT: Fast Computer Tool for the Analysis of On-Board Antennas," *IEEE Antennas and Propagation Magazine,* Vol. 41, No. 2, pp. 94–98.

Rademacher P. (1997), "Ray Tracing: Graphics for the Masses," ACM Crossroads, URL: http://www.cs.unc.edu/~rademach. (2002-12-02).

Rosquist K. (2001), Personal communication, Stockholm University (SCFAB), Department of Physics.

Spiegel M. R. (1994), *Mathematical Handbook of Formulas and Tables,* Schaum's Outline Series, McGraw-Hill, 32nd printing.

Struik D. J. (1988), *Lectures on Classical Differential Geometry,* 2nd ed., New York, Dover.

Zhou Y. and Ling H. (2003), "Evaluation of the Multiplaten Z-Buffer Algorithm for Electromagnetic Ray Tracing in High-Frequency Electromagnetic Scattering Computations," in *Proceedings of 2003 IEEE AP-S/URSI Symposium,* Columbus, OH, June 22–27.

Willmore T. J. (1984), *An Introduction to Differential Geometry,* Oxford University Press, fourth impression.

6

ANTENNAS ON SINGLY CURVED SURFACES

Singly curved surfaces, especially circular cylindrical surfaces, are common in conformal antenna applications. In this chapter, basic antenna characteristics of antenna elements on singly curved surfaces are treated. Both waveguide-fed apertures and microstrip-patch antennas are discussed. The aim is to give a basic overview of the mutual coupling among the elements versus surface geometry and size, and the effect of the mutual coupling on the radiation characteristics. Note, however, that conformal array characteristics are not discussed in this chapter.

6.1 INTRODUCTION

Antennas mounted on singly curved surfaces are an important class of conformal arrays for applications in which a large (azimuthal) angular coverage is required. These types of antennas have been used in many experimental radar and communication systems. An overview of different types of antennas is found in [Provencher 1972, Mailloux 1994, Hansen 1981, 1998]. Other examples are [Rai et al. 1996, Kanno et al. 1996, Gniss 1998, Löffler et al. 1999, Visser and Gerini 1999, Martín Polegre et al. 2001]. Other reference sources include the *IEEE Transactions on Antennas and Propagation,* Special issue on conformal antennas (January 1974), Proceedings of the Array Antenna Conferences held in 1970 and 1972 at the Naval Electronics Laboratory Center, San Diego, USA, and

Conformal Array Antenna Theory and Design. By Lars Josefsson and Patrik Persson
© 2006 Institute of Electrical and Electronics Engineers, Inc.

Proceedings from the European Workshops on Conformal Antennas (www.ewca-home.org). The singly curved surface can also be used as an approximation of the shape of an aircraft wing or fuselage.

In this chapter, we will give an overview of the performance of radiating elements located on singly curved surfaces. Radiating elements on doubly curved surfaces are discussed in Chapter 7. The focus is on the mutual coupling and its influence on the radiating characteristics. We will also point out the differences between antennas on singly curved surfaces and antennas on planar surfaces. The emphasis is on antenna element characteristics when a single element is fed, both isolated and in the array environment. Note that the characteristics of conformal array antennas will be discussed in Chapter 8

Arrays on singly curved surfaces (in most cases on circular cylinders) have been the subject of research for many years, especially the radiating element characteristics. Both dipoles and apertures have been used as elements; some references are [Carter 1943; Gladman 1970; Hessel and Sureau 1970; Borgiotti and Balzano 1970, 1972; Sureau and Hessel 1969, 1971, 1972; Hessel 1972; Steyskal 1977; Herper et al. 1985a,b; Persson and Josefsson 2001; Persson and Rojas 2003; Thors and Rojas 2003; Thors et al. 2003; Holter 2005; Persson and Josefsson 2005; Zmak and Sipus 2005]. In some of these examples "infinite" cylindrical arrays are discussed and analyzed with a unit cell approach, and in other cases an element-by-element approach is used; compare Chapter 4. Studies of the microstrip patch antenna on a circular cylinder include [Munson 1974; Ashkenazy et al. 1985; Nakatini et al. 1986; Jayakumar et al. 1986; Ashkenazy et al. 1988; Sohtell 1989; Habashy et al. 1990; Da Silva et al. 1991; Tam et al. 1995; Wu 1995; Huang and Chang 1997; Jin et al. 1997; Herscovici et al. 1997; Wong 1999; Löffler et al. 1999; Ertürk et al. 2003, 2004; Raffaelli et al. 2005].

The main part of this chapter deals with the canonical case of waveguide-fed aperture antennas mounted on various singly curved, perfectly electric conducting (PEC) surfaces. The aperture antenna is a simple element and can be analyzed easily with the methods described in Chapter 4. Thus, the complexity of the antenna itself is minimized and the focus is directed to the influence of the curved surface on the antenna characteristics. A dielectric coating will also be added, demonstrating the influence of a radome covering the antenna. In most of the examples, a circular cylinder is used (as when a radome is included), but we will also extend the discussion to other types of singly curved PEC surfaces such as elliptic and hyperbolic cylinders. Finally, microstrip-patch antennas are treated. These antennas are also relatively simple elements, but can have a complex feeding structure. They require that a dielectric layer (the dielectric substrate supporting the patch element) be included in the analysis. As mentioned in Chapter 4, this is, in general, a difficult problem. Hence, only circular cylinders are considered.

The mutual coupling results are presented in terms of the amplitude and phase of scattering matrix, or in terms of the self and mutual impedances. Two cases are distinguished. The first is the mutual coupling between two elements only, referred to as isolated coupling since no other elements are involved. The other elements of the array are assumed to be absent (short-circuited in case of apertures). If all elements are present, we use the term "array mutual coupling." The influence of the mutual coupling is also demonstrated by comparing radiation patterns of isolated and embedded elements.

We will first discuss aperture antennas flush mounted on singly curved, noncoated PEC surfaces (Sections 6.2–6.3). A special case of such surfaces is discussed in Section 6.4, in which faceted cylinders are considered. Coated surfaces are discussed in Section 6.5. Microstrip-patch elements on cylinders are discussed in Section 6.6. We conclude with a discussion of antennas on conical surfaces in Section 6.7.

6.2 APERTURE ANTENNAS ON CIRCULAR CYLINDERS

6.2.1 Introduction

Aperture antennas flush mounted on a conducting circular cylinder represent an important class of conformal antennas. The analysis of this canonical problem provides insight into many of the characteristics of singly curved conformal antennas. In the examples to follow, the circular cylinder is assumed to be infinitely long. The apertures have a rectangular shape with either axial or circumferential orientation. The aperture field is given by a combination of rectangular waveguide modes. In general, several waveguide modes should be included, but in many cases a single mode approximation is enough. We will also demonstrate the influence of the higher-order modes, and show when they are important if accurate results are required.

The geometry and all relevant parameters are defined in Figure 6.1. The radius of the cylinder is denoted by R and equals 5 λ in most of the cases to be presented. The apertures are described by the width and height, w and h. s is the geodesic length between the centers of the apertures. The angle α is also defined in Figure 6.1.

The apertures are linearly polarized (excited by the fundamental TE_{10} mode). Figure 6.1 shows a typical case with $w > h$ and axial (vertical) polarization. Axially oriented apertures, in which $w < h$, represent a case with circumferential (horizontal) polarization. We will also consider cross polarized cases, that is, coupling with one aperture of each kind.

6.2.2 Theory

The results in this chapter concerning aperture antennas are found by using an UTD–MoM method, except when otherwise stated. This method is based on solving the

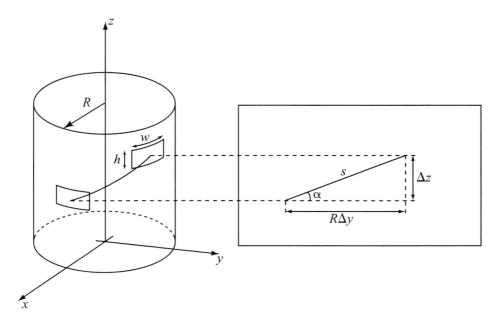

Figure 6.1. Geometry for the study of the characteristics of azimuthally oriented aperture antenna elements on circular cylinders—vertical polarization.

integral equation (IE) for the aperture electric field $\overline{E}_{ap}$. The integral to be solved is schematically given as

$$\overline{H}_{\tan}^{\text{ext}}(\overline{E}_{ap}) = \overline{H}_{\tan}^{\text{int}}(\overline{E}_{ap}) \tag{6.1}$$

The internal field is represented by waveguide modes and the external field is found by determining the fields on the surface using UTD, as described in Chapter 4; see Equation (4.20). The unknown aperture field is then found by using a standard MoM approach. Once found, the mutual admittance between elements 1 and 2 can typically be expressed as

$$Y_{21} = \frac{1}{V_1 V_2} \int_{S_2} \overline{E}_{ap}^2 \times \overline{H}^2(\overline{E}_{ap}^1) d\overline{S} \tag{6.2}$$

Note that a geodesic distance of approximately 0.5 λ or larger is needed for the UTD solution to be valid. For smaller distances, a planar approximation is used. If we take out two radiators from a large array with the other radiators short-circuited in case of apertures, Equation (6.2) gives an element in the mutual admittance (impedance) matrix. From this matrix, we can derive the scattering matrix:

$$\mathbf{S} = (\mathbf{I} - \mathbf{Y})(\mathbf{I} + \mathbf{Y})^{-1} \quad \text{or} \quad \mathbf{S} = 2(\mathbf{I} + \mathbf{Y})^{-1} - \mathbf{I} \tag{6.3}$$

If isolated coupling is considered, $\mathbf{Y}$ is a 2 × 2 element matrix calculated as a function of the spacing or any other parameter of interest. The array mutual coupling is obtained by considering all elements in the array; thus, $\mathbf{Y}$ is an $N \times N$ element matrix, where N is the number of elements in the array. If higher-order waveguide modes are used in the analysis, the matrix is generally of the size $MN \times MN$, where M is the number of modes used; see also Equation (4.22).

The far-field radiation pattern can be obtained directly once the aperture field is found with or without the effect of mutual coupling.

An important observation is that the IE is referred to the curved aperture plane, as discussed in Section 4.5.2. However, the waveguide modes used in the IE formulation are valid for a planar surface in a cross section of the waveguide; see Figure 4.14. This gap is often disregarded; however, we have taken it into account by adding a phase correction to the aperture field at the planar surface.

6.2.3 Mutual Coupling

6.2.3.1 *Isolated Mutual Coupling.* We start by analyzing the isolated coupling, that is, the mutual coupling between two elements only. The radius of the cylinder is 5 λ, and the aperture dimensions are 0.27 λ × 0.65 λ. When changing from the E-plane case to the H-plane case, both apertures are rotated 90°.

The first results in Figures 6.2 and 6.3 show the E- and H-plane couplings, respectively, in the azimuth direction ($\alpha = 0°$), and as a function of the geodesic length s. The total distance covered in the figures corresponds to an angular sector of approximately 180° of the circular cylinder. A comparison with the corresponding planar case is also included.

We observe that the H-plane coupling is weaker than the E-plane coupling (as it is in the corresponding planar case). The ripple is caused by the interference of two waves traveling in opposite directions around the cylinder. The amplitude of the coupling in the

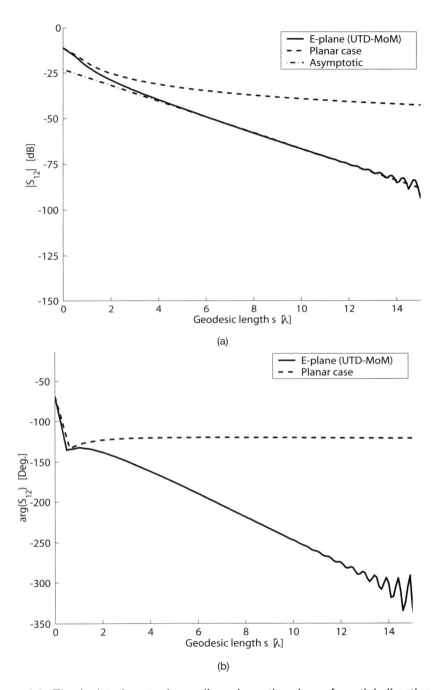

(a)

(b)

Figure 6.2. The isolated mutual coupling along the circumferential direction of a cylinder while using circumferentially polarized elements, that is, the isolated coupling in the E plane (solid line). The result is compared with the corresponding planar case (dashed line). Note that the distance in wavelengths corresponds to an angular interval of approximately 180° at the cylinder. (*a*) Amplitude. The result is also compared with an asymptotic formula in Table 6.1 (dash–dotted line). (*b*) Phase.

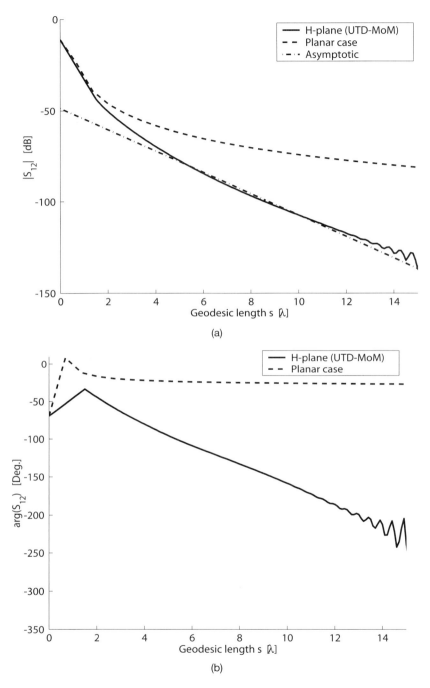

(a)

(b)

Figure 6.3. The isolated mutual coupling along the circumferential direction of a cylinder while using axially polarized elements, that is, the isolated coupling in the H plane (solid line). The result is compared with the corresponding planar case (dashed line). Note that the distance in wavelengths corresponds to an angular interval of approximately 180° at the cylinder. (*a*) Amplitude. The result is also compared with an asymptotic formula in Table 6.1 (dash–dotted line). (*b*) Phase.

planar case decays as $1/s^n$ (n is given in Table 6.1 and depends on the polarization). Obviously, the surface field along the circumferential direction of the cylinder experiences a faster decay due to continuous leaking (cf. Section 4.4.2), resulting in an exponential decay due to the curvature. Table 6.1 shows how the amplitude of the field (the mutual coupling) decays asymptotically for the different cases.

Figures 6.2 (a) and 6.3 (a) show the results using the cylindrical surface asymptotic formulas in Table 6.1 (with appropriate constants) for the E- and H-plane couplings. The dominant term is shown. Higher-order terms are needed for a highly accurate evaluation of the mutual coupling, in particular for the cylindrical H-plane. In general, the asymptotic behavior (for both polarizations) of the cylindrical case is given as

$$U \propto e^{-jks} \sum_{n=1}^{\infty} (D_n)^2 \, e^{-\Omega_n s} \tag{6.4}$$

where U is the field component, and D_n and Ω_n are appropriate constants [Keller 1956, McNamara et al. 1990]. Thus, the asymptotic behavior of the cylindrical case is the same for both the E-plane and H-plane couplings, which is in contrast to the planar case. Using the dominant term only [cf. Table 6.1 and Figures 6.2 (a) and 6.3 (a)] we find that the asymptotic expressions for the circular cylinder case predict the correct decay for moderate to large distances s, whereas the asymptotic expressions for the planar case are accurate for small distances as well (not shown).

The phase data in Figures 6.2 and 6.3 are shown with the phase delay corresponding to the geometrical separation subtracted. For the planar case, the phase follows almost a horizontal line, indicating that the wave is propagating as a TEM wave in free space. For the circular cylinder, there is an additional phase delay. One interpretation of this is that the wave is propagating at another radius. Analyzing the phase data gives an equivalent geodesic length $\tilde{s}$ corresponding to a radius of $\tilde{R} \approx R + 0.2 \, \lambda$; thus, the surface field can be interpreted as traveling approximately 0.2λ above the curved surface. This is an interesting result when discussing the physics of the ray-based model. In this model, the surface field is assumed to travel tightly along the surface; but a different radius could have been used, as indicated in Appendix 4A.2. In fact, from Keller's surface diffraction analysis [Eq. (6.4)], the approximate solution for the H_z component (circumferential polarization) gives

$$H_z \propto e^{-(jk+\Omega_n)s} \tag{6.5}$$

where

$$\Omega_n = \frac{\alpha_1'}{R} \left(\frac{\pi R}{\lambda} \right)^{1/3} e^{j\pi/6} \tag{6.6}$$

Table 6.1. Asymptotic decay of the amplitude of the isolated coupling in the E and H planes

	Planar surface	Cylindrical surface
E plane	Amplitude $\approx C_E \dfrac{1}{s}$	Amplitude $\approx C_1^E e^{-C_2^E s}$
H plane	Amplitude $\approx C_H \dfrac{1}{s^2}$	Amplitude $\approx C_1^H e^{-C_2^H s}$

and α_1' is a zero of the Airy function, ≈ 1.02. We see already from this solution that there will be an extra phase delay since Ω_n is complex. Inserting the appropriate numbers gives the same 0.2 λ extra as derived above.

Next, we will study the dependence of the mutual coupling on the cylinder radius. The following radii are tested: $R = 2 \lambda$, 5λ, 10λ, 20λ, and 50λ. The apertures have the same size as in the previous examples. Figure 6.4 shows the E-plane couplings (for $\alpha = 0°$) and Figure 6.5 shows the H-plane couplings (also for $\alpha = 0°$). In both cases, the results are shown versus the geodesic length between the apertures. The corresponding planar result ($R \rightarrow \infty$) is included for comparison. Analysis shows that the phase data closely follow Equations (6.5) and (6.6).

As indicated in the figures, the circular cylinder solution approaches the planar solution when the radius is increased to infinity, as expected. Another interesting result is that the planar solution seems to be surprisingly accurate for small distances, for both polarizations. This is a feature often used when analyzing arrays on curved surfaces. By using the planar solution, approximate values of the mutual coupling can be found easily for neighboring elements as well as between elements along the axis; see also Section 6.5.3.2. However, the effect of the curvature of the surface is quite strong for larger distances. For a 50 λ radius, the planar solution and the cylinder solution deviate considerably after a distance of approximately 6–8 λ (both in the E and H planes). Similar examples can be found in the literature; see, for example, [Borgiotti and Balzano 1970, Hessel 1972, Hansen 1981].

So far, only the circumferential direction ($\alpha = 0°$) of the cylinder has been considered. For a complete picture of the behavior of the mutual coupling, other directions must also be analyzed. Therefore, a range of α values, namely $\alpha = 0°$, $\alpha = 20°$, $\alpha = 40°$, $\alpha = 60°$, $\alpha = 80°$ and $\alpha = 90°$, were investigated, with the results shown in Figures 6.6 and 6.7. The aperture dimensions are the same as before and the aperture orientations are kept fixed for each figure. Hence, in Figure 6.6 the aperture polarization for $\alpha = 0°$ corresponds to E-plane coupling, but as α increases, the coupling becomes mixed. When α reaches 90°, we have, instead, H-plane coupling along the axis of the cylinder. Figure 6.7 shows the opposite scenario, starting with the H-plane coupling with $\alpha = 0°$, and reaching E-plane coupling along the axis of the cylinder when $\alpha = 90°$. The maximum geodesic length s is kept constant, corresponding to half the perimeter of a cylinder with a 5 λ radius ($5\pi\lambda$). The planar case is included, to be compared with the cylinder results for $\alpha = 90°$.

These results show that the transition from E-plane coupling in the azimuth to H-plane coupling along the axial direction is different from the opposite transition—going from H-plane coupling to E-plane coupling. The general behavior is governed by two factors: the changing radius of curvature along the geodesic (which goes toward infinity when α reaches 90°), and the changing orientation of the surface magnetic field of the source at different α angles; see Figure 6.8.

The different decay rates shown in Figure 6.6 are due to the above-mentioned two factors. When α is increased the radius of curvature along the geodesic increases, resulting in a successive slower decay of the surface field. At the same time, the amplitude of the surface field from the aperture remains quite stable for most α values, expect when reaching the axial direction; see Figure 6.8 (a), where the field decays quickly. The effect of the curvature dominates in the circumferential direction, whereas the decay of the surface field dominates in the axial direction.

Figure 6.7 shows an opposite trend compared to Figure 6.6; the mutual coupling increases continuously when going from the H plane to the E plane. In this situation, the effects of the radius of curvature and the surface field work in the same direction when the angle α is changing; see also Figure 6.8 (b).

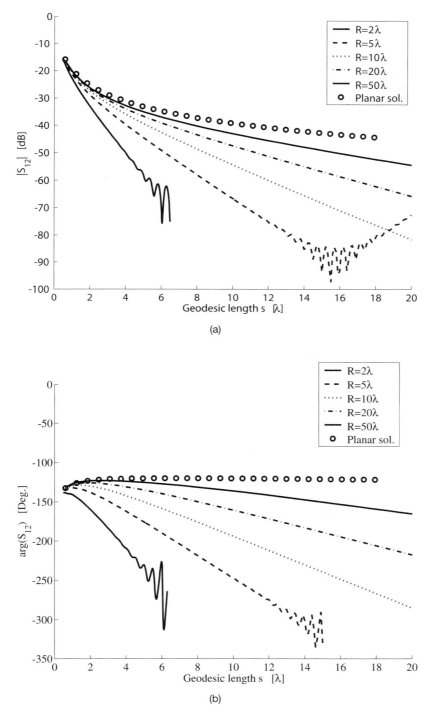

(a)

(b)

Figure 6.4. The isolated *E*-plane mutual coupling along the circumferential direction of cylinders with different radii. (*a*) The amplitude. (*b*) The phase, with the free space wave contribution subtracted.

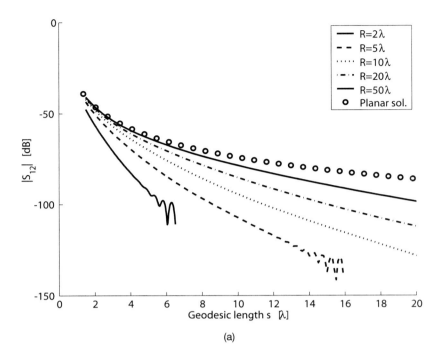

(a)

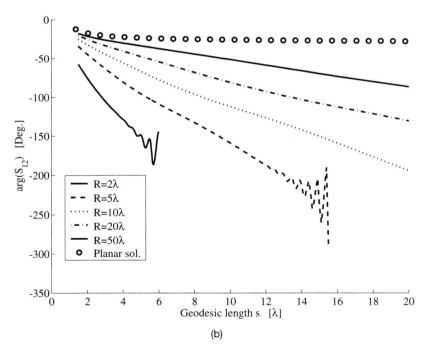

(b)

<u>Figure 6.5.</u> The isolated *H*-plane mutual coupling along the circumferential direction of cylinders with different radii. (*a*) The amplitude. (*b*) The phase, with the free space wave contribution subtracted.

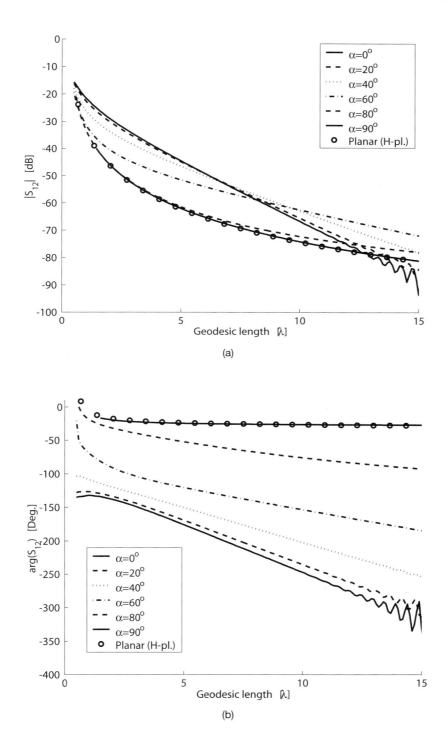

(a)

(b)

Figure 6.6. The isolated mutual coupling as function of φ for a given radius (5λ) and geodesic length. Here, $\alpha = 0°$ corresponds to *E*-plane coupling and $\alpha = 90°$ corresponds to *H*-plane coupling since the apertures do not rotate when changing α. (a) Amplitude. (b) Phase.

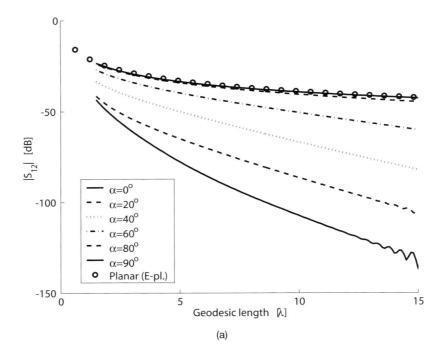

(a)

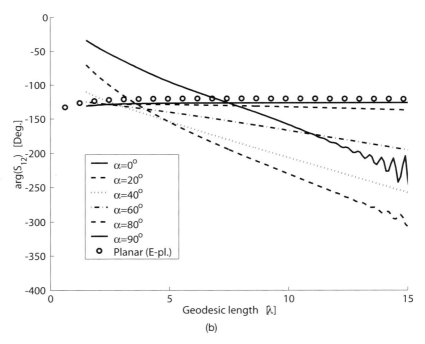

(b)

Figure 6.7. The isolated mutual coupling as function of α for a given radius (5λ) and geodesic length. Here, $\alpha = 0°$ corresponds to H-plane coupling and $\alpha = 90°$ corresponds to E-plane coupling since the apertures do not rotate when changing α.

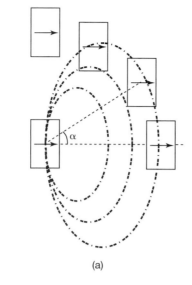

(a)

---·---·---·---·-- Magnetic field

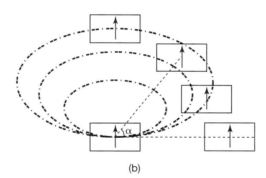

(b)

Figure 6.8. A two-dimensional sketch of the planar magnetic field of the source element and its interaction with another aperture. (*a*) *E*-plane type coupling along the horizontal line. (*b*) *H*-plane type coupling along the horizontal line.

Figures 6.6 and 6.7 show that the planar solution agrees quite well with the cylinder solution along the axial direction. An interesting question arises as to whether the cylinder-axial solution recovers the planar solution or not, that is, does the finite radius of curvature along the binormal direction influence the result or not? This can be investigated by studying the cylinder solution along the axial direction according to UTD, and compare it with the planar solution. In [Pathak and Wang 1981] the surface field due to an infinitesimal magnetic current moment $d\bar{p}_m$ is given by

$$d\bar{\bar{H}}_m(Q|Q') \approx d\bar{p}_m \cdot [\hat{b}'\hat{b}H_b + \hat{t}'\hat{t}H_t + (\hat{b}'\hat{t} + \hat{t}'\hat{b})H_{bt}] \qquad (6.7)$$

where H_b is the transverse component and H_t is the longitudinal component with respect to the geodesic [see Figure 4.12 (*b*)]. H_{bt} represents higher-order terms. Evaluating this expression when $\alpha \to 90°$ for a fixed radius R and a large distance s gives the result that H_t and H_{bt} do recover the planar solution, whereas H_b is obtained as

$$H_b^{\text{cylinder}} \underbrace{A\,\frac{e^{-jks}}{ks}}_{H_b^{\text{planar}}} + \underbrace{B\,\frac{1}{kR}\,\frac{e^{-jks}}{\sqrt{ks}}}_{\to\, 0 \text{ when } R \to \infty} \tag{6.8}$$

Obviously, the finite radius of curvature in the binormal direction influences the transverse component in such a way that it becomes stronger than the planar solution. Thus, the axial solution does not exactly recover the planar solution. The reason is that the cylinder acts more as a waveguide, whereas at a planar surface the energy is diffracted over a plane, resulting in reduced amplitude. See also the discussion in Section 6.6.3.2.

6.2.3.2   Cross-Polarization Coupling.
In the previous examples, only the co-polar coupling was considered. It is also of interest to investigate the cross-polar coupling since this can be a source of cross-polarized radiation in dual-polarized apertures. More details on controlling the polarization are found in Chapters 7 and 8. The first example in Figure 6.9 is equivalent to Figures 6.6 and 6.7 except for the cross-polarized case. Since there is no cross-polar coupling for $\alpha = 0°$ and $\alpha = 90°$, these cases have been left out. The parameters are the same as before. In the example shown, the source is circumferentially polarized with the receiving aperture axially polarized. The variation in decay rates can be explained by the same reasoning as was used for the results in Figures 6.6 and 6.7.

The next example in Figure 6.10 shows the E-plane, H-plane, and cross-polar mutual couplings as a function of the axial distance z with a fixed angular separation in the azimuth, here 30°. The aperture dimensions and the size of the cylinder are the same as in the previous examples. In the cross-polar case, the source is circumferentially polarized and the receiving aperture is axially polarized. The results follow our expectations based on the earlier discussion.

6.2.3.3   Array Mutual Coupling.
The next step is to study the array mutual coupling, that is, the coupling among the elements in the array environment with all elements present and terminated in matched loads. We choose the same geometry as in the previous examples, that is, the radius of the cylinder is 5 λ and the dimensions of the apertures are 0.27 λ × 0.65 λ. In the first example, we have a single-row array covering an arc of approximately 120°. Along this arc, 18 elements are placed equidistantly with circumferential polarization. Figure 6.11 shows a comparison between the isolated and array mutual couplings for this case. The effect of the elements in between is clearly visible in both amplitude and phase, although the phase is more sensitive to the presence of other elements than the amplitude. The corresponding planar case (isolated and array mutual couplings) is also included for comparison.

It was found earlier (see Table 6.1), that the amplitude of the isolated coupling in the cylinder case decays as $C_1 e^{-C_2 s}$ (E plane) for moderate to large distances. The analysis here shows that the amplitude of the array coupling has the same type of decay, but with different constants C_1 and C_2. Thus, both the isolated and the array mutual couplings have an exponential decay in the single-row cylinder case. See also Section 7.2.2.2, where a similar result is found for the doubly curved surface. An example for a 3 × 18 element array is given next, where we also compare theoretical results with measured results.

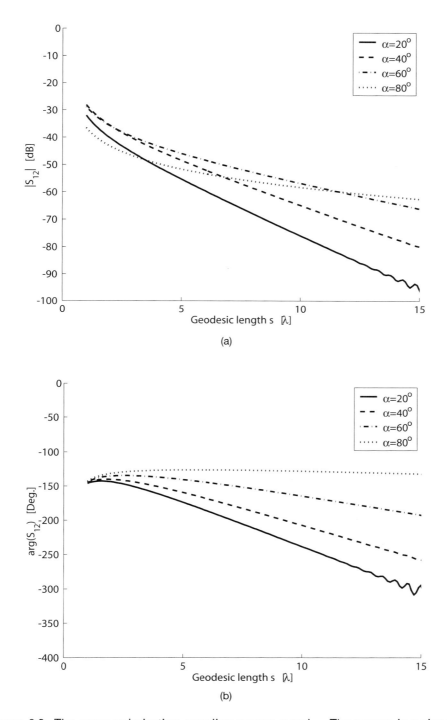

Figure 6.9. The cross-polarization coupling versus s and α. The source is a circumferentially polarized aperture and the receiver is axially polarized. (*a*) The amplitude. (*b*) The phase.

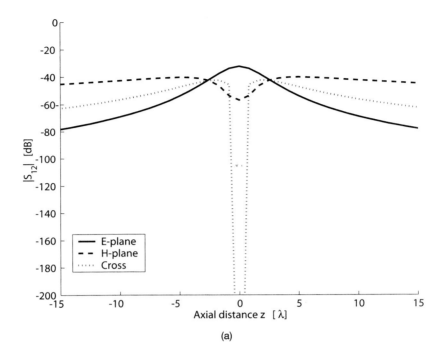

(a)

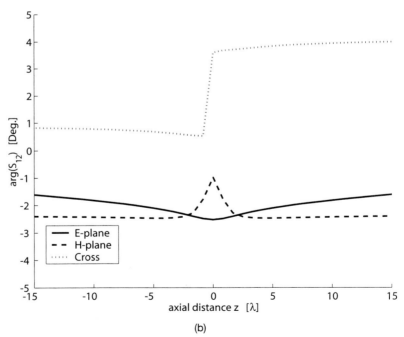

(b)

<u>Figure 6.10.</u> The isolated mutual coupling versus the axial distance *z* for a fixed angular separation $\Delta\varphi = 30°$. Both the *E*-plane, *H*-plane, and cross-polar couplings are shown. (*a*) The amplitude. (*b*) The phase.

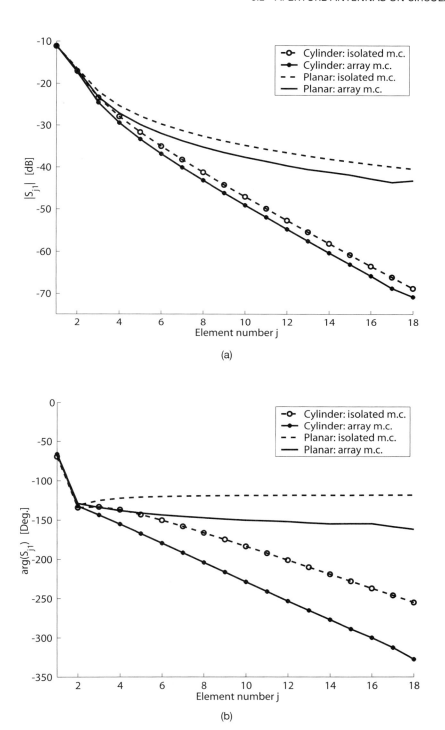

Figure 6.11. A comparison between the array and isolated mutual coupling for a single-row array on a cylinder for the *E* plane (lines with circles). The corresponding planar case is also shown for a single row of 18 elements. Single-mode results.

Including higher-order modes in the calculations increases the accuracy of the aperture field solution and, hence, also other results such as the mutual coupling. We will compare the calculated results with measured data [Lindgren and Josefsson 1998] obtained from an experimental antenna built by Ericsson Microwave Systems AB in Mölndal, Sweden, as a test bed for singly curved conformal antennas. The surface is a circular cylinder with three rows of rectangular apertures; see Figure 6.12. The main dimensions are:

Element apertures	39 mm × 16 mm
Element grid	Rectangular, 3 × 18 elements extending over about 120°
Element spacings	37.1 mm in the azimuth
	41 mm in the axial direction
Polarization	Circumferential
Cylinder radius	300 mm
Frequency interval	4.3 → 6.8 GHz
Cut-off frequency of second-order mode	$f_{c,20} \approx 7.6$ GHz
Frequency used here	$f = 5.65$ GHz

In the calculations, no surface rays encircling the cylinder were taken into account since the cylinder is truncated at the rear for practical reasons. Absorbers were placed on all edges to minimize edge effects. We used the following numbering of the apertures: the aperture in the lower-left corner of Figure 6.12 is aperture number 1; aperture number 54 is located in the upper-right corner.

Figure. 6.12. The experimental antenna. Courtesy of Ericsson Microwave Systems AB, Göteborg, Sweden. See also color insert, Figure 2.

The first example comparing theoretical and measured data in Figure 6.13 shows the isolated mutual coupling (E plane) between two elements in one row as a function of the angular distance, for $f = 5.65$ GHz. During the measurements, all apertures, except the two being measured, were covered with conducting tape. As can be seen in Figure 6.13, the agreement is very good (cf. Figure 6.2 where a similar case was considered). Only a single-mode approximation was used in the simulations; if higher-order modes are used, the results do not change.

In the following examples (Figures 6.14 to 6.17), the array mutual coupling is presented for $f = 5.65$ GHz. Additional results for other frequencies can be found in [Persson and Josefsson 2001]. The first result, Figure 6.14, shows the result for the array mutual coupling in the E plane for the first row of elements ($1 \rightarrow 18$) and for the second row of elements ($19 \rightarrow 36$) using a single-mode approximation. We observe some differences between the first and second row depending on the different element surroundings. Furthermore, even with a single-mode approximation, the mutual coupling diagrams show good agreement between the measured and the calculated results, at least down to about –60 dB. For elements far away, the agreement is not so good but the coupling levels are very low.

In order to improve the accuracy even further, higher-order waveguide modes were included in the analysis. With the four lowest TE modes (in increasing cutoff order—TE_{10}, TE_{20}, TE_{01}, and TE_{11}) only small differences were observed, but when taking also the TM_{11} mode into account, a significant improvement was obtained (see Figure 6.15). Now the simulated results show good agreement with measurements down to coupling levels as low as –80 dB. The agreement of the phase is also improved. Using even more modes (up to 20 modes were tried) improved the results very little [Persson and Josefsson 2001].

The importance of the TM_{11} mode is probably explained by the fact that this mode is the first mode with a field component parallel to the direction of propagation. Obviously, this component is important for a precise modeling of the aperture field, and also for the planar case [Bonnedal 1996]. See also Section 11.4.2, where scattering versus higher-order modes is discussed.

For the H plane (along the axis of the cylinder) only limited data are available since only two data points can be taken for the three rows. Figure 6.16 shows the H-plane coupling (single mode) for two different columns in the array—an edge column and the center column. The agreement is slightly worse compared to the E-plane results, but by using higher-order modes, the results improve. However, five modes are not enough in this case; at least 10 modes are needed for good agreement with measured data (see Figure 6.17). The need for additional higher order modes in the H plane is due to the size of the aperture in this plane; see also Section 11.4.2.

6.2.4 Radiation Characteristics

In this section, the focus is on element patterns, both isolated and embedded. The isolated-element pattern shows the radiation properties of a single element, including the effect of the curved surface. The embedded-element pattern carries more information, including the effect of the surrounding elements and a particular element grid. The array pattern, when all elements are fed according to a particular excitation function, represents an even more specific case, to be discussed in Chapter 8.

For the embedded-element patterns, only a single element is fed. The other elements in the array are present but terminated in matched loads. They are being excited by the mutual coupling only. Thus, mutual coupling is an important parameter for the embedded radiation patterns.

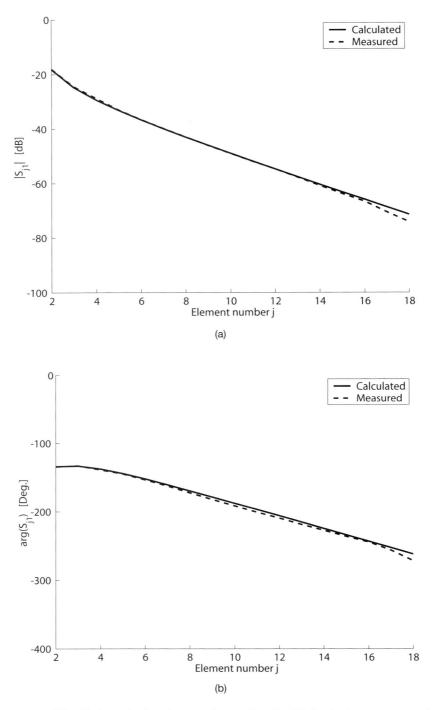

(a)

(b)

Figure 6.13. The *E*-plane isolated mutual coupling (solid line) along one row in the experimental antenna compared to measurements (dashed line). Here, only a single-mode approximation is used and *f* = 5.65 GHz. (*a*) Amplitude. (*b*) Phase.

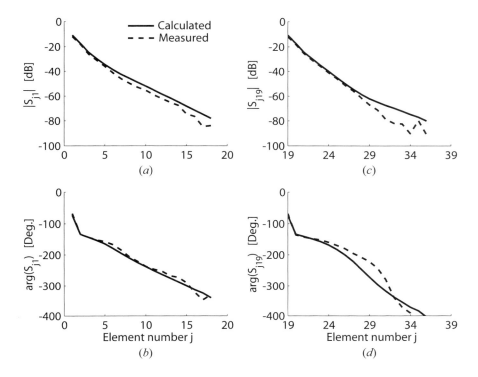

<u>Figure 6.14.</u> The array mutual coupling in the *E* plane for the first row (elements 1 → 18) and the second row (elements 19 → 36). Single-mode approximation, *f* = 5.65 GHz. Solid line: calculated, dashed line: measured. (*a*) Amplitude, first row. (*b*) Phase, first row. (*c*) Amplitude, second row. (*d*) Phase, second row [Persson and Josefsson 2001].

The radiation characteristics, in particular the embedded-element patterns in cylindrical arrays, have been the subject of careful research in past. One reason for that is that the embedded-element patterns can be significantly different from those in a planar array. Some of the references are [Sureau and Hessel 1969, Borgiotti and Balzano 1970, Munger et al. 1971, Sureau and Hessel 1971, Herper et al. 1985a,b, Hansen 1981, Mailloux 1982]. A summary of the results obtained is found in [Hessel 1972, Mailloux 1994, Hansen 1998].

Most of the results presented here are obtained using the UTD–MoM method presented earlier, and some results are taken from references and from measurements. In the calculated results, we will use the same radiating elements as before, that is, waveguide-fed rectangular apertures of size $0.27 \lambda \times 0.65 \lambda$ with the waveguides terminated in matched loads, except when otherwise stated. The size of the cylinder will be varied as well as the spacing between the elements. In most cases, a rectangular grid arrangement will be used, although a triangular grid is considered as well. Note that the phase variation shown here is always referenced to the element center.

6.2.4.1   Isolated-Element Patterns. Figures 6.18 and 6.19 show the amplitude and phase patterns in the azimuth of a single rectangular aperture excited in the fundamental TE_{10} mode with phase compensation for the curved aperture (see Section 6.2.2).

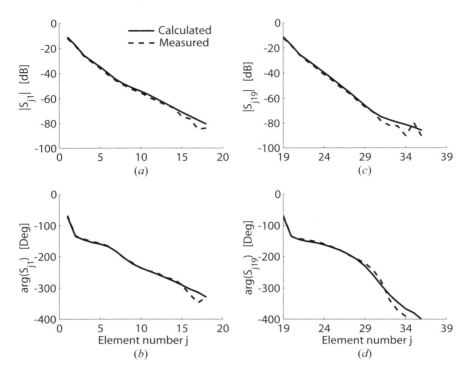

Figure 6.15. The array mutual coupling in the *E* plane for the first row (elements 1 → 18) and the second row (elements 19 → 36). Five-mode approximation, *f* = 5.65 GHz. Solid line: calculated, dashed line: measured. (*a*) Amplitude, first row. (*b*) Phase, first row. (*c*) Amplitude, second row. (*d*) Phase, second row [Persson and Josefsson 2001].

For Figure 6.18, the aperture is circumferentially polarized and the patterns are *E*-plane patterns. For 6.19, the aperture is axially polarized and the patterns are *H*-plane patterns. The results are calculated for different cylinder radii, including one case for an infinite planar ground plane.

The main difference between the patterns for different radii is found in the shadow and transition regions, as expected. In the lit region, only minor changes are observed and most of the energy (for both polarizations) is radiated into the upper hemisphere. The radiated field appears to be not as sensitive as the mutual coupling to changes in the radius of the cylinder (cf. Figures 6.4 and 6.5).

6.2.4.2 Embedded-Element Patterns. As shown earlier, the magnitude of the mutual coupling among the elements on a curved surface is less than for the corresponding planar case. However, the effects of the mutual coupling on the embedded-element patterns can still be considerable for the curved case. The embedded-element pattern gives an indication of the performance of the fully excited array and serves, therefore, as a valuable tool in the array design process. We will here study the embedded-element patterns of circular arrays with the elements uniformly spaced and covering the complete (or almost complete) circumference of a cylinder. Thus, no edge effects are present in the

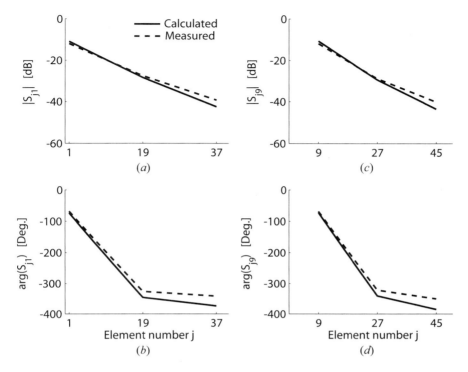

<u>Figure 6.16.</u> The array mutual coupling in the *H* plane for an edge column and for the center column. Single-mode approximation, *f* = 5.65 GHz. Solid line: calculated, dashed line: measured. (*a*) Amplitude, edge column. (*b*) Phase, edge column. (*c*) Amplitude, center column. (*d*) Phase, center column.

azimuth and all azimuth element patterns are identical. This resembles the infinite planar array case, and we will compare patterns for the two cases.

If the array does not cover the full circumference, the symmetry is lost and each element will have its own embedded pattern. Such a truncated array can be a suitable model for some realistic applications and will be discussed in Section 6.3.3 in connection with arrays on other types of singly curved surfaces.

In the main part of this section, all embedded-element patterns are calculated in the azimuth plane for circumferentially polarized elements (*E*-plane patterns). If a similar study is done for axially polarized elements, the difference between the isolated-element pattern shown in Section 6.2.4.1 and the embedded-element pattern in the broadside region in the azimuth is almost negligible due to the weak *H*-plane coupling.

Finally, dipole elements are compared with aperture elements. At the same time, elevation patterns are discussed, both for rectangular and triangular grids.

In the first result, shown in Figure 6.20, the gain of the *E*-plane embedded-element pattern for a single ring of circumferentially polarized rectangular apertures is shown for different cylinder radii. The element spacing is kept constant, equal to 0.6 λ. The patterns are also compared with those of the corresponding planar case. As may be seen, there are no major differences between the pattern for the planar case and the pattern for a cylinder with a large radius in the lit region. Both patterns exhibit a clearly visible drop-off near

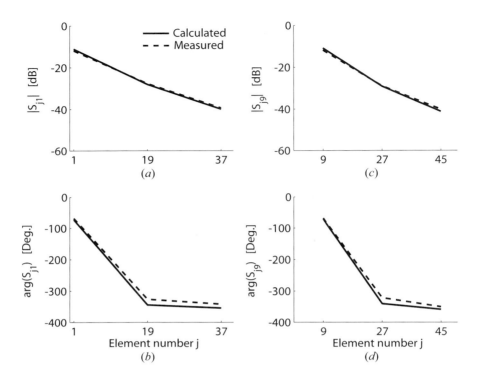

<u>Figure 6.17.</u> The array mutual coupling in the *H* plane for an edge column and for the center column. Ten-mode approximation, *f* = 5.65 GHz. Solid line: calculated, dashed line: measured. (*a*) Amplitude, edge column. (*b*) Phase, edge column. (*c*) Amplitude, center column. (*d*) Phase, center column.

$\phi = 42°$. This drop-off in the planar case is due to an endfire grating lobe at $\phi = 41.8°$. Its counterpart in the cylindrical array is a "cylindrical endfire grating lobe" [Mailloux 1994]. It should be noted that the cylindrical grating lobe is not as distinct as in the planar case due to the curved surface. However, the phenomenon is caused by the same mechanisms in both cases.

Since the grating lobe is more diffuse in the cylindrical case, the pattern dips are not as deep as in the (infinite) planar case. We have in the cylinder case a finite number of elements participating, the number decreasing with the radius of the cylinder. The mutual coupling amplitude is also less for the curved case.

The decay of the element pattern in the shadow region ($|\phi| > 90°$) is linear on a decibel scale, indicating that the angular pattern is primarily due to a single creeping wave contribution [Sureau and Hessel 1971, Harper et al. 1985a]. However, the rate of decay for the embedded-element pattern is greater due to the additional loss mechanism introduced by the terminated elements. The ripple in the region $\phi > |150°|$ is due to the interference between two creeping waves traveling in opposite directions, as in the isolated case.

Another interesting observation is the ripple in the broadside (lit) region for the circular cylinder, which is not observed in the (infinite) planar case. This ripple is caused by the finiteness of the curved array, and is also observed in finite planar arrays [Amitay et al. 1972] but is not as pronounced as in the cylinder case.

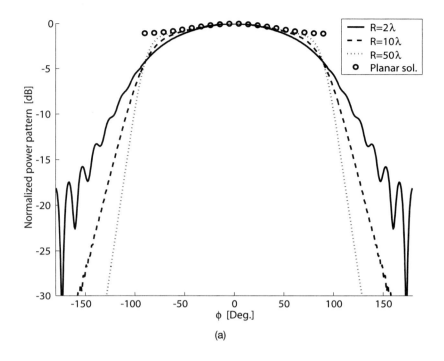

(a)

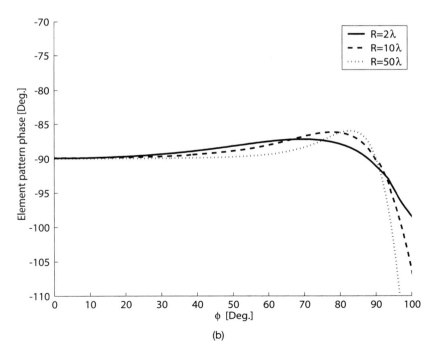

(b)

<u>Figure 6.18.</u> The *E*-plane isolated-element pattern in the azimuth versus the radius of the cylinder. (*a*) Amplitude. The planar case is also included (circles). (*b*) Phase.

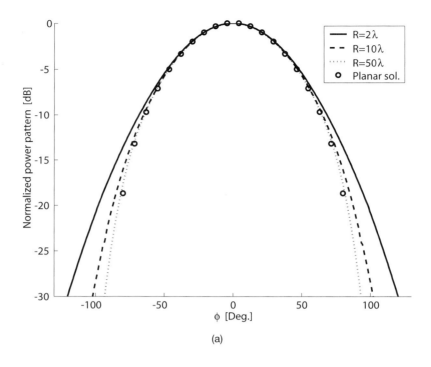

(a)

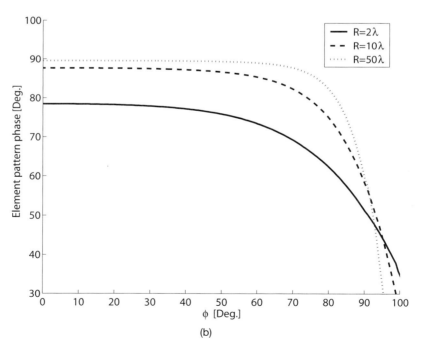

(b)

<u>Figure 6.19.</u> The *H*-plane isolated-element pattern in the azimuth versus the radius of the cylinder. (*a*) Amplitude. The planar case is also included (circles). (*b*) Phase.

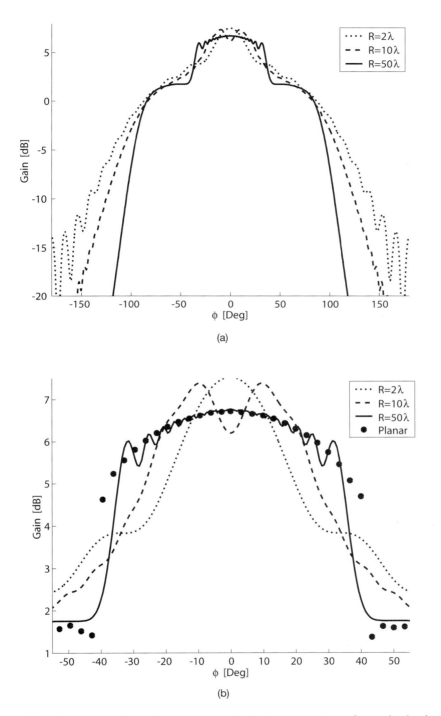

Figure 6.20. The gain of the *E*-plane embedded-element pattern for a single-ring array in the azimuth versus the radius of the cylinder for a fixed-element spacing ($d = 0.6\ \lambda$). (*a*) Overview for $R = 2\lambda$, 10λ, and 50λ. (*b*) The corresponding planar case is also shown (dots) in a close-up view.

The ripple in the broadside region will be discussed next by examining the embedded-element pattern dependence on the element spacing for a fixed radius. To do this, embedded-element patterns are calculated for four values of element spacing, namely $d = 0.4\ \lambda$, $0.5\ \lambda$, $0.6\ \lambda$, and $0.7\ \lambda$. The radius is fixed and equals $5\ \lambda$. The gain of the E-plane embedded-element patterns is shown in Figure 6.21 (a) together with the single element gain for comparison. The corresponding phase variation (referenced to the element) is shown in Figure 6.21 (b).

Studying the amplitude pattern, a progressive narrowing of the embedded-element patterns is found when increasing the element spacing, similar to the planar case. This is due to the endfire grating lobe (EGL), which occurs (in the planar case) at ϕ_{EGL} = not visible ($d = 0.4\ \lambda$), $90°$ ($d = 0.5\ \lambda$), $41.8°$ ($d = 0.6\ \lambda$), and $25.4°$ ($d = 0.7\ \lambda$). As seen here, no deep nulls appear at the "EGL" angles; instead, shallow dips are found. The reason for this is that the curvature reduces the influence of the mutual coupling, and the number of elements is not very large. Furthermore, there is a ripple in the broadside region, which is not visible in the planar case. This ripple is reduced when decreasing the element spacing and becomes negligible for $d = 0.4\ \lambda$. Thus, the gain of the antenna embedded element is (sometimes strongly) dependent on the spacing. The phase data in Figure 6.21 (b) show the phase variation in the fed-element pattern with respect to the observation angle. One observes that the phase varies only by a few degrees up to the "EGL" drop-off.

The ripple discussed above is not caused by an edge effect since there is no discontinuity in the angular direction; instead, it is due to the mutual coupling among the elements on this nonplanar geometry. The embedded-element pattern is given by a superposition of the isolated-element patterns of all elements when excited through mutual coupling. When added together, the interference between the different contributions will result in an oscillating behavior. As we see in Figure 6.21 (a), the interference may add out of phase ($d = 0.6\ \lambda$) or add in phase ($d = 0.7\ \lambda$). The ripple is reduced by reducing the element spacing.

To provide another way of looking at the sometimes very fast changes of the element pattern when changing the geometry slightly, two additional examples are considered. Figure 6.22 shows the gain (E plane) for two cases with very different results obtained by only changing the element spacing for a fixed radius of $6\ \lambda$. In Figure 6.22 (a), the element spacing is $0.6\ \lambda$ and in Figure 6.22 (b) it is increased by one-tenth of a wavelength to $0.7\ \lambda$. Figure 6.22 shows the gain and element excitations (amplitude and phase). The excitation (obtained from the mutual coupling analysis) is shown for the elements located within $\pm 90°$ with respect to the actively fed element denoted as element number "0." The phase of the elements is shown with respect to the fed element. Together with the original excitation, a projected excitation is obtained. This projected excitation is the excitation of the elements on the curved surface when projected on a planar surface in front of the cylinder. Hence, this equivalent planar array becomes an array with nonuniform element spacing. The new excitations are found by adding a phase corresponding to the distance between the element position on the curved surface and the new position on the planar surface.

Studying the excitation data, now from a planar array point of view, we can more easily explain why the embedded-element patterns look like they do. For the smaller element spacing [Figure 6.22 (a)], there is a large change in phase among the elements closest to the excited element. For these elements, the element spacings are almost equal on the projected plane. If a planar analysis is applied, we get a pattern similar to the embedded-element pattern shown in Figure 6.22 (a). The contributions from the elements closest to the

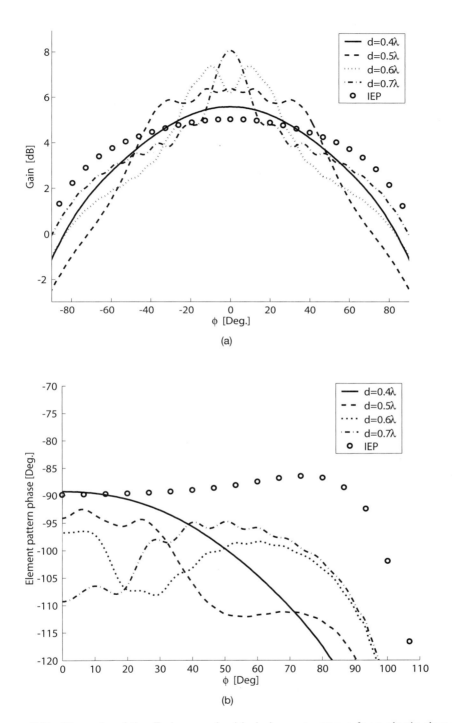

Figure 6.21. The gain of the *E*-plane embedded-element pattern for a single-ring array versus the element spacing for a fixed radius ($R = 5\lambda$). The isolated element pattern (IEP) is also shown for comparison (circles). (*a*) Gain. (*b*) Phase.

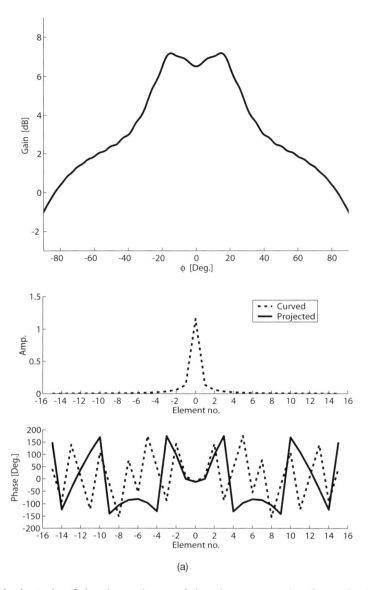

Figure 6.22. A study of the dependence of the element spacing for a single-row array located at a circular cylinder with a 6λ radius. (*a*) *E*-plane gain and element excitation for a 0.6λ element-to-element distance. *(continued)*

fed element add out of phase. For the second case, with increased element spacing, the change in phase among the elements is smaller. Thus, we get a single peak in broadside since the contributions from the elements add in phase. With this analysis, we get more insight into the wide variety of embedded-element patterns found for circular arrays. However, at the end it is still an effect of the curved surface, thus unique for a nonplanar array.

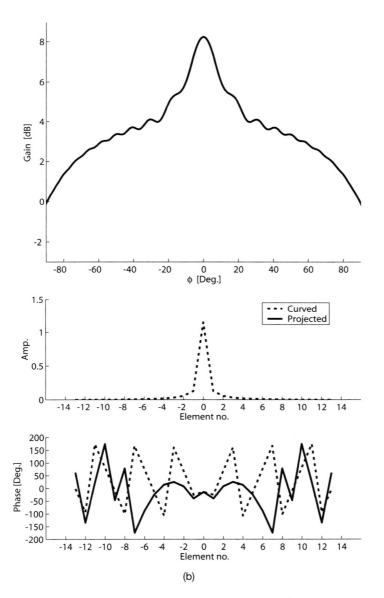

(b)

Figure 6.22. (*cont.*) (*b*) *E*-plane gain and element excitation for a 0.7λ element-to-element distance. The projected excitation (solid line) in both cases is the excitation of the elements when projected on a planar screen in front of the array.

In all discussions above, a single ring of apertures was assumed. Shown next are calculated results for a 13-ring array. In this example, the element distance in the azimuth is 0.6 λ, the axial distance between the rows is 0.68 λ, and the radius is 5 λ. In total, this equals an array of 13 × 52 elements. Figure 6.23 shows the embedded-element pattern of an element in the edge row and the embedded-element pattern of an element in the middle row. Also included is the embedded-element pattern of the single-row array studied earlier.

Adding rows will only change the results slightly, as seen here. The results are shown as normalized power patterns.

The behavior of the embedded-element patterns for cylindrical arrays has been discussed in the past. Several papers, for example [Munger et al. 1971, Hessel 1972, Herper et al. 1985a,b], show embedded-element patterns with rather peculiar patterns, similar to those shown here.

As an example, Figure 6.24 shows embedded-element patterns for a 60° sector cylindrical array developed at the Naval Electronics Laboratory Center [Hessel 1972]. The array is a stacked-ring array of open-ended rectangular waveguides in a conducting cylinder with 32 elements in each column. The total number of elements is 1344, forming a triangular grid. The array radius is 26.3 λ, the waveguide aperture is 0.13 λ by 0.68 λ, and the vertical spacing is 0.72 λ. The results in Figure 6.24 (a) and (b) are for 0.73 λ azimuth spacing and 0.79 λ azimuth spacing, respectively. Measured and calculated results are shown when an element near the center was excited with all other elements terminated. The asymmetry of the measured patterns reflects errors in the experimental array and end effects. These results are very similar to the results presented earlier in this section. The pattern becomes narrower as the grating lobes move into visible space, and the ripple in the main lobe is due to the mutual coupling among the elements. In the paper by Hessel, as well as in many of the references mentioned earlier, the ripple is described in terms of the interference between a (direct) space wave and creeping waves. This is another way of describing the ripple in the broadside region based on the asymptotic solution in which different wave types can be identified.

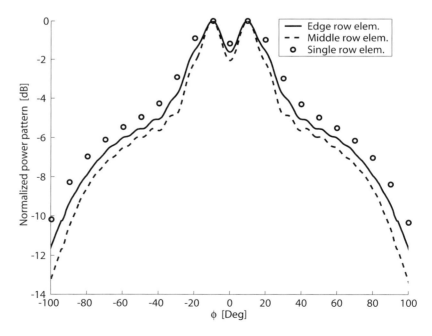

Figure 6.23. The embedded-element pattern for different elements in a 13 × 52 elements array (solid line: edge-row element, dashed line: middle-row element). The radius equals $R = 5\lambda$ and the element spacing in the azimuth and axial directions is 0.6λ and 0.68λ, respectively. The single-row result is also shown for comparison (circles).

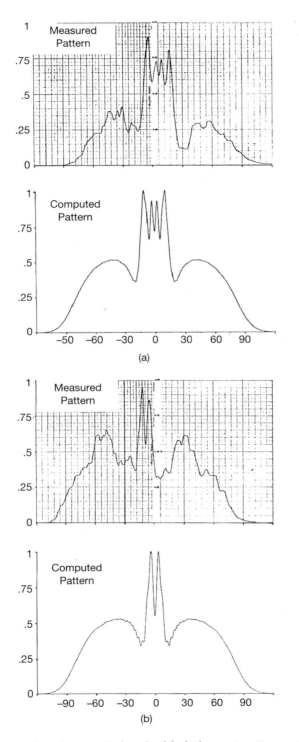

Figure 6.24. Measured and computed embedded-element patterns at broadside. (*a*) Azimuth spacing 0.73λ. (*b*) Azimuth spacing 0.79λ [Munger et al. 1971].

Measured results from a smaller array are shown in Figure 6.25. The results are obtained from the experimental antenna built at Ericsson Microwave Systems AB discussed earlier [Lindgren and Josefsson 1998]; see Section 6.2.3.3 and Figure 6.12. Figure 6.25 shows the *E*-plane embedded-element pattern for the center element in the middle row (out of three rows with 18 elements in each row). Results for two frequencies are shown: 4.9 GHz and 5.65 GHz.

The calculated pattern is due to a closed, infinitely long, circular cylinder. The measured patterns are for a truncated cylinder in both azimuth and axial directions (cf. Figure 6.12). Furthermore, some disturbances due to the measuring range were also detected. Thus, some ripple is seen in the measured results, which is not visible in the calculated results. These results illustrate once again the special features of the embedded-element patterns of a cylindrical antenna.

It has been mentioned in some references, for example [Hessel 1972, p. 284; Mailloux 1994, p. 219] that element patterns with angle- and frequency-dependent ripple make it difficult to synthesize low-sidelobe patterns. However, if the ripple is caused by the mutual coupling among the elements, it can be accounted for in the synthesis, provided the synthesis procedure includes the mutual coupling. This is discussed in more detail in Chapter 10. The frequency dependence, grating lobe effects, and manufacturing tolerances may still cause problems in the practical situation.

Finally, element patterns of a periodic array of axial dipole radiators ($L = \lambda/2$) placed coaxially on a conducting circular cylinder are considered briefly. We will use this type of element to discuss the characteristics of the elevation pattern, but also to investigate the differences between a rectangular and triangular grid. First, a comparison of the dipole element and the aperture element is made.

The dipole array model consists of an infinite number of equispaced stacked rings of axial dipoles in a rectangular grid; see Figure 6.26 (*a*). The dipoles are placed at a distance *s* ($s = 0.25\ \lambda$) above the circular ground plane and the results are obtained by a modal solution applied to a radial unit cell with phase-shift walls [Herper et al. 1985a]. Figure 6.26 (*b*) shows the voltage element gain pattern of the dipole array for $ka = 105$, $b = 0.6\ \lambda$, and $d = 0.8\ \lambda$. The results for two elevation angles are shown: $\theta = 90°$ and $\theta = 60°$. Also included are the results for a cylindrical array of apertures [Sureau and Hessel 1972]. The size of the aperture is $0.32\ \lambda \times 0.75\ \lambda$ with circumferential polarization. Thus, the circumferential plane corresponds for both types to the direction of the strongest mutual coupling.

As seen, there is a larger peak-to-peak amplitude in the ripple for the aperture array. Also, the grating lobe effect of the aperture array appears as a more pronounced dip instead of a pattern drop-off as in the dipole array. This is explained by the fact that the mutual coupling effects in the axial dipole array are weaker than in the corresponding circumferentially polarized aperture array. With these differences and similarities in mind, we will discuss the effect of different grid arrangements for the dipole array. The results are taken from [Herper et al. 1985a].

The first result is for a rectangular lattice. Figure 6.27 (*a*) shows the voltage element gain pattern (element pattern normalized to the unit cell gain) as a function of the elevation angle. Here, $ka = 120$, $b = 0.6\ \lambda$, and $d = 0.7\ \lambda$. For the rectangular lattice, the grating lobe drop-off broadens and shifts toward larger values of ϕ when θ decreases. Furthermore, as we approach the axial direction (decreasing θ), the *E*-plane gain is reduced and the ripple is reduced due to weaker coupling

The grating lobe effect can be explained in quasiplanar terms when the *m* and *n* lobes with $u_{mn}^2 + v_{mn}^2 < 1$ are visible. Note, however, that for planar arrays there is a very clear

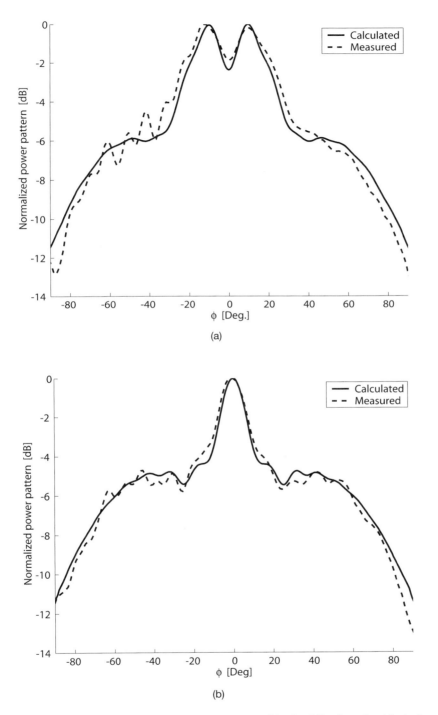

Figure 6.25. Computed (solid line) and measured (dashed line) embedded-element patterns (*E* plane) for a three-row array with 18 elements in each row. The result is for the center element in the middle row. (*a*) *f* = 4.9 GHz. (*b*) *f* = 5.65 GHz.

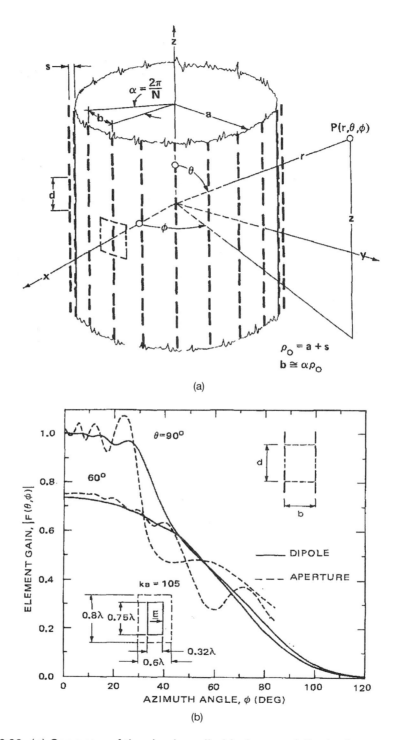

(a)

(b)

Figure 6.26. (*a*) Geometry of the circular cylindrical array of dipoles in a rectangular lattice. (*b*) Voltage element gain patterns between dipole and rectangular aperture radiators. $ka = 105$, $b = 0.6\lambda$, and $d = 0.8\lambda$ [Herper et al. 1985a].

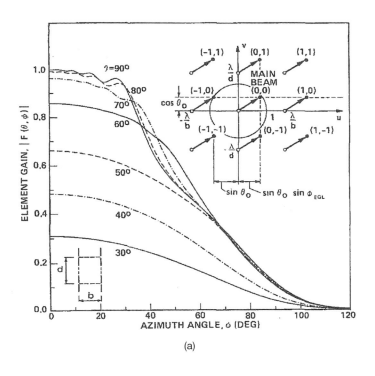

(a)

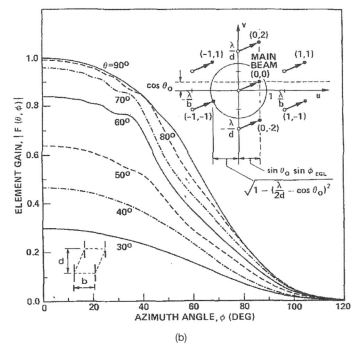

(b)

Figure 6.27. Voltage element gain pattern and corresponding local grating lobe diagram for $ka = 120$, $b = 0.6\lambda$, and $d = 0.7\lambda$. Parameters: elevation angle $\theta = 30°$, $40°$, ..., $90°$. (a) Rectangular grid. (b) Triangular grid [Herper et al. 1985a].

distinction between visible and invisible space. With curved arrays, there is no such clear distinction, as discussed in Section 8.3.2. Nevertheless, a local planar grating lobe diagram is used here and it is shown for the cylindrical (rectangular grid) array in Figure 6.27 (*a*). This diagram leads to the relation

$$\sin \phi_{GL} = \frac{1}{\sin \theta} \left[\frac{m}{|m|} \sqrt{1 - \left(\cos \theta + n \frac{\lambda}{d} \right)^2} - m \frac{\lambda}{b} \right]$$

(6.9)

$$m = 0, \pm 1, \pm 2, \dots, \text{ and } n = 0, \pm 1, \pm 2, \dots$$

From Equation (6.9), we see that the *H*-plane grating lobe ($m = -1$, $n = 0$) enters real space at $\phi_{GL} = 42°$ for $\theta = 90°$, and ϕ_{GL} shifts toward larger values when θ decreases. This is also seen in Figure 6.27 (*a*).

Figure 6.27 (*b*) shows the result for a triangular grid, derived from the rectangular grid studied previously by displacing every second ring by half the circumferential spacing. A grating lobe diagram for this lattice is also shown in the figure. From this we get

$$\sin \phi_{GL} = \frac{1}{\sin \theta} \left[\frac{m}{|m|} \sqrt{1 - \left(\cos \theta + n \frac{\lambda}{2d} \right)^2} - m \frac{\lambda}{b} \right]$$

(6.10)

$$m = 0, \pm 1, \pm 2, \dots, \text{ and } n = 0, \pm 1, \pm 2, \dots, \text{ where } m + n \text{ is even}$$

Here, the *H*-plane grating lobe $(-1, -1)$ causes narrowing of the element pattern with decreasing θ, in contrast to the rectangular grid reference array. Furthermore, an increased sharpness of the grating lobe drop-off and an increased level of the associated off-broadside ripple with decreasing θ is found. This is due to the effect of the grating lobe entering real space with decreasing θ, interfering with the main lobe.

The *E*-plane gain performance is found to be similar in both cases, since in both cases the axial grating lobe is located at $-\lambda/d$.

So, the lattice design of large, cylindrical, uniformly spaced arrays can be based on the local planar grating lobe diagram. See also Chapter 8.

6.3 APERTURE ANTENNAS ON GENERAL CONVEX CYLINDERS

6.3.1 Introduction

Arrays on circular cylinders have been used quite a lot in conformal antenna applications. However, other types of singly curved surfaces are better suited for some applications, for example, the aircraft wing shapes shown in Figures 1.2 and 1.3. For these more general cylinder surfaces, there are very few results available in the literature. The main difference from the previous examples in Section 6.2 is that the radius of curvature is varying along the geodesic, resulting in a varying decay rate of the surface field. This will influence the characteristics of the antenna, as we will demonstrate.

Structures with a varying radius of curvature can be analyzed with a generalized version of the UTD–MoM approach used previously. This method is very useful and can be applied easily; the only limitation is the restriction on the radius of curvature in the UTD formulation. As mentioned in Section 4.4.2, the surface must be smoothly curved and large in terms of wavelength. From a practical (engineering) point of view, this limitation is seldom a problem since most surfaces are large and smooth enough. However, as will be mentioned later (see also [Persson 1999]), quite sharp surfaces can be analyzed with the UTD–MoM approach.

In this section, we will present results for three types of surfaces: the elliptic, parabolic, and hyperbolic cylinders. The antenna elements are the same as used in the previous sections, that is, rectangular apertures with dimensions 0.27 λ × 0.65 λ. The angle α is defined as before (see Figure 6.1).

6.3.2 Mutual Coupling

We focus on the effect of the curvature on the mutual coupling, especially the effect of a varying curvature. Only the isolated coupling is discussed since the curvature effects are most easily seen in this parameter. The analysis is also limited to the *E*-plane single-mode case.

6.3.2.1 *The Elliptic Cylinder.* The first geometry to be discussed is the elliptic cylinder. This generalization of the circular cylinder solution is very interesting. Changing the ellipticity makes it possible to study a wide variety of shapes. Note that the ellipticity is denoted by the term *a*/*b*, where *a* is the major axis and *b* is the minor axis. Different elliptic cylinders are shown in Figure 6.28 where a cross denotes the position of the transmitting aperture. The isolated mutual coupling is calculated from that

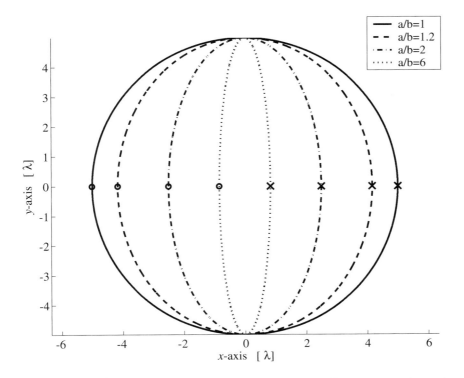

Figure 6.28. The elliptic geometry and the position of the first aperture (cross) and the final position of the second aperture (circle).

aperture to an aperture located at different positions along the surface, in the counter-clockwise direction. A ring denotes the final position of the second aperture. The circular cylinder considered earlier ($R = 5\ \lambda$) is included as a reference. The results for $\alpha = 0°$ are shown in Figure 6.29.

As expected, the decay of the isolated mutual coupling is reduced when the surface is flat but increases again when passing through the sharpest part of the surface. The UTD solution is pushed to its limit for some of the surfaces with very sharp edges. In the results shown here, the local radii of curvature at the sharpest point are (starting with the circular cylinder) $r = 5\ \lambda$, $3.47\ \lambda$, $1.25\ \lambda$, and $0.14\ \lambda$. A radius of curvature as small as $\approx 1\ \lambda$ is very close to the commonly accepted limit for the UTD formulation. The result for the even smaller radius shown here has a reasonable behavior, but the accuracy cannot be ascertained. For a circular cylinder we have verified results for apertures (with the same dimension as here) when the radius of the cylinder equals $1\ \lambda$ [Persson and Josefsson 1999]. At 5 GHz, this equals a cylinder with a radius of 0.06 m. However, since no reference results have been found, it is difficult to verify the result for the most pointed surface presented in Figure 6.29. The important conclusion is that the UTD solution gives satisfactory results for many surfaces that are interesting from a practical engineering point of view.

The ripple at large separations is caused by the interference of the two waves traveling in opposite directions around the elliptic cylinder. The extent of the interference region is larger when the elliptic cylinder gets sharper. This is explained by the smaller curvature (more flat surface) on the rear side causing the amplitudes to decay more slowly

We have also considered the *H*-plane coupling. A similar behavior is found, although the amplitudes are reduced in general. Furthermore, if the mutual coupling is studied with respect to the angle α, the same type of behavior is found as was found for the circular cylinder.

6.3.2.2 *The Parabolic Cylinder.*

We assume that the surface of the parabolic cylinder is extended to infinity in all directions, that is, no edge effects are included in the calculations. The three shapes shown in Figure 6.30 have been studied. The start and end positions of the apertures are chosen so that the geodesic length to the vertex always equals $5\ \lambda$. Thus the total distance traveled is $10\ \lambda$. As before, the starting point is denoted with a cross, and the final position is shown as a circle. The results for the isolated *E*-plane coupling are shown in Figure 6.31. In the examples shown here, the local radius of curvature at the vertex ranges between $1.15\ \lambda$ and $11.3\ \lambda$. There is no ripple since no encircling rays are possible. Otherwise, the results are similar to the results for the elliptic cylinder.

6.3.2.3 *The Hyperbolic Cylinder.*

The hyperbolic surface gives results similar to those for the parabolic cylinder. Here, the local radius of curvature at the sharpest edge ranges between $1.5\ \lambda$ and $10\ \lambda$. The geometry for different hyperbolic surfaces is shown in Figure 6.32 and the results follow in Figure 6.33, where the isolated coupling in the *E*-plane is shown.

6.3.3 Radiation Characteristics

The varying radius of curvature on general cylinders implies that the radiation patterns of the elements on the cylinder depend on the element locations. We will illustrate this for elements on different cylinders. We will first investigate cases with elements all around

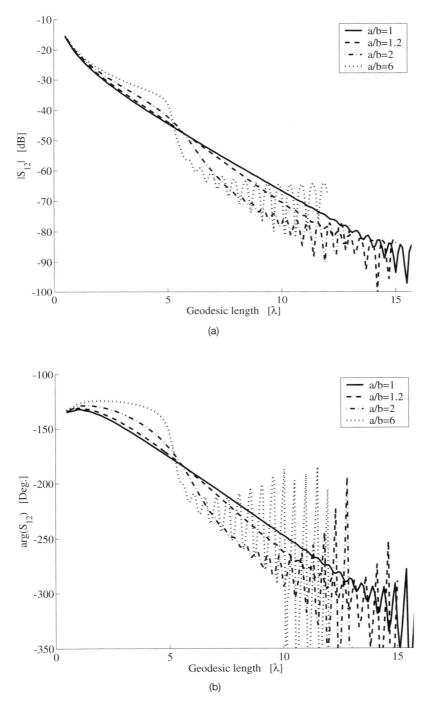

Figure 6.29. The isolated mutual coupling in the circumferential direction (*E* plane) for apertures at elliptic cylinders. The curves are coded according to Figure 6.28, and the circular cylinder is also included for comparison. (*a*) The amplitude. (*b*) The phase.

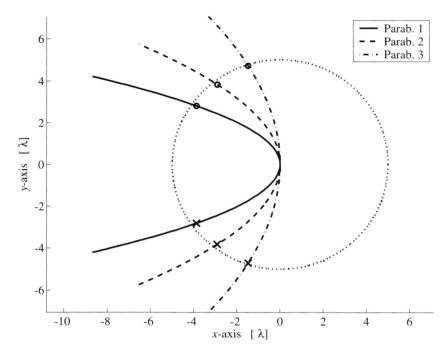

Figure 6.30. The parabolic geometry and the position of the first aperture (cross) and the final position of the second aperture (circle). The circular cylinder ($R = 5\lambda$) is also shown for geometrical reference.

the structure and no end effects, that is, an elliptic cylinder is used in these examples. Then, end effects will be studied for several geometries.

6.3.3.1 The Elliptic Cylinder. For each of the surfaces in Figure 6.28, except the ellipse with $a/b = 6$, aperture elements are placed in one row covering the circumference of the elliptic cylinder. In these examples, the element distance is 0.5 λ. Figure 6.34 shows the element locations when $a/b = 1.2$.

Figure 6.35 (*a*) shows the *E*-plane embedded-element patterns for the different elliptic geometries when the excited element is located in the flat part, and Figure 6.35 (*b*) shows the results for the elements in the most strongly curved part. Figure 6.36 shows the corresponding results when the element distance is increased to 0.6 λ.

As for the circular case, the ripple in the broadside region gets stronger when the element spacing is increased. Furthermore, the patterns are relatively insensitive to a moderate change of the radius of curvature, and the most strongly curved part has to be quite sharp before any large differences appear.

6.3.3.2 End Effects. In this part, we will discuss the end effects for a sector array, in particular the embedded-element patterns for different element locations in the array. We have chosen four typical surfaces for our investigation: planar, circular cylinder, elliptic cylinder, and hyperbolic cylinder. The planar surface is included as a reference. Additional results are found in Section 6.6.4, where microstrip-patch elements are used. In the examples shown here, the following parameters apply:

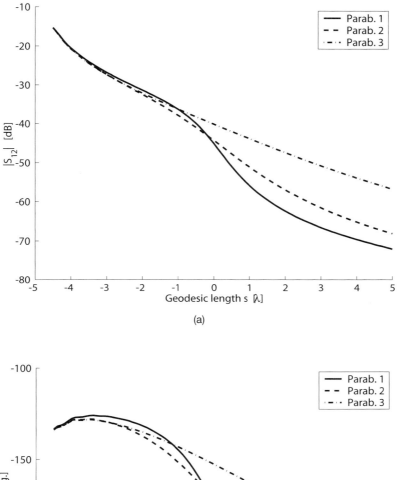

(a)

(b)

<u>Figure 6.31.</u> The isolated mutual coupling in the circumferential direction (*E* plane) for apertures at parabolic cylinders. (*a*) The amplitude. (*b*) The phase.

Element-to-element distance	$0.5\,\lambda$
Number of elements	31
Embedded-element patterns shown	Element numbers 1, 6, 11, and 16
Polarization	E-plane patterns
Aperture size	$0.27\,\lambda \times 0.65\,\lambda$

Figure 6.37 shows the planar case to be used as a reference for the following results. As may be seen, most element patterns look the same and it is only near the end of the array that some differences appear. This has been discussed in [Holter and Steyskal 2002a,b] where the size requirement for finite phased-array models is investigated. The general conclusion is that at any frequency, a finite array should be about $5\,\lambda \times 5\,\lambda$ in size. The central elements then approximately simulate an infinite-array environment, and the edge elements approximate the edge elements of a large finite array. This represents a generalization of the customary 10×10 element model, which is often used as an engineering "rule of thumb" in the normal narrow-band case with $0.5\,\lambda$ element spacing.

The first nonplanar antenna array to be considered is an arc array on a circular cylinder. The embedded-element patterns for elements 1, 6, 11, and 16 (out of 31 elements) are shown in Figure 6.38 together with the geometry. In this example, the radius is $5\,\lambda$. Except for the edge element, the patterns have almost the same shape, compare with the planar case.

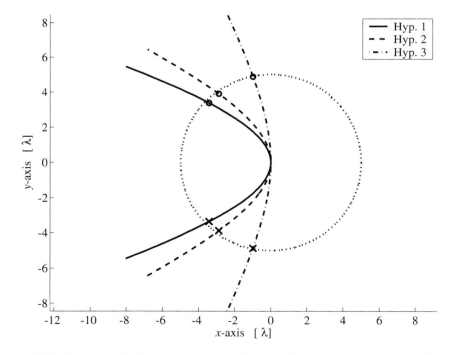

Figure 6.32. The hyperbolic geometry and the position of the first aperture (cross) and the final position of the second aperture (circle). The circular cylinder ($R = 5\lambda$) is also shown for geometrical reference.

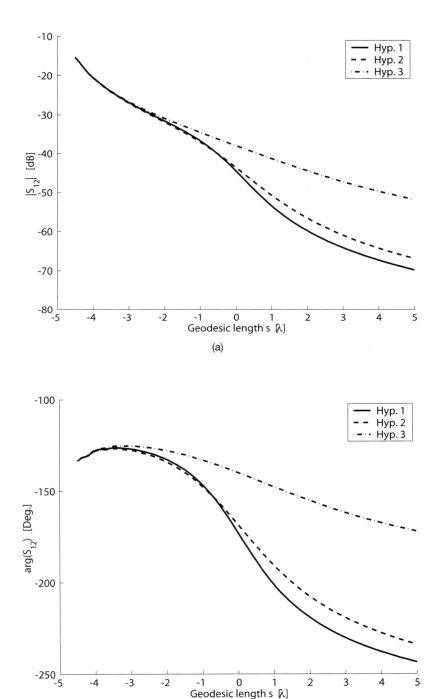

Figure 6.33. The isolated mutual coupling in the circumferential direction (*E* plane) for apertures at hyperbolic cylinders. (*a*) The amplitude. (*b*) The phase.

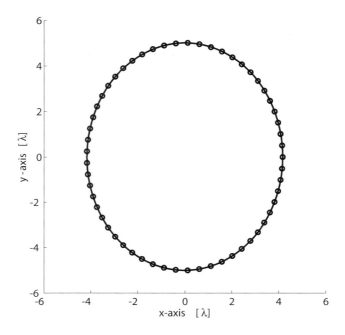

Figure 6.34. The elliptic cylinder array with an element-to-element distance of 0.5λ, *a/b* = 1.2.

The last two examples show a larger impact of the curved surface. In these examples, the elliptic cylinder (Figure 6.39) and the hyperbolic cylinder (Figure 6.40) are considered. Here, the embedded-element patterns are not as equal to each other as in previous examples. This is to be expected since the radius of curvature is varying, which affects the mutual coupling among the elements, thus changing the embedded-element patterns. Note that the ϕ coordinate is referenced to the ellipse and hyperbolic cylinder, respectively; it is not relative to the element angular locations. Hence, the patterns do not overlap as in the previous examples.

6.4 APERTURE ANTENNAS ON FACETED CYLINDERS

6.4.1 Introduction

Faceted conformal antenna arrays are an interesting extension of nonplanar antennas in which a curved surface is approximated by planar facets. Each facet can be viewed as a subarray and can be obtained from standardized building blocks, thus reducing the cost in building conformal antennas. Such antennas have already been proposed for use in use in satellites [Polegre Martín et al. 2001]. In [Jamnejad et al. 2002], options are discussed for future phased arrays for the JPL/NASA Deep Space Network, and multifaceted arrays are one option. Some results for faceted antennas are found in, for example [Agrawal and Powel 1986], where an antenna consisting of 16 facets of 2 × 4 dipole arrays is discussed. General ray-tracing tools for planar and curved facets are found in, for example [Saez de Adana et al. 1998, Hussar et al. 2000].

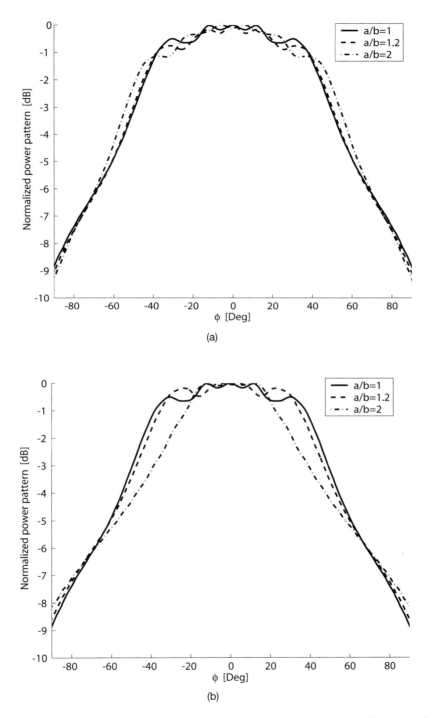

Figure 6.35. The embedded-element pattern versus the shape of elliptic surfaces with an element-to-element distance of 0.5λ. ϕ is a relative angle and equals zero at the element position. (*a*) The element on the flattest part is excited. (*b*) The element on the most strongly curved part is excited.

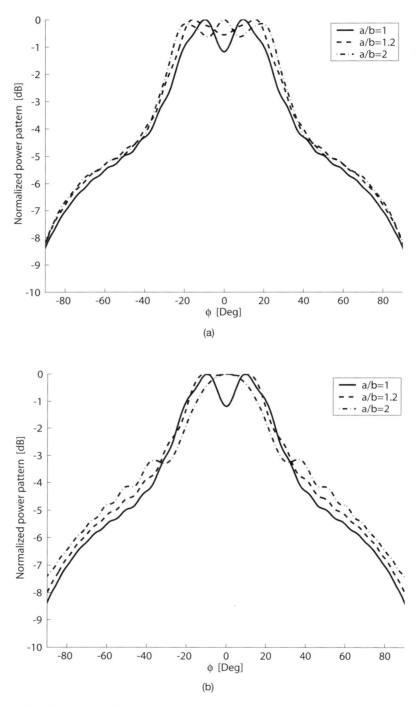

Figure 6.36. The embedded-element pattern versus the shape of elliptic surfaces with an element-to-element distance of 0.6λ. ϕ is a relative angle and equals zero at the element position. (*a*) The element on the flattest part is excited. (*b*) The element on the most strongly curved part is excited.

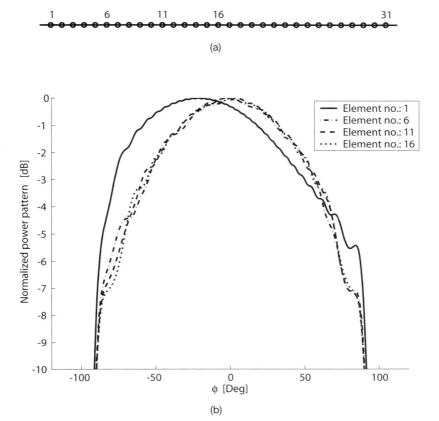

Figure 6.37. Embedded-element patterns in the *E* plane due to apertures in a sector array containing 31 elements, infinite ground-plane. The element-to-element distance equals 0.5λ. (*a*) The geometry. (*b*) The results.

In this section, we will consider a faceted circular cylinder and a faceted elliptic cylinder. Both the isolated mutual coupling and isolated element patterns are analyzed with waveguide-fed apertures as radiators. The results are compared with results obtained for a corresponding smooth surface. The facets are supposed to be subarrays and must always be large enough to support the radiating elements. Thus, there is no sense in making the facets smaller until a limiting smooth case is obtained.

The results for the faceted surfaces are calculated using a wedge UTD–MoM approach. The general approach is similar to the one used in the previous examples. Edge diffraction at grazing incidence is included but not multiple reflections [Marcus 2004]. We will find that the results for the faceted surfaces considered here agree closely with the results for the corresponding smooth surfaces.

The faceted cylindrical antennas considered here consist of rectangular facets with waveguide-fed rectangular apertures (0.33 λ × 0.73 λ). The longest side of each aperture is parallel to the axis of the cylinder and the aperture is centered on the facet, resulting in horizontal or *E*-plane polarization. The faceted cylinders have been chosen so that their circumference is identical to that of a corresponding smoothly curved cylinder. The radius

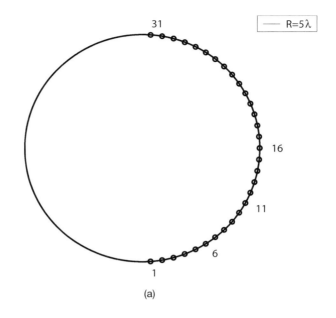

(a)

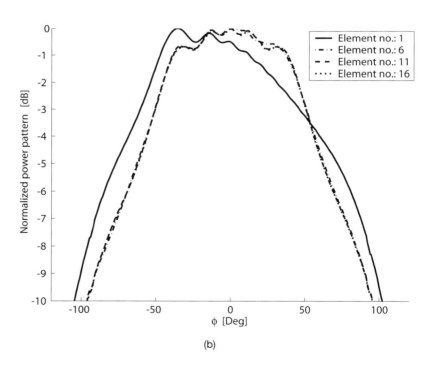

(b)

<u>Figure 6.38.</u> Embedded-element patterns in the *E* plane due to apertures in a sector array containing 31 elements, circular cylinder (*R* = 5 λ). The element-to-element distance equals 0.5 λ. φ is a relative angle and equals zero at the element position. (*a*) The geometry. (*b*) The results.

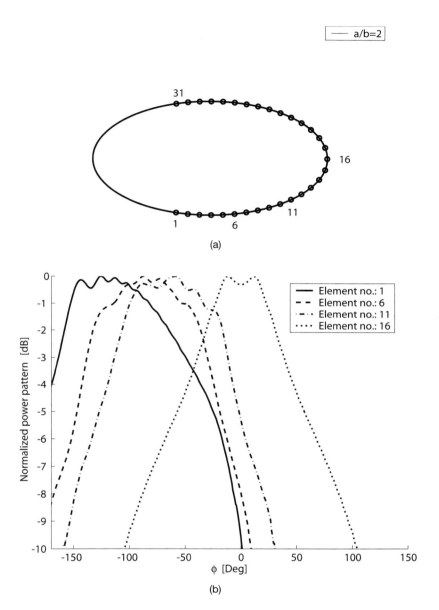

Figure 6.39. Embedded-element patterns in the *E* plane due to apertures in a sector array containing 31 elements, elliptic cylinder (*a*/*b* = 2, *a* = 5 λ). The element-to-element distance equals 0.5 λ. (*a*) The geometry. (*b*) The results.

of the circular cylinder is 10 λ and the number of facets is 20, which means that the smallest facet is at least 3 λ across. This is large enough to retain the properties of the faceted antenna but small enough to make a meaningful comparison with the results from the smooth cylinder. The faceted circular cylinder is seen in Figure 6.41 (*a*). Figure 6.41 (*b*) shows the faceted elliptic cylinder (*a*/*b* ≈ 2) where 16 segments have been used. See also [Marcus et al. 2003].

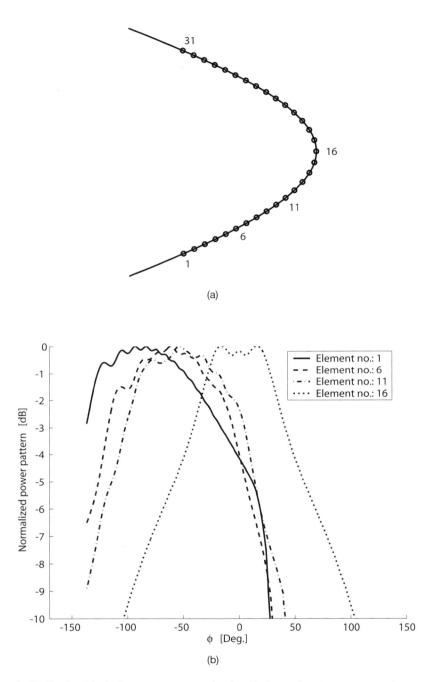

(a)

(b)

<u>Figure 6.40.</u> Embedded-element patterns in the *E* plane due to apertures in a sector array containing 31 elements, hyperbolic cylinder (*b/a* = 0.3, *a* = 16.67λ). The element-to-element distance equals 0.5λ. (*a*) The geometry. (*b*) The results.

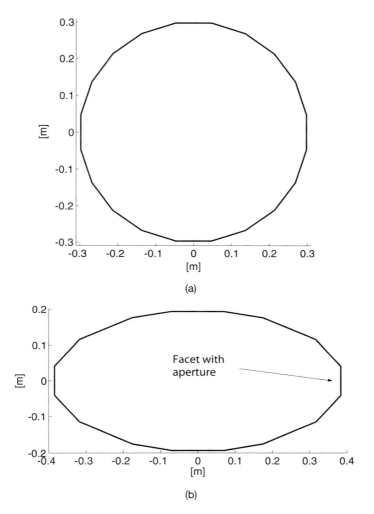

Figure 6.41. Faceted geometries, $f = 10$ GHz. (*a*) The circular cylinder with a radius of 10λ; the number of facets is 20. (*b*) The elliptic cylinder, the facet with the source aperture is shown. The ellipticity is $a/b \approx 2$ and the number of facets is 16 [Marcus et al. 2003].

6.4.2 Mutual Coupling

The isolated mutual coupling is calculated for two configurations for both the circular cylinder and the elliptic cylinder. The first case is for $\alpha = 0°$ (see Figure 6.1), that is, the apertures are located at the same height on the cylinder. In the second case $\alpha = 40°$, which means that the centers of the apertures are located on a spiral with the inclination 40°. For the elliptic surface, the transmitting aperture is located at the facet of the most strongly curved part of the surface; see Figure 6.41 (*b*). The isolated mutual coupling is calculated from that aperture to an aperture located at the center of the facets.

Figure 6.42 shows the isolated mutual coupling for the faceted circular cylinder. The isolated mutual coupling is expressed in terms of the mutual admittance; see Equation

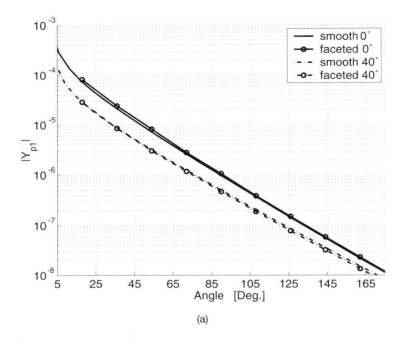

(a)

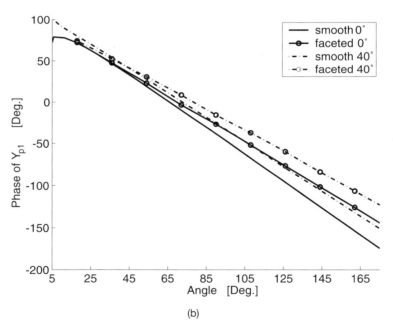

(b)

Figure 6.42. The isolated mutual coupling (admittance) (Ω^{-1}) for a faceted circular cylinder (lines with circles), $\alpha = 0°$ and $40°$. Results for a smooth circular cylinder are also included for comparison [Marcus et al. 2003]. (*a*) Amplitude. (*b*) Phase.

(6.2). To get the *S*-matrix, Equation (6.3) can be used together with the self-admittance of the waveguide aperture on an infinite ground plane, $Y_{11} = (1.13 - j1.53) \cdot 10^{-3} \ \Omega^{-1}$, and the characteristic admittance for the TE_{10} mode of the waveguide, $Y_c = 1.94 \cdot 10^{-3} \ \Omega^{-1}$. The phase in Figure 6.42 (*b*) is the phase when the free-space contribution is subtracted. The faceted elliptic cylinder is considered in Figure 6.43. As may be seen, the faceted and smoothly curved results agree quite well in both cases.

6.4.3 Radiation Characteristics

Here, the isolated element pattern is discussed for an aperture located on one of the facets [see Figure 6.41 (*b*) for exact location on the elliptic surface]. The results are shown as relative amplitude and relative phase with the phase center located at the center of the aperture.

Figure 6.44 shows the result for the circular cylinder. The ripple in the amplitude is expected since the pattern of the faceted cylinder is composed of diffracted fields from each of the cylinder's edges, whereas the smooth cylinder diffracts from all points along the surface. The phase has essentially the same behavior for both cases, and the ripple in the phase corresponds to that of the amplitude.

Finally, the results for the elliptic cylinder are considered. Figure 6.45 shows the relative amplitude and relative phase in the *E* plane. The ripple in amplitude is higher than for the circular cylinder since the wedge angles close to the source facet are smaller compared to the circular case. This gives higher diffracted field amplitudes. The phase has corresponding oscillations.

We noted before that the mutual coupling results were almost the same for the smoothly curved and the faceted cylinders. We see here also that the radiation patterns agree well for the two cases. Hence, the edges of the facets do not affect the results very much in the typical cylinder cases studied here.

6.5 APERTURE ANTENNAS ON DIELECTRIC COATED CIRCULAR CYLINDERS

6.5.1 Introduction

For mechanical protection, the array antenna is often covered by a dielectric radome. The introduction of this external dielectric layer changes the characteristics of the antenna. Now, more energy will be trapped and propagate along the surface and interact with the antenna elements. In this section, we will analyze the effects of this coating on aperture antennas on PEC circular cylinders. As discussed in Chapter 4, this is sometimes a difficult problem if the surface is electrically large, in particular when surfaces other than the circular cylinder are considered.

In this section, we will concentrate on the main differences compared to a noncoated surface. Most of the results are for a single layer only. In practice, it may be difficult to avoid an air gap between the PEC surface and the coating, and we will discuss this also.

The circular cylinder geometry used before will be used again (see Figure 6.1) but now a dielectric coating with thickness *t* is added. The same type of aperture elements as considered in Section 6.2 will be used and the radius of the cylinder will also be the same. To repeat, this means rectangular apertures with dimensions $0.27 \ \lambda \times 0.65 \ \lambda$ and the radius of the PEC cylinder is 5 λ.

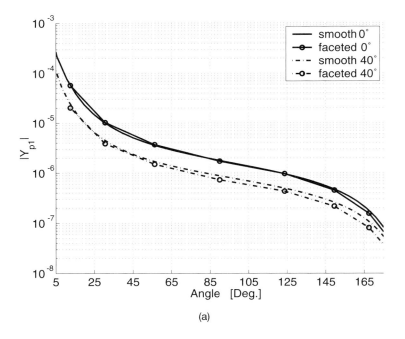

(a)

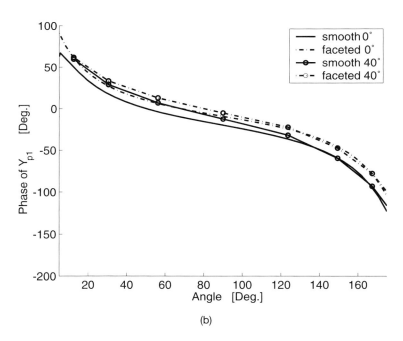

(b)

<u>Figure 6.43.</u> The isolated mutual coupling (admittance) (Ω^{-1}) for a faceted elliptic cylinder (lines with circles), $\alpha = 0°$ and $40°$. Results for a smooth elliptic surface is also included for comparison. The transmitting aperture is located at the most strongly curved part of the surface [Marcus et al. 2003]. (*a*) Amplitude. (*b*) Phase.

COLOR FIGURES

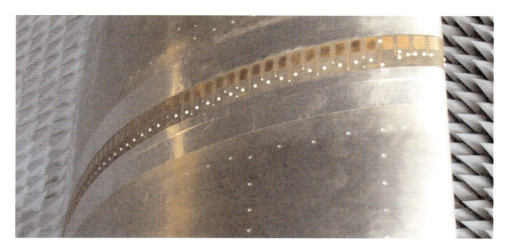

Figure 1. Microstrip array on wing shape.(Courtesy of Air Force Research Lab./Antenna Technology Branch, Hanscom AFB, USA.)

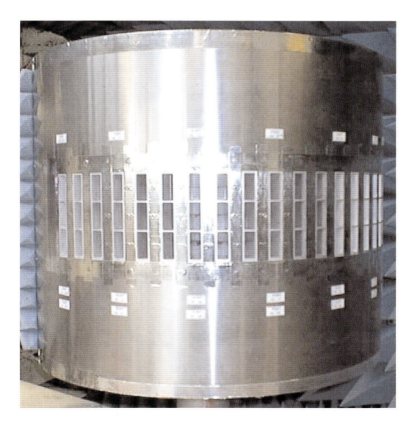

Figure 2. Waveguide array on cylinder. (Courtesy of Ericsson Microwave Systems AB, Göteborg, Sweden.)

Conformal Array Antenna Theory and Design. By Lars Josefsson and Patrik Persson
© 2006 Institute of Electrical and Electronics Engineers, Inc.

Figure 3. Waveguide array on paraboloid. (Courtesy of Ericsson Microwave Systems AB, Göteborg, Sweden.)

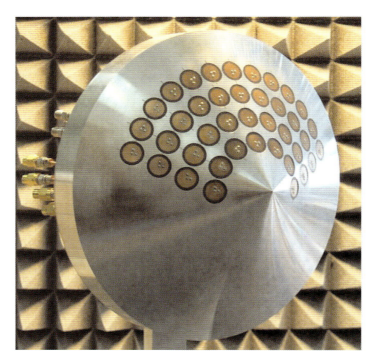

Figure 4. Conical microstrip array. (Courtesy of FGAN Research Establishment for Applied Science, Wachtberg-Werthhoven, Germany.)

Figure 5. Conical microstrip array. (Courtesy of Alcatel-Space and ESA.)

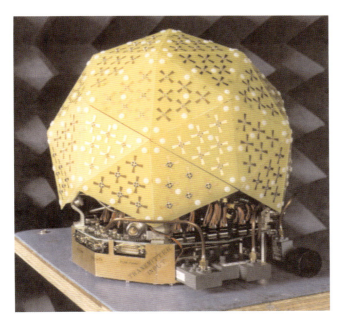

Figure 6. Faceted spherical array. (Courtesy of Roke Manor Research Ltd., Roke Manor, Romsey, Hampshire, UK.)

Figure 7. Broadband circular array. (Courtesy of Anaren Inc., Syracuse, NY, USA.)

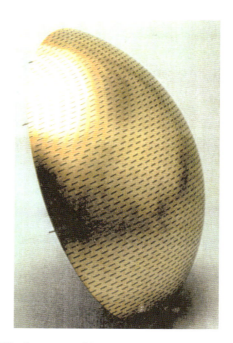

Figure 8. Spherical FSS slot array. (Courtesy of Ericsson Microwave Systems AB, Göteborg, Sweden.)

Figure 9. Elliptical microstrip array. (Courtesy of FGAN Research Establishment for Applied Science, Wachtberg-Werthhoven, Germany.)

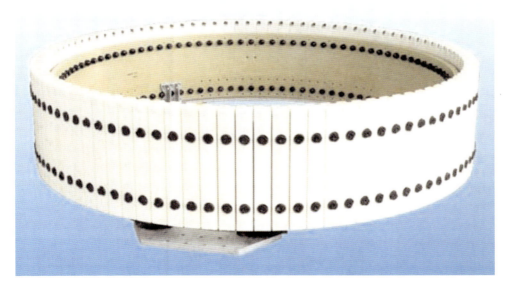

Figure 10. Cylindrical sonar array. (Courtesy of Atlas Elektronik GmbH, Submarine Systems, Bremen, Germany.)

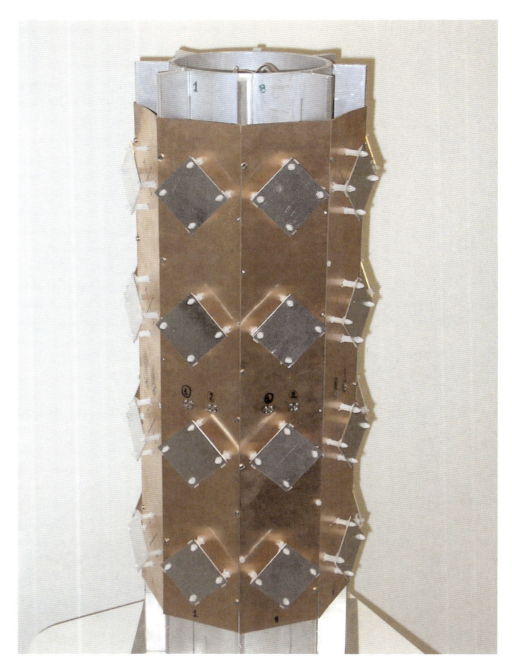

Figure 11. Cylindrical microstrip array. (Courtesy of Ericsson AB, Göteborg, Sweden.)

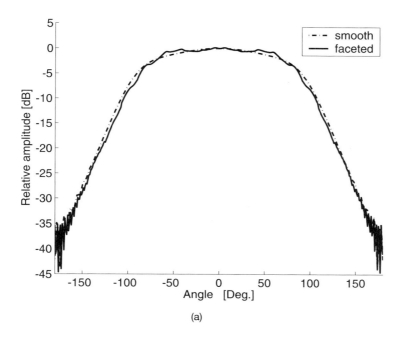

(a)

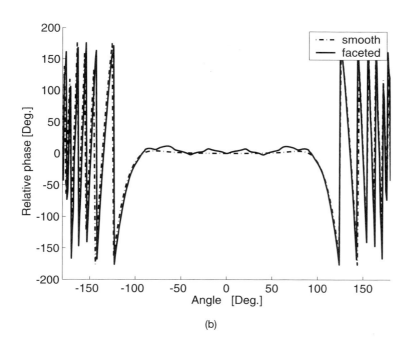

(b)

Figure 6.44. The isolated-element pattern for a faceted circular cylinder (solid line). The radius of the cylinder is 10λ and the number of facets is 20. Results for a smooth circular cylinder are also included for comparison (dash–dotted line) [Marcus et al. 2003]. (*a*) Amplitude. (*b*) Phase.

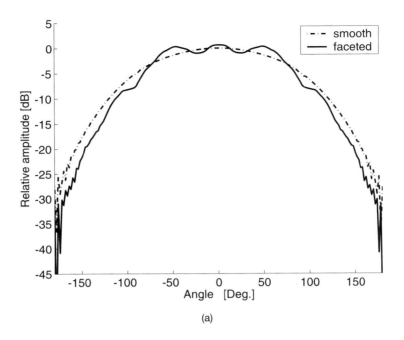

(a)

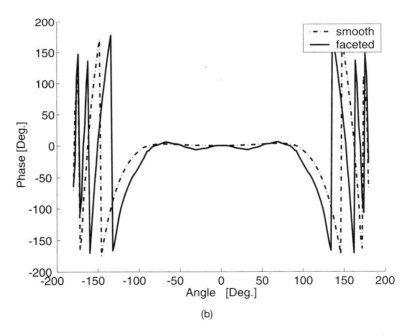

(b)

<u>Figure 6.45.</u> The isolated-element pattern for a faceted elliptic cylinder (solid line). The ellipticity is $a/b \approx 2$ and the number of facets is 16. Results for a smooth elliptic cylinder are also included for comparison (dash–dotted line) [Marcus et al. 2003]. (*a*) Amplitude. (*b*) Phase.

<u>Figure 6.46.</u> Experimental antenna with dielectric coating. Courtesy of Ericsson Microwave Systems AB, Göteborg, Sweden.

We will also use measured data from the experimental antenna described in Section 6.2 as a reference solution [Lanne and Josefsson 1999a,b]. The antenna consists of 3 × 18 rectangular waveguides mounted on the surface of a circular sector (see Figure 6.12), but now with a dielectric radome added as shown in Figure 6.46. The dielectric layer has relative permittivity $\varepsilon_r = 2.32$, thickness $t = 3.99$ mm, and $\tan \delta = 2.47 \cdot 10^{-4}$. The apertures are numbered as before (see Section 6.2.3.3), that is, row by row, from left to right, starting with aperture 1 in the lower-left corner of the array. As may be seen in Figure 6.46, the cylinder is truncated at the rear for practical reasons. Thus, absorbers were placed on the edges to minimize edge effects during the measurements.

6.5.2   Mutual Coupling

The calculated results presented here are obtained with the G1DMULT code [Sipus et al. 1998] instead of using the asymptotic solution [Persson and Rojas 2003]. The reason for using G1DMULT is that we would like to include an air gap between the PEC surface and the single-layer coating. This can be handled by G1DMULT for moderate cylinder radii. The asymptotic solution,* on the other hand, can handle larger radii, but the method is limited to just one single dielectric layer. We will first study single-layer cases and introduce the effect of an air gap later.

6.5.2.1   *Isolated Mutual Coupling.*   The first examples illustrate how the coating affects the isolated mutual coupling in the E plane along one row, that is, $\alpha = 0°$. Figure 6.47 shows the amplitude and phase of the isolated mutual coupling along the surface

---

*For a single layer, the two methods have been compared (see [Persson and Rojas 2003]) and the difference between them is measured in tenths of a dB.

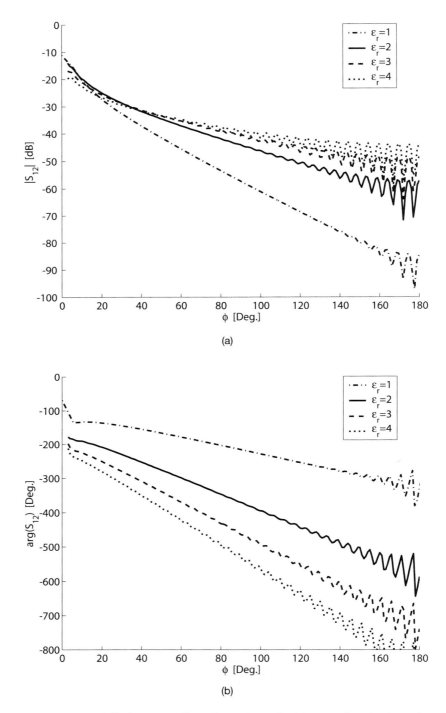

Figure 6.47. Isolated *E*-plane coupling along the cylindrical surface in the azimuth (α = 0°). Here, the coupling is studied with respect to the relative permittivity of the lossless coating; a single-waveguide mode is used. (*a*) The amplitude. (*b*) The phase with the free-space contribution subtracted.

while changing the relative permittivity ($\varepsilon_r = 2$, 3, and 4) of the coating. The coating is assumed to be lossless with a thickness of 4 mm and the frequency is 5 GHz. The next example (see Figure 6.48) shows the dependency of the thickness ($t = 1$ mm, 3 mm, and 5 mm) while keeping the relative permittivity constant ($\varepsilon_r = 2$) and assuming a lossless coating. The frequency is again 5 GHz, which means that the thickness becomes 0.02 λ_d, 0.0707 λ_d, and 0.1178 λ_d (λ_d is the wavelength in the dielectric). The noncoated cylinder result is also included for comparison. Note that the results are all single-mode results and no air gap is included.

From the results, it is clearly visible that the mutual coupling between the two elements is stronger when the coating is present. This is to be expected since the coating will reduce the scattered field and the field along the surface becomes trapped along the surface due to the coating. If the coating gets thicker or if the permittivity is increased, the field becomes more and more trapped, as seen here. As a consequence, the interaction between the encircling waves is much more pronounced. The phase data are shown with the free space contribution subtracted, but the phase is still decaying as function of the distance. This was also observed for the noncoated cylinder (see Figure 6.2) and it was shown that this could be interpreted as if the wave propagates slightly above the surface. Here, a similar analysis shows that the wave propagation cannot be explained only by considering a wave propagating in free space above the surface. If so, the wave would propagate roughly half a wavelength above the surface, which is not likely. Instead, the surface wave along a nonplanar coated PEC surface is made up of two components interacting with each other: one free-space wave (where most of the energy is propagating) and one wave propagating in the coating [Anderson and Hass 1970, Lanne and Josefsson 1999a]. This can be examined further by a detailed analysis of the pole locations for the geometry; see Appendix 1, Section 4A.4.

The calculated isolated coupling (no air gap) along one row has also been compared against measured data. The agreement is good, as seen in Figure 6.49. Using higher-order modes does not produce any significant improvement of the result (up to 12 modes were tried). Another factor to consider is the air gap between the PEC cylinder and the dielectric layer. It was not possible to avoid this air gap due to practical reasons. The average distance of the air gap equals 0.85 mm and we will discuss the effect of the air gap next.

To do this we use a lossless coating with thickness $t = 4$ mm and $\varepsilon_r = 2$. Figure 6.50 shows the isolated coupling versus the thickness of the air gap, and two cases have been considered. The following thicknesses of the air gap are used: 1 mm, and 1.5 mm. In free-space wavelengths ($f = 5$ GHz), these thicknesses equal 0.0167 λ_0 and 0.025 λ_0. For comparison, the result without an air gap and the measured result (with an air gap) are also shown. The effect of the air gap is visible, although it is quite small when considering the isolated coupling. Higher-order modes in the calculations do not change the results.

6.5.2.2 *Array Mutual Coupling.*
We noted in Section 6.2 for the noncoated case that the array mutual coupling is more sensitive to higher-order modes than the isolated coupling. We can expect the same thing here for the coated case. The air gap should also be more critical. The calculated results shown here will be compared with measured data [Lanne and Josefsson 1999a] obtained from the experimental antenna built by Ericsson Microwave Systems AB in Mölndal, Sweden, and presented earlier in this chapter. The main dimensions are the following:

Element apertures	39 mm × 16 mm
Element grid	Rectangular, 3 × 18 elements, extending over about 120°

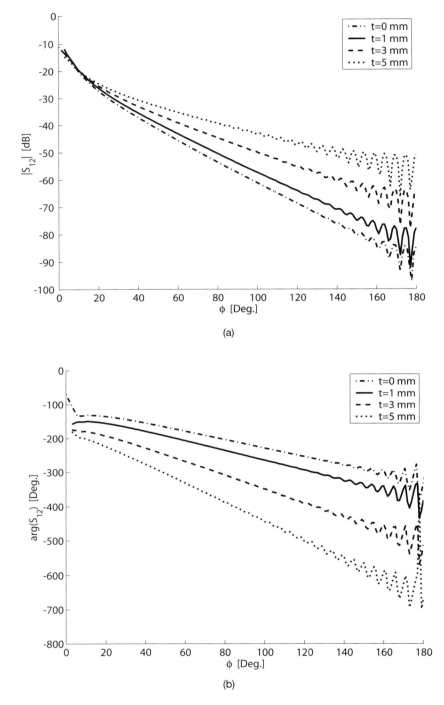

(a)

(b)

Figure 6.48. Isolated *E*-plane coupling along the cylindrical surface in the azimuth ($\alpha = 0°$). Here, the coupling is studied with respect to the thickness of the lossless coating; a single-waveguide mode is used. (*a*) The amplitude. (*b*) The phase with the free-space contribution subtracted.

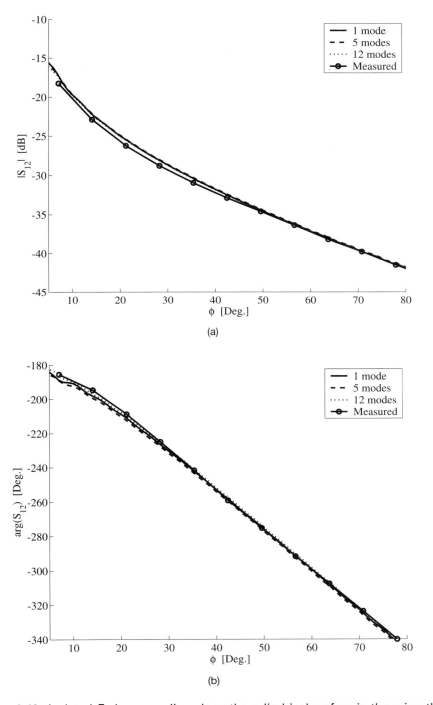

Figure 6.49. Isolated *E*-plane coupling along the cylindrical surface in the azimuth (α = 0°). Here, the coupling is studied with respect to the number of modes used when ε_r = 2.32 and *t* = 3.99 mm. Measured data are also included (line with circles). (*a*) The amplitude. (*b*) The phase with the free-space contribution subtracted.

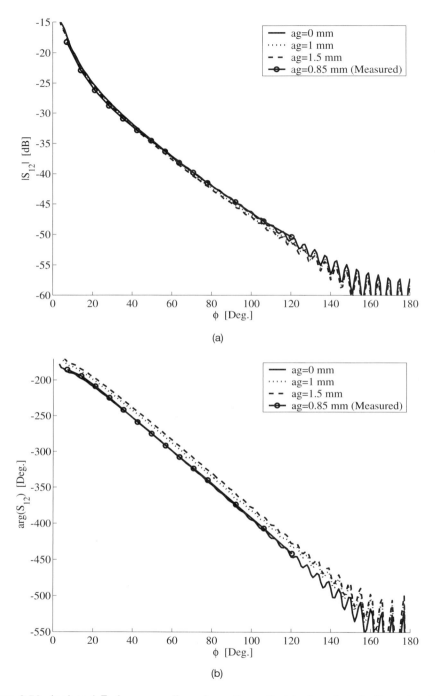

Figure 6.50. Isolated *E*-plane coupling along the cylindrical surface in the azimuth (α = 0°). Here, the coupling is studied with respect to the thickness of the air gap (ag) between the PEC surface and the coating. Measured data are also included (line with circles). (*a*) The amplitude. (*b*) The phase with the free-space contribution subtracted. *(continued)*

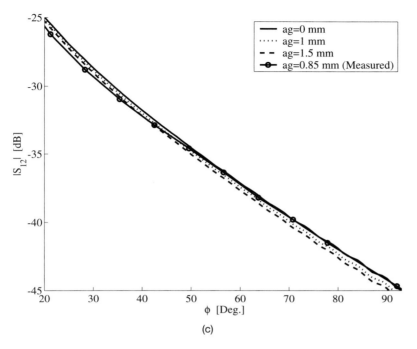

Figure 6.50. (*cont.*) (*c*) Close-up view of the amplitude result.

Element spacings	37.1 mm in the azimuth
	41 mm in the axial direction
Polarization	Circumferential
Cylinder radius	300 mm
Frequency	$f = 5.65$ GHz
Coating	Relative permittivity $\varepsilon_r = 2.32$
	Thickness $t = 3.99$ mm
	$\tan \delta = 2.47 \cdot 10^{-4}$

The first example in Figure 6.51 shows the array mutual coupling (amplitude and phase) for the first row of the array calculated using 1, 5, and 12 waveguide modes. The coating is assumed to be lossy and there is no air gap included. The similar example for the noncoated surface was shown in Figure 6.15, where 5 modes were used.

Comparing the coated (Figure 6.51) and the noncoated (Figure 6.15) cases, we notice that the coating has a large effect on the array mutual coupling. For example, the local amplitude minimum at element number 11 is not seen in the noncoated case. Another observation is that the result (both amplitude and phase) for the coated surface is oscillating. This behavior is not noted in any of the earlier examples, not even for the isolated coupling with a coating (cf. Section 6.5.2.1). As mentioned earlier, the wave propagation along the coated surface can be divided into two parts; the oscillations may be due to the interference between these two parts. Furthermore, the apertures probably also take part in this phenomenon since the oscillations are visible only when studying the mutual coupling among terminated elements coated with dielectric material [Anderson and Hass 1970, Lanne and Josefsson 1999a].

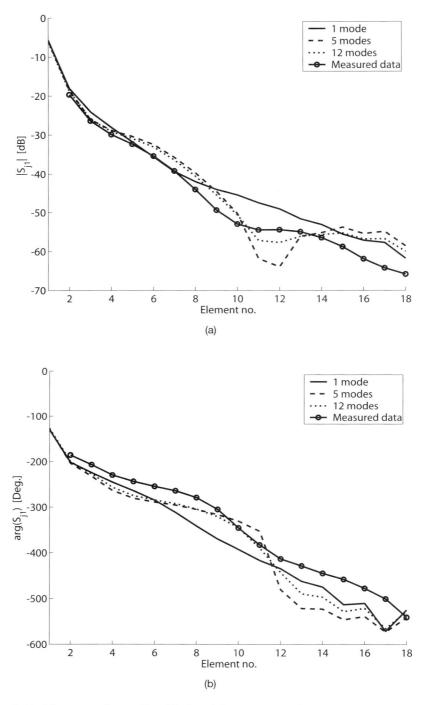

(a)

(b)

Figure 6.51. The mutual coupling (*E* plane) for an array of apertures with a lossy dielectric coating. The calculated results are obtained using 1, 5, and 12 modes; the air gap between the PEC cylinder and the coating is not included. The results are compared with measured results (line with circles). (*a*) Amplitude. (*b*) Phase.

In Section 6.2, it was also shown that five waveguide modes were needed for good/excellent accuracy (without coating). Using the same number of modes for the coated surface does not, however, give the same close agreement with the measurements. Using up to 12 modes does not give much improvement.

To improve the calculated results, it is necessary to include the air gap in the analysis. Doing this, the results shown in Figure 6.52 are obtained. The agreement with the measured data improves considerably when higher-order waveguide modes are included. It seems that five modes give satisfactory results, 12 modes improve the results even more [Sipus et al. 2001]. However, it is still not possible to obtain an agreement as good as obtained in the noncoated case. An explanation can be that the dielectric coating and the spacing could not be controlled very tightly in the measurement setup. However, we conclude that if a coated conformal array antenna has a similar air gap, it must be included in the analysis if good accuracy is required.

6.5.3 Radiation Characteristics

In this section, we will give some examples of how the dielectric coating affects the radiation characteristics of the antenna. The results shown here have been obtained with an asymptotic method [Thors and Rojas 2003]. This asymptotic method can be used for the analysis of thin dielectrics up to a maximum thickness of approximately $0.15 \lambda_d$ (λ_d is the wavelength in the dielectric). This is not a serious limitation from a practical point of view. However, an air gap cannot be included using this model. We will show typical examples of how the coating affects the radiation characteristics, but the treatment is not as detailed as in Section 6.2.4, since the main features have already been discussed.

6.5.3.1 *Isolated-Element Patterns.* The isolated-element pattern is presented for two polarizations to illustrate the effect of the coating. The geometry is the same as before, but with some differences in the parameter setup. To simplify the problem, the aperture is approximated with a magnetic current moment located in the center of the aperture. Although this is a simplification of the real case, we will get a good impression of what to expect when adding the coating. Both a $\hat{z}$- and a $\hat{\varphi}$-oriented dipole are considered; see Figure 6.53 and Figure 6.54, respectively. Furthermore, the frequency (f_0) is 6.8 GHz and the radius of the PEC cylinder is $6.8 \lambda_0$. The coating is assumed to be lossless, its thickness is $0.0357 \lambda_0$, and the permittivity ε_r is chosen in the interval $1.1 \rightarrow 2.94$. Note that no air gap is included. The isolated element pattern is calculated as a function of the azimuth angle φ for an elevation angle $\theta = 50°$. The noncoated result is included for comparison, obtained with the UTD solution in [Pathak et al. 1981].

From these results, it is evident that the largest effect of an increasing dielectric constant is that more energy is guided along the dielectric layer, causing higher radiation levels in the shadow region. Also, thicker layers increase the radiation levels in the shadow region.

It is well known that the cross-polarized far field from a $\hat{z}$-directed magnetic dipole situated on a PEC cylinder is zero. However, if a dielectric cladding is introduced, rather high cross-polarization levels are obtained as a consequence of coupling within the dielectric material; see Figure 6.53 (*b*). To check that zero cross-polarization is obtained as $\varepsilon_r \rightarrow 1$, calculations were made for a dielectric constant of $\varepsilon_r = 1 + \Delta\varepsilon$, where $\Delta\varepsilon$ is a small number. In the limit when $\varepsilon_r \rightarrow 1$, the radiation patterns should approach the well-defined radiation patterns from a magnetic dipole situated on a PEC cylinder without dielectric. As expected, when the relative permittivity is lowered, the results approach the UTD solution [Pathak et al. 1981].

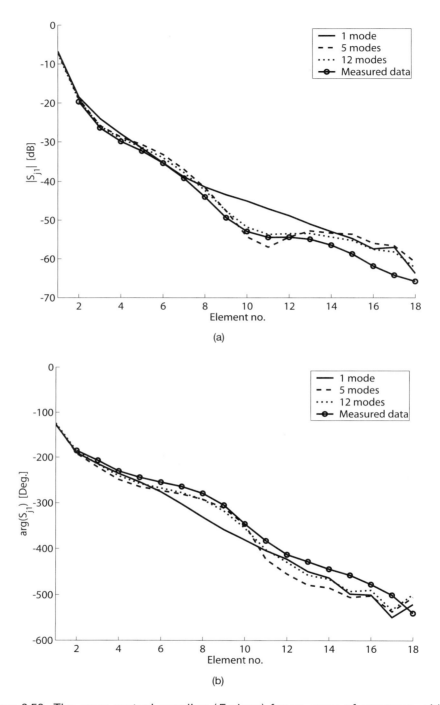

Figure 6.52. The array mutual coupling (*E* plane) for an array of apertures with a lossy dielectric coating. The calculations were made using 1, 5, and 12 modes, including a small air gap (0.85 mm) between the PEC cylinder and the coating. The result is compared with the measured results (line with circles). (*a*) Amplitude. (*b*) Phase.

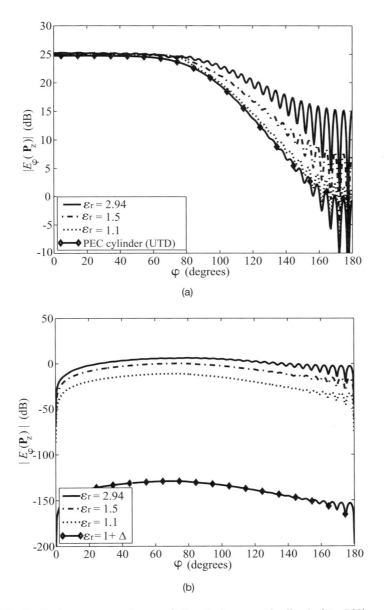

Figure 6.53. Radiation patterns from a $\hat{z}$-directed magnetic dipole ($\theta = 50°$) versus the relative permittivity of the lossless coating. The noncoated result is also included for comparison (line with diamonds). (*a*) $|E_\varphi|$. (*b*) $|E_\theta|$ ($\Delta\varepsilon = 10^{-6}$) [Thors and Rojas 2003]. Copyright 2003 American Geophysical Union. Reproduced/modified by permission of American Geophysical Union.

6.5.3.2   Embedded-Element Patterns. The embedded-element patterns presented are taken from [Thors et al. 2003], providing an overview of the effect of the dielectric coating. All results are obtained with the asymptotic method. The mutual coupling among the elements is found from the asymptotic method presented in [Persson and Rojas 2003]. However, we lack an asymptotic solution for the mutual coupling within the parax-

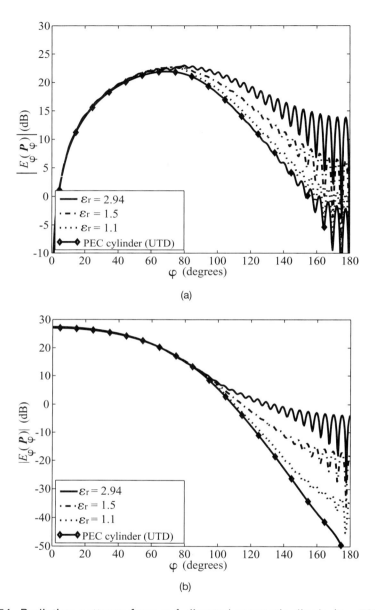

Figure 6.54. Radiation patterns from a $\hat{\varphi}$-directed magnetic dipole ($\theta = 50°$) versus the relative permittivity of the lossless coating. The noncoated result is also included for comparison (line with diamonds). (a) $|E_\varphi|$. (b) $|E_\theta|$. Copyright 2003 American Geophysical Union. Reproduced/modified by permission of American Geophysical Union.

ial region. To overcome this, a planar solution is assumed for this region, as indicated in Figure 6.55. This solution is accurate enough if the radius of the cylinder is relatively large, and if the apertures are oriented in such a way that the weak H-plane coupling is along the cylinder axis (see Figure 6.55). For this case, the E-plane coupling is much stronger and will contribute more to the coupling in the paraxial region than the direct H-plane coupling.

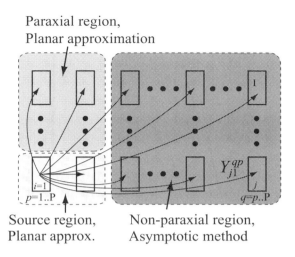

Figure 6.55. The admittance values are calculated with the source aperture in the lower-left corner of the array. For computational purposes, the rectangular grid array is divided into three regions: source region, paraxial region, and nonparaxial region [Thors et al. 2003].

In this section, the theoretical results are presented for a rectangular grid array consisting of 4×32 elements with element spacings $\Delta z = 41$ mm and $R\Delta\varphi = 20$ mm ($\approx 0.38 \lambda$ at $f = 5.65$ GHz). This gives an array covering an arc of about $120°$ on the cylinder. The rectangular waveguide apertures have $h = 39$ mm and $w = 16$ mm, that is, azimuthal polarization. The radiation properties are studied in the frequency band 4.0–7.3 GHz where only the fundamental mode propagates; the cutoff frequency for the second-order mode is $f_{c,20} \approx 7.6$ GHz. Five waveguide modes are used in the analysis and the cylinder radius is taken to be $R = 0.3$ m. The thickness of the coating is 2.4 mm and its relative permittivity $\varepsilon_r = 2.0$.

In Figure 6.56, normalized E-plane embedded-element patterns are shown at two different frequencies both with and without a dielectric coating for the element in the middle of the array. Since the element-to-element distance is small, the embedded-element patterns are smooth, but when a dielectric coating is applied, the mutual coupling increases and the neighboring elements have a larger effect on the radiated pattern. This effect is more pronounced at higher frequencies where the coating becomes electrically thicker.

Figure 6.57 shows the corresponding H-plane embedded-element patterns. The effect of the dielectric layer is clearly visible in the oscillations. The magnitude of these oscillations increases when the frequency increases since the dielectric layer becomes electrically thicker and the energy is guided more efficiently along the cylinder surface.

Since the array is finite, different elements will have different environments, which will affect the embedded-element pattern. This is illustrated in Figure 6.58 where the E-plane element patterns are shown for the first element (no. 33) of the second row, both with and without a dielectric coating. As shown in Section 6.3.3.2, elements further away from the array center have more unsymmetrical patterns. The center of the 33rd element is located at $\varphi \approx -52.9°$ but due to mutual coupling effects the lobe direction is shifted further away from the array center, especially at lower frequencies.

Finally, embedded-element patterns have also been calculated and compared to mea-

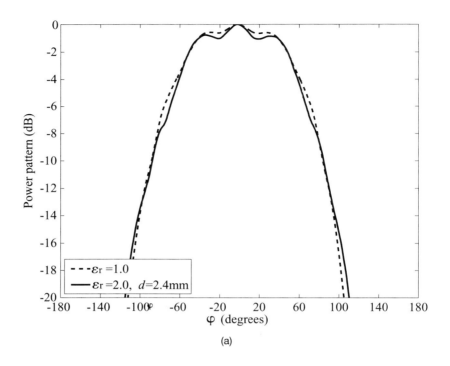

(a)

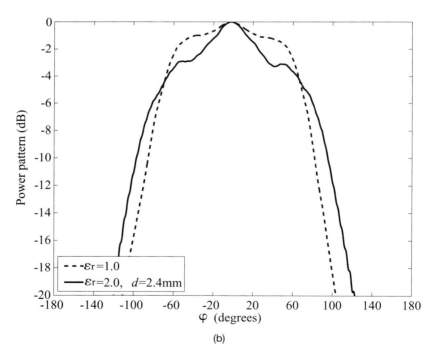

(b)

<u>Figure 6.56.</u> Normalized *E*-plane embedded-element pattern for the element in the middle of a 4 × 32 elements array (solid line). The noncoated result is also included for comparison (dashed line). (*a*) *f* = 5.65 GHz. (*b*) *f* = 6.8 GHz [Thors et al. 2003].

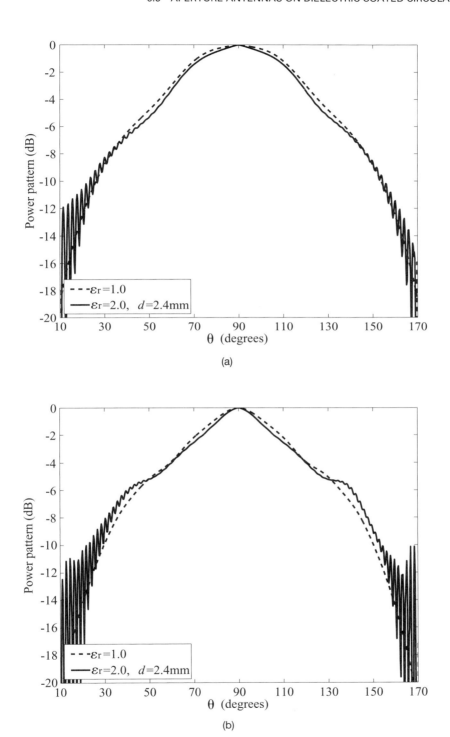

Figure 6.57. Normalized *H*-plane embedded-element pattern for the element in the middle of a 4 × 32 elements array (solid line). The noncoated result is also included for comparison (dashed line). (*a*) *f* = 5.65 GHz. (*b*) *f* = 6.8 GHz [Thors et al. 2003].

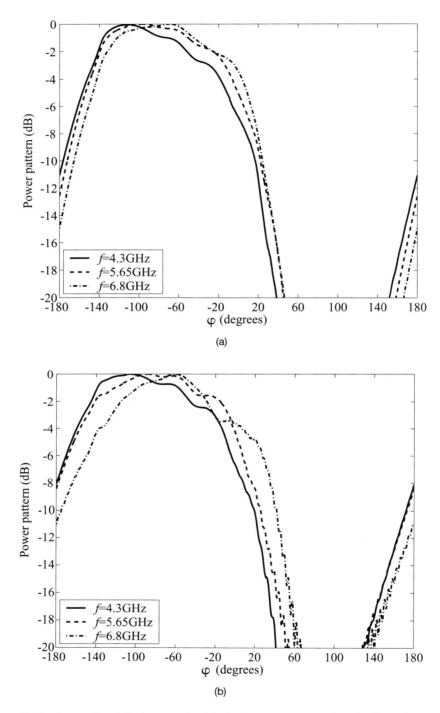

Figure 6.58. Normalized *E*-plane embedded-element patterns for the first element of the second row of the 4 × 32 elements array. The results are shown versus the frequency. (*a*) Without dielectric coating. (*b*) With dielectric coating (ε_r = 2.0, *d* = 2.4 mm) [Thors et al. 2003].

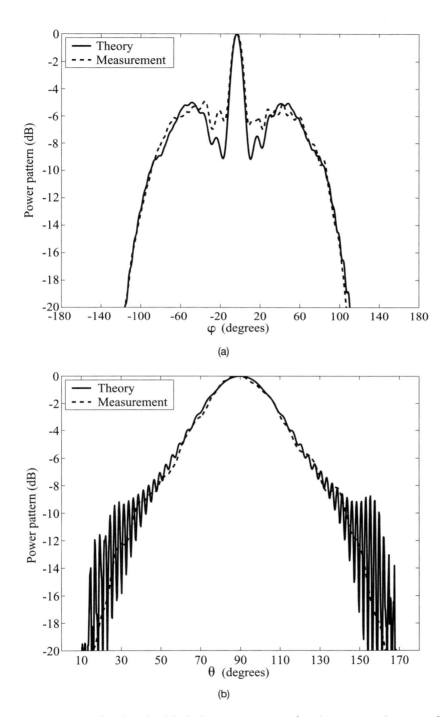

Figure 6.59. Normalized embedded-element patterns for the center element of the experimental antenna shown in Figure 6.46. The patterns are calculated at $f = 5.4$ GHz (solid line) and the results are compared with measured results (dashed line). The relative permittivity of the coating is $\varepsilon_r = 2.32$ and its thickness is 3.99 mm. (a) E plane. (b) H plane. [Thors et al. 2003].

surements. In Figure 6.59, the normalized E- and H-plane patterns are shown for the experimental antenna shown in Figure 6.46. The result is for the element located in the center of the array calculated at $f = 5.4$ GHz.

As shown, the agreement between theory and measurements is quite good both in the E and the H planes. The small differences that can be observed in the E plane around $\varphi = \pm 20°$ is due to the air gap in the measurements, as has been verified in [Sipus and Kildal 2003]. In [Thors et al. 2003], it was shown that the corresponding E-plane radiation pattern for an isolated element in this array is very broad. However, when the element is placed in an array environment, the radiation pattern is changed considerably due to the mutual coupling. The oscillations in the theoretical H-plane results are a consequence of interference between rays creeping around the cylinder. These oscillations are not observed in the measurement since the experimental antenna was truncated in the azimuthal direction, as shown in Figure 6.46.

As shown in Figure 6.59 (a), the E-plane embedded-element pattern is very oscillatory. The origin of the ripples can be attributed to the mutual coupling and the rather large azimuthal element spacing in the experimental antenna ($R\Delta\varphi \approx 0.67$ λ at $f = 5.4$ GHz). This was discussed in Section 6.2.4. The most practical way to avoid the problem is to reduce the element spacing. Similar oscillations are not observed in Figure 6.59 (b). This can be understood by realizing that the H-plane coupling is much weaker than the E-plane coupling.

6.6 MICROSTRIP-PATCH ANTENNAS ON COATED CIRCULAR CYLINDERS

6.6.1 Introduction

Microstrip antennas are widely applied in mobile communication systems, radar systems, and aerospace applications [James and Hall 1989] because of their low weight, low profile, and ease of manufacturing. Because of their ability to conform to nonplanar structures, they are often used in conformal arrays. The surface on which the elements are mounted influences the radiation properties and it is important to be able to predict this effect; however, the complex geometries can lead to difficult analytical problems. In this case, it is the curved coated surface that must be solved and, as indicated (see Chapter 4), this is in general a difficult problem.

Research on microstrip patches has been focused on the development of suitable analysis methods. Some references are [Nakatini et al. 1986; Tam et al. 1995; Ertürk et al. 2000, 2002, 2003; Raffaelli et al. 2005]; others can be found in Sections 4.3.3 and 4.4.3. In addition, there are several microstrip-patch antennas built to verify the calculated results, or to study them just by measurements. Some examples are [Sohtell 1987, Ashkenazy et al. 1988, Gniss 1998, Löffler et al. 1999].

The analysis of conformal microstrip antennas reveals no dramatic differences compared to the results discussed in Section 6.5 (apertures with coated surfaces). Therefore, the results shown in this section have been limited to a few typical cases.

The microstrip antennas considered here are probe-fed rectangular antennas. They are mounted on the outer surface of a cylindrically shaped dielectric substrate with an inner radius R, thickness t, and relative permittivity $\varepsilon_r > 1$. A perfectly conducting, cylindrically shaped ground plane backs the substrate and the cylinder is assumed to be infinite in the z direction. A typical geometry is shown in Figure 6.60.

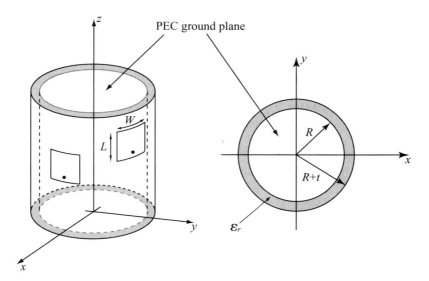

<u>Figure 6.60.</u> Geometry of two probe-fed rectangular microstrip antennas located on a circular cylinder.

6.6.2 Theory

Most of the calculated results shown here are obtained by solving an integral equation. The integral equation is set up with a spectral-domain Green's function that incorporates the effect of the size of the cylinder as well as the parameters of the dielectric substrate. The Green's function can be calculated in several ways. For small cylinders with a single layer, a closed-form expression can be used [Habashy et al. 1990, Tam et al. 1995]. For large single-layered cylinders, an asymptotic solution is possible [Ertürk and Rojas 2003]. For multilayer cylindrical structures, a general numerical algorithm such as G1DMULT [Sipus et al. 1998] is needed.

The electric field integral equation (EFIE) is found by enforcing the boundary conditions on the curved surface of the patch. We get

$$\hat{n}_x[\overline{E}^i(\overline{r}) + \overline{E}^s(\overline{r})] = 0 \qquad (6.11)$$

where $\overline{E}^i(\overline{r})$ is the incident field on the patch produced by the probe, and $\overline{E}^s(\overline{r})$ is the field due to the currents of the patch. The equation is solved for the unknown currents of the patch by applying the method of moments. The matrix equation is shown in Equation (4.26), where the moment method impedance matrix is derived. Once found, the scattering matrix is calculated by

$$\mathbf{S} = (\mathbf{Z} - \mathbf{Z}_0)(\mathbf{Z} + \mathbf{Z}_0)^{-1} \qquad (6.12)$$

where $\mathbf{Z}_0$ is the characteristic impedance of the feeding transmission line.

The radiation from a rectangular microstrip-patch antenna can be computed from the current distribution on the patch. An alternative approach is to view the two opposing patch ends as slot apertures and compute the radiation from an assumed dual-slot field distribution, see, for example, [Dahele et al. 1987]. We will use this method as well for comparison.

6.6.3 Mutual Coupling

Before discussing the mutual coupling among patch elements in an array, we will first look at the resonant frequencies and the input impedance of a single element. Both calculated and measured data are included.

6.6.3.1 Single-Element Characteristics.

By using a planar solution, an approximate value of the resonance frequency can be found. The cavity model of a probe-fed patch antenna can be used if the coating thickness is much smaller than one wavelength and the radius is electrically large. Thus, the resonant frequency of the mnth TM mode is given by [Dahele et al. 1987]

$$f_{mn} = \frac{c}{2\sqrt{\varepsilon_r}} \left[\left(\frac{m}{W} \right)^2 + \left(\frac{n}{L} \right)^2 \right]^{1/2} \tag{6.13}$$

where $W = 2R\phi_0 + t/\sqrt{\varepsilon_r}$ is the effective circumferential length of the patch and $L = z_m + t/\sqrt{\varepsilon_r}$ is the effective axial length. The actual dimensions of the patch are $2R\phi_0$ and z_m. An approximate formula for the input impedance can also be found using the cavity model [Luk et al. 1989].

Although the cavity model can be used to get an approximate estimation of the resonance frequency and input impedance, it does not incorporate the effect of the curvature and will not give an exact answer. To include the curvature, a more rigorous model must be used. In the next example, we will show how the resonant frequency and the input impedance depend on the curvature. The results are obtained using the G1DMULT code [Sipus et al. 1998]. The following parameters apply for the substrate coating: $t = 2.3$ mm, $\varepsilon_r = 2.85$, and $\tan \delta = 0.0073$. The dimensions of the patch are $L = 44.4$ mm and $W = 66.6$ mm, yielding resonance at 2.1 GHz. The position of the probe has been adjusted to achieve 50 Ω input impedance for the isolated patch (the distance of the probe from the patch center is 13.5 mm for vertical polarization and 9.5 mm for horizontal polarization). Furthermore, an air gap (approximately 0.4 mm) between the PEC cylinder and the coating is included; see also Section 6.6.3.2. Figure 6.61 shows the input impedance for this patch when mounted on a circular cylinder with different radii ($R_{PEC} = 2\ \lambda$, $R_{PEC} = 5\ \lambda$, and $R_{PEC} = 10\ \lambda$). Both axially and circumferentially polarized patches are considered.

Figure 6.61 shows that the curvature does not affect the resonant frequency very much. The axially polarized patch is more or less unaffected, whereas a slight change in resonant frequency is found for the circumferentially polarized patch. The amplitude of the input impedance is also more or less unaffected by the change in curvature. It should be pointed out, however, that the resonance resistance is sensitive to the air gap between the dielectric substrate and the metallic cylinder, in particular for cylinders with small radii. Tam and coworkers [1995] reported a large difference between the calculated (rigorous formulation but no air gap) and measured (with air gap) resonance resistance for a cylinder with $R \approx 0.26\ \lambda$ amounting to around 50%. For the resonant frequency, however, only a small disagreement (less than 1%) was observed.

Note that for doubly curved surfaces there is a larger effect of the radius, as discussed in Section 7.3.2.1.

6.6.3.2 Isolated and Array Mutual Coupling.

The first results in this section give an overall view of the isolated mutual coupling between two patches on a cylinder. To simplify the problem, the patches are represented by (electric) current modes and the results are given in terms of mutual impedances. A current mode is defined by a piecewise sinusoid along the direction of the current and by a constant value along the direc-

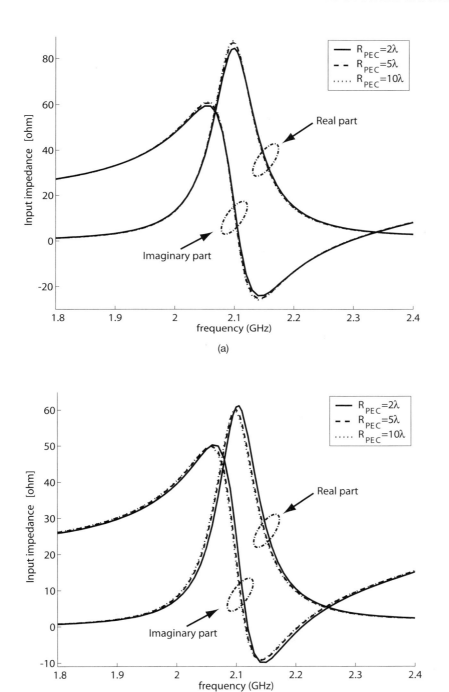

Figure 6.61. The input impedance of a single-patch antenna mounted on a circular cylinder versus the radius. (*a*) Axial polarization. (*b*) Circumferential polarization.

tion perpendicular to the current. Results for coupling between two patch elements (in terms of S-parameters) are also given and compared with measured results. The mutual coupling in an array environment is discussed briefly at the end.

In the current-mode representation, each current element has the dimensions of $0.05 \lambda_0$ (along the direction of the current) by $0.02 \lambda_0$. The results are obtained using an asymptotic Green's function [Ertürk and Rojas 2000, 2002] and the eigenfunction solution is included as well. The cylinder is defined by the following parameters: $R = 3 \lambda_0$, $t = 0.06 \lambda_0$, and $\varepsilon_r = 3.25$.

Figure 6.62 shows the mutual impedance between two identical z-directed current modes for different distances s and angles α (cf. Figures 6.1 and 6.60). As expected, the coupling is stronger for $\alpha = 90°$ and weaker for $\alpha = 0°$. The corresponding case with φ-directed current modes is shown in Figure 6.63. These results follow the same trends as the examples shown earlier in this chapter (cf. Figures 6.6 and 6.7).

The mutual impedance between φ- and z-directed current modes (cross-polarization) is shown in Figure 6.64. This coupling is weak at $\alpha = 0°$ and $90°$, since the cross-polar component of the surface field exhibits a $\sin 2\alpha$ type pattern, similar to the planar case as shown in [Marin and Pathak 1992].

The coupling in terms of the S-parameter, $S_{12} = S_{21}$, versus the edge-to-edge spacing between two identical patches, for two polarization cases, is shown next. The results are compared with measured results [Raffaelli et al. 2005]. The radius of the cylinder is 60 mm, and the dimensions for the patches and substrate are as for Figure 6.61. The parameters are repeated for convenience: $\varepsilon_r = 2.85$, $\tan \delta = 0.0073$, $t = 2.3$ mm, $L = 44.4$ mm, and $W = 66.6$ mm. Unfortunately, an air gap (approximately 0.4 mm) between the PEC cylinder and the coating could not be avoided. This air gap is included in the calculated results, and it was important in order to obtain accurate results (see also Section 6.5.2).

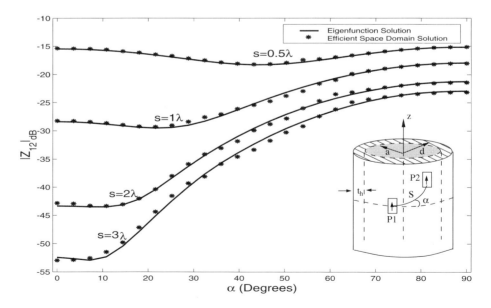

Figure 6.62. Mutual coupling, Z_{12}, between two identical z-directed current modes for a coated cylinder with $R = 3\lambda_0$, $t = 0.06\lambda_0$, and $\varepsilon_r = 3.25$ [Ertürk 2000]. $a = R$ and $d = R + t$.

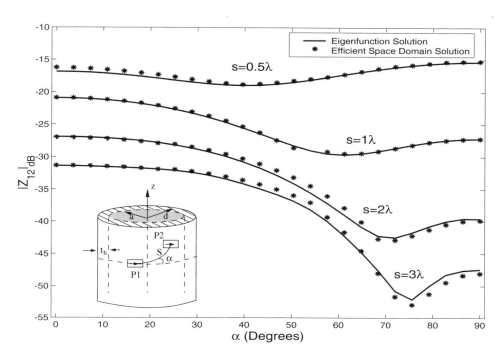

Figure 6.63. The mutual impedance Z_{12} between two φ-directed current modes for a coated cylinder with $R = 3\lambda_0$, $t = 0.06\lambda_0$, and $\varepsilon_r = 3.25$ [Ertürk 2000]. $a = R$ and $d = R + t$.

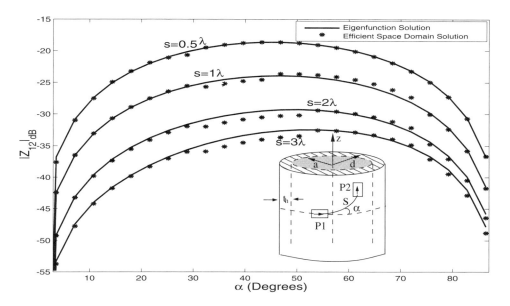

Figure 6.64. Mutual impedance Z_{12} between φ- and z-directed current modes (cross-polarization) for a coated cylinder with $R = 3\lambda_0$, $t = 0.06\lambda_0$, and $\varepsilon_r = 3.25$ [Ertürk 2000]. $a = R$ and $d = R + t$.

Figure 6.65 (*a*) shows the isolated mutual coupling between two patches mounted next to each other in the azimuthal direction. Thus, vertical polarization represents in this case *H*-plane coupling, which is weaker and decays faster than the *E*-plane coupling (as expected). At large spacings, some effect of the interference of creeping waves traveling in both directions around the cylinder is seen. The results for two patches mounted next to each other in the axial direction are shown in Figure 6.65 (*b*). As expected, the coupling decreases much faster for the horizontally polarized patches (that is, in the *H* plane) than for the vertically polarized ones. Note that the magnitude of S_{12} decays slowly for the vertical polarization. This is due to the small cylinder diameter ($0.84 \lambda_0$), making the cylinder act as an open waveguide; compare the Gobau line [Gobau 1964]. On the other hand, if the radius is increased, the results will move toward the planar case [Tam et al. 1995] for both polarizations.

The agreement between calculations and measurements in Figure 6.65 is good. The discrepancy between the measured and calculated results can be explained by the simple

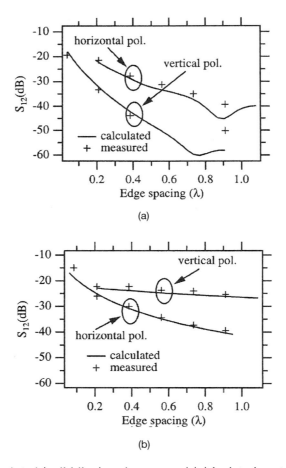

(a)

(b)

Figure 6.65. Calculated (solid line) and measured (+) isolated mutual couplings between two patches, for vertical and horizontal polarizations, versus edge spacing in wavelengths. (*a*) Separated in the horizontal direction. (*b*) Separated in the vertical direction [Raffaelli et al. 2005].

model of the probe and by a nonprecise value of the air gap between the metal and the dielectric layer in the experimental antenna. Note that six current modes were used in the calculations.

The mutual coupling in an array environment is seen in Figure 6.66. The array is a 5 × 10 patch array (five elements in the φ and 10 in the z direction). The distance between the elements is 78.76 mm both in the horizontal and vertical directions. They are mounted on the same cylinder as previously described. The mutual coupling between two patches separated in the z direction, both isolated and in the array environment, is shown. Both vertical and horizontal polarizations are considered.

6.6.4 Radiation Characteristics

6.6.4.1 Isolated-Element Patterns. The radiation patterns of a microstrip-patch antenna depend on the radius of the cylinder as well as the parameters of the coating. The general behavior can be expected to follow the previous results for covered aperture elements in Section 6.5. This also applies to the embedded-element patterns. However, for a specific design simulations are needed to predict the exact features of the antenna and make sure that no unexpected phenomena (resonances, etc.) appear. Single-element patterns versus the radius, as well as embedded-element patterns, can be found in [Sohtell 1987, Ashkenazy et al. 1988, Luk et al. 1989, Löffler et al. 1999, Railton et al. 2003, Raffaelli et al. 2005].

We present here the radiation pattern in the azimuthal plane of one single patch for horizontal and vertical polarizations. The patch antenna is the same as used earlier, that is, $R = 60$ mm, $\varepsilon_r = 2.85$, $\tan \delta = 0.0073$, and $t = 2.3$ mm. The dimensions of the patch are $L = 44.4$ mm and $W = 66.6$ mm, yielding a resonance at 2.1 GHz. The calculated results are compared with measurements and the agreement is very good; see Figure 6.67 (*a*) and (*b*). Note that an air gap between the PEC cylinder and the coating was unavoidable and is included in the calculations [Raffaelli et al. 2005]. In addition to the results of the full-wave analysis shown here, we will also use the cavity model for comparison [Dahele et al. 1987]. The results obtained with the cavity model are shown in Figure 6.67 (*c*) and (*d*). This shows that the cavity model gives accurate results for this example. Thus, it can be

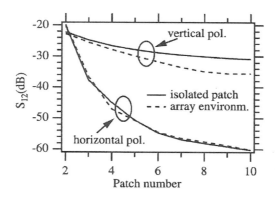

Figure 6.66. Mutual coupling in the z direction for horizontally and vertically polarized patches in a 5 × 10 elements array (dashed line) compared to the mutual coupling between two isolated patches (solid line) [Raffaelli et al. 2005].

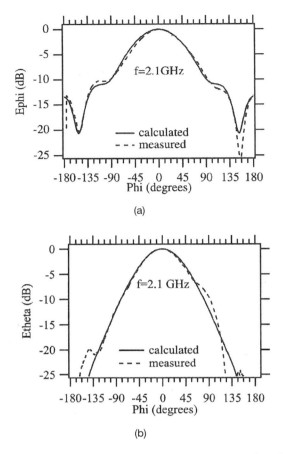

Figure 6.67. Calculated (solid line) and measured (dashed line) radiation patterns of one patch on a cylinder at 2.1 GHz. (*a*) Horizontal polarization, measured and calculated results [Raffaelli et al. 2005]. (*b*) Vertical polarization, measured and calculated results [Raffaelli et al. 2005]. *(continued)*

used for simple examinations of what to expect when the thickness of the coating is much smaller than one wavelength.

6.6.4.2 Embedded-Element Patterns.

Embedded-element patterns of patches are expected to behave as they do for aperture antennas coated with a dielectric. In the literature, only a few examples are found; see, for example, [Knott 2002, Railton et al. 2003]. Shown here are examples taken from [Knott 2002]; extracts are found in [Knott and von Winterfeld 1999, 2001]. The examples show embedded-element patterns of both a circular and an elliptic sector array. Both measured and calculated results are shown. Figure 6.68 shows the geometry of these microstrip-patch antenna arrays. The radius of the circular cylinder is 3.3 λ and the element-to-element distance is 0.5 λ, with the elements distributed within a 270° sector (that is, 32 elements). The dielectric substrate is RT Duroid 5880 with $\varepsilon_r = 2.2$ and the thickness is 1.57 mm. The elliptic cylinder is defined by $a/b = 2$ and the same type and number of elements is used (with an element-to-element distance of 0.5 λ within a 180° sector). The calculated results have been obtained by an approximation

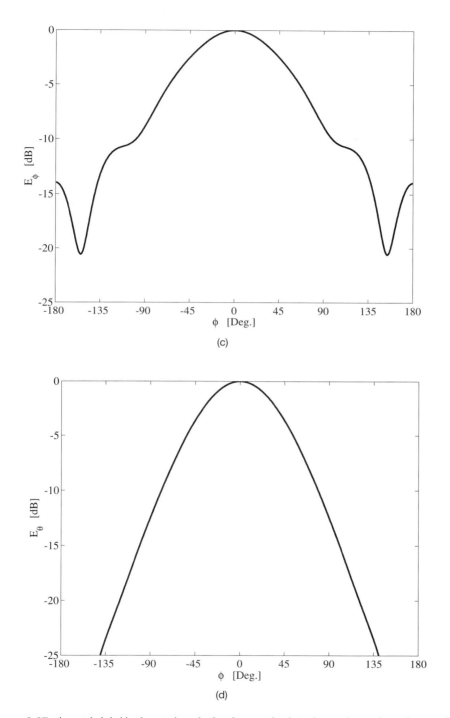

(c)

(d)

Figure 6.67. (*cont.*) (*c*) Horizontal polarization, calculated results using the cavity model. (*d*) Vertical polarization, calculated results using the cavity model.

(a)

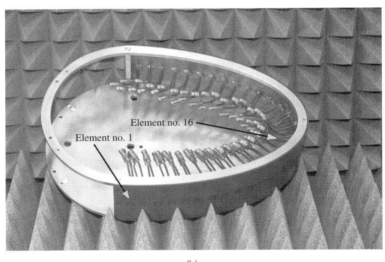

(b)

<u>Figure 6.68.</u> Experimental patch antenna arrays. (Courtesy of FGAN Research Establishment for Applied Science, Wachtberg-Werthhoven, Germany). (*a*) Circular cylinder sector array. (*b*) Elliptic cylinder sector array. See also color insert, Figure 9.

method using a combination of electric and magnetic short dipole radiators (Huygens sources) to allow for the calculation of mutual coupling effects [Knott 2002]. It should be pointed out that the model is an approximation and a perfect agreement is not expected.

The results shown here are for the embedded-element pattern of a centrally located element (element no. 16) as well as for an edge element. Both horizontal and vertical polarization are considered. The isolated element pattern (calculated) is also shown for comparison. Figure 6.69 shows the results for the circular cylinder, and Figure 6.70 shows the results for the elliptic cylinder [element no. 16 is shown in Figure 6.68 (*b*)].

The results for the horizontal polarization agree well with measured data in all cases. However, the vertical polarization results do not. The reason for this is that the **S** matrix was only calculated for the horizontal polarization case but was used in the calculation of

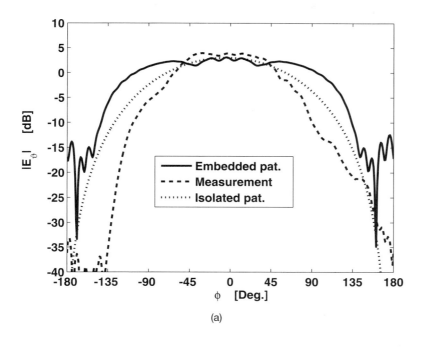

(a)

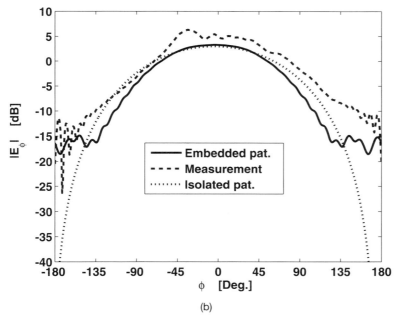

(b)

Figure 6.69. Embedded-element patterns of the circular cylinder (solid line). Measured results (dashed line) as well as isolated element patterns (dotted line) are shown [Knott 2002]. (*a*) Centrally located element (element no. 16), vertical polarization. (*b*) Centrally located element (element no. 16), horizontal polarization. (*continued*)

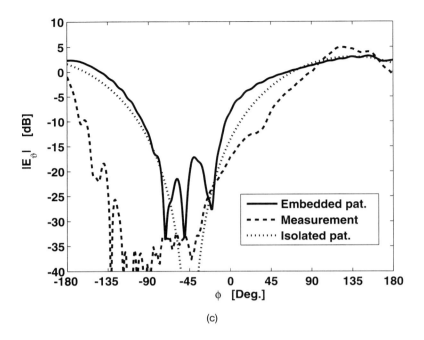

(c)

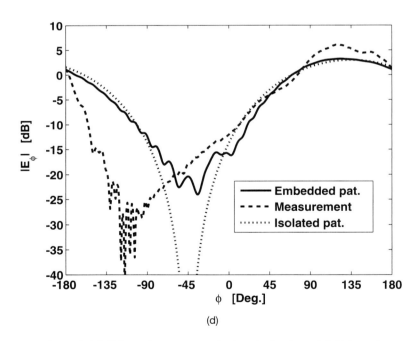

(d)

<u>Figure 6.69.</u> (*cont.*) (*c*) Edge element, vertical polarization. (*d*) Edge element, horizontal polarization.

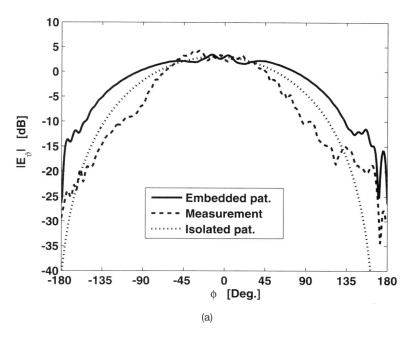

(a)

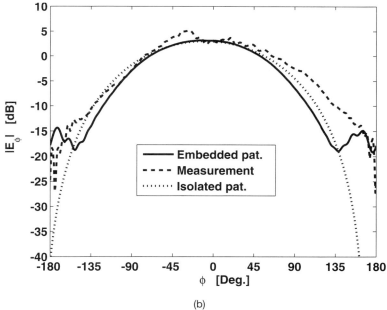

(b)

Figure 6.70. Embedded-element patterns of the elliptic cylinder (solid line). Measured results (dashed line) as well as isolated element patterns (dotted line) are shown [Knott 2002]. (*a*) Centrally located element (element no. 16), vertical polarization. (*b*) Centrally located element (element no. 16), horizontal polarization. (*continued*)

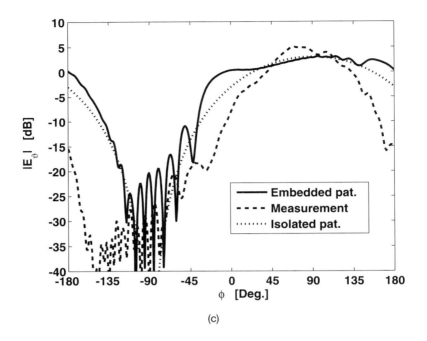

(c)

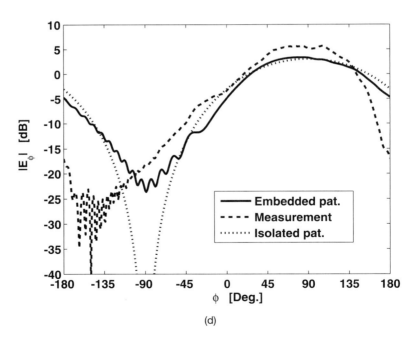

(d)

Figure 6.70. (*cont.*) (*c*) Edge element, vertical polarization. (*d*) Edge element, horizontal polarization.

the vertical polarization case as well. If the proper **S** matrix were calculated the results for the vertical polarization would have been more accurate.

Comparing these results with the aperture case considered in Section 6.3.3.2, the same type of behavior is found. The patterns exhibit the expected effects of the mutual coupling, such as the ripple in the main beam, filling of pattern nulls, and, for the edge elements, the main lobe shift toward the array center. However, the effect is not as visible as in Section 6.3.3.2 due to a weaker coupling among the patch elements.

6.7   THE CONE

6.7.1   Introduction

Like the cylinder, the cone is characterized by a singly curved surface. The conical surface can be of particular interest for applications in the noses of streamlined airborne vehicles, rockets, and missiles. The conical surface offers wide-angle coverage and good aerodynamic performance. A small conical angle, however, poses problems with the installation of electronics in the tip region and the radiation performance in the forward direction is poor. The conical array has been studied in several reports, as discussed in Section 8.6.4. Here, the emphasis is on antenna element characteristics when a single element is fed, both isolated and in the array environment.

Figure 6.71 shows a slotted cone geometry with an equivalent cylinder. The importance of the tip was considered in an early study [Held and Hasserjian 1958]. In that work, circumferential and radial slots were considered. The conclusion of the study was that the

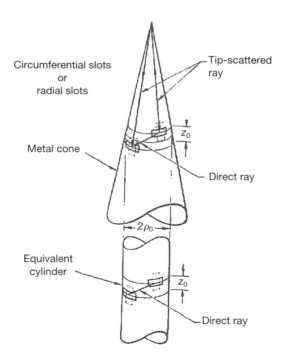

Figure 6.71. Slotted cone and equivalent cylinder [Golden et al. 1974].

scattering from the tip could be disregarded for slots further than 3 λ from the tip. From that position, calculations based on an equivalent cylinder could be utilized to simplify the calculations, especially for slender cones. The cylinder radius was chosen to $R' = (c_1 + c_2) \sin \theta_0/2$, where c_1 and c_2 are the radial distances from the apex to the apertures. However, later studies have shown that the tip can have a significant effect on the characteristics of cone arrays and antennas, even for large distances from the array.

In this section, we will show some results characterizing antennas on cones. Mostly, aperture antennas on PEC cones are considered and the results are taken from the literature. The cone is assumed to be infinitely long. A lot of work was done in the 1970s in this field. However, from recent literature we can see that there has been a renewed interest in conical arrays. Slot arrays have been considered in [Holter 2005, Zmak and Sipus 2005], and microstrip antennas are discussed in [Descardeci and Giarola 1992, Meeks and Wahid 1996, Shavit 1997, Smith et al. 2001].

6.7.2 Mutual Coupling

6.7.2.1 Aperture Antennas.
The Green's function for an infinite PEC cone can be derived and expressed exactly in terms of a doubly infinite series, with the aid of spherical Bessel functions and associated Legendre functions [Chan et al. 1977]. However, the numerical evaluation of such a series is quite tedious. An alternative method is to use approximation techniques, that is, high-frequency methods. With this approach the mutual admittance (Y_{12}) has two dominant components: the direct ray contribution (Y_{12}^d) going from aperture 1 to aperture 2, and the other (Y_{12}^t) from the rays diffracted at the tip of the cone (cf. Figure 6.71). The scattering from the tip can be found using diffraction coefficients [Pridemore-Brown 1972]. This high-frequency approach has been used in [Golden et al. 1974, Lee 1978, Hansen 1981] to study the mutual admittance of apertures in cones and the results shown here are taken from [Lee 1978].

In the first example, calculated and measured data (S_{12}) are shown versus frequency for a fixed position of two circumferential apertures. The size of the waveguide is 22.86 mm × 10.16 mm, the angular separation $\phi_0 = 60.8°$, the half cone angle $\theta_0 = 12.2°$, and the local radius at the aperture position is 45.53 cm. Shown in Figure 6.72 are three sets of data: (i) Experimental data taken from [Golden et al. 1974], (ii) Calculated results in which Y_{12}^d (the direct ray contribution) is obtained using the cone solution [Lee 1978], and (iii) calculated results in which Y_{12}^d is obtained using the exact modal solution of an equivalent cylinder [Golden et al. 1974]. Y_{12}^t is included [both in sets (ii) and (iii)] by using the asymptotic expressions for the diffracted field given in [Pridemore-Brown 1972].

The result shows a couple of interesting features. The peaks and valleys are caused by the interference between Y_{12}^d and Y_{12}^t; thus, both components are of comparable amplitude at this angular distance. At smaller distances, the direct signal would be expected to be dominant, whereas the tip diffraction can be the dominant one at larger separations (cf. Figures 6.73–6.75). The equivalent cylinder solution seems to give accurate results, even though the cone solution is slightly better compared with measured results. The use of an equivalent cylinder approximation has been discussed in [Lee and Mittra 1977]. It was shown that the equivalent cylinder seems to be good enough for a small-angled cone, $\theta_0 \approx 15°$. However, with $\theta_0 = 30°$ an error in Y_{12}^d of 2.5 dB was noted.

The next three figures show the mutual admittance for two circumferential slots (0.2 λ × 0.5 λ) versus the angular separation ϕ_0 and the radial separation ($c_1 - c_2$) for a cone with $\theta_0 = 15°$. Both Y_{12} and Y_{12}^d are shown; thus, the difference between the two, the tip dif-

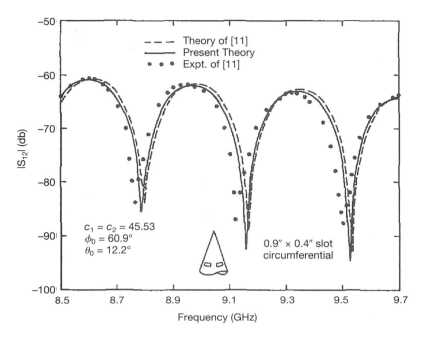

Figure 6.72. Coupling coefficient S_{12} between two circumferential slots on a cone as a function of frequency [Lee 1978]. Reference [11] is [Golden et al. 1974].

fraction, is clearly visible. In the first example (Figure 6.73), the circumferential slots are at the same altitude. In this case, the direct coupling is weak (*H*-plane coupling); hence, tip diffraction is affecting the results even at small angular separations.

Figure 6.74 shows the mutual admittance versus the angular separation when the apertures are separated by a small radial distance (2λ). In this situation, the tip contribution is almost negligible for $\phi_0 < 65°$. However, when the two apertures are widely separated in the radial direction (6λ), with one aperture near the tip, the tip contribution gets stronger. This is seen in Figure 6.75.

The last example shows the mutual admittance as a function of the orientation of the apertures. The two apertures are separated by 1λ along the radial direction and the apertures are rotated by angles ω_1 and ω_2, as seen in Figure 6.76. The largest coupling is found, as expected, when both apertures are circumferential, that is, $\omega_1 = \omega_2 = 90°$. If they are radially oriented ($\omega_1 = \omega_2 = 0°$), the amplitude is reduced. The minimum value is, of course, found when the two apertures are orthogonal.

The results shown here are mainly for circumferential apertures (radial electric fields) in which the tip is directly illuminated. Thus, the effect of the apex is clearly visible. For radial apertures, there is no radial field component and therefore no tip contribution to the mutual admittance. This has been verified by calculated results and comparisons with measurements [Golden et al. 1974].

6.7.2.2 *Microstrip-Patch Antennas.* Microstrip-patch elements on a conical surface is a configuration of great potential interest. However, no mutual coupling data for this case have been found in the published literature. This may be due to the complex-

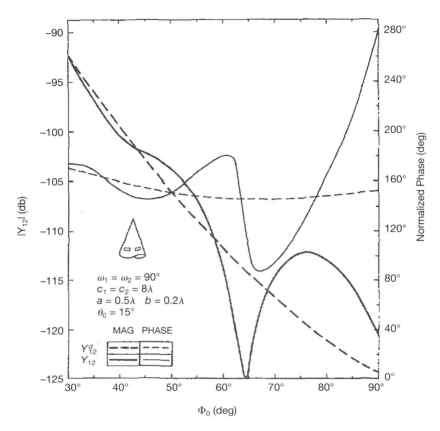

Figure 6.73. Mutual admittance between two circumferential slots on a cone at the same altitude versus the angular separation [Lee 1978].

ity of analyzing the near fields on a coated cone. However, input impedance results are discussed in, for example [Meeks and Wahid 1996] for the special case of a wraparound microstrip antenna. A rectangular patch is discussed in, among others, [Descardeci and Giarola 1992, Shavit 1997]. In these references, a cavity model is used, and tip diffraction is not included. Figure 6.77 shows the input impedance for a wraparound microstrip antenna on a conical surface. The input impedance is calculated as a function of the feed position and the cone half-angle θ_s. The following parameters apply: $\varepsilon_r = 2.32$, $t = 0.15$ cm, $r_a = 0.10$ m, and $r_b = 0.129$ m.

6.7.3   Radiation Characteristics

6.7.3.1 Aperture Antennas. Many authors have studied the radiation patterns from apertures on conical surfaces; some references are [Bailin and Silver 1956, Felsen 1957, Wait 1969, Pridemore-Brown 1972, Pridemore-Brown and Stewart 1972, Pridemore-Brown 1973, Balzano and Dowling 1974]. The solutions are obtained using an integral representation or a high-frequency approach. Shown here are some results

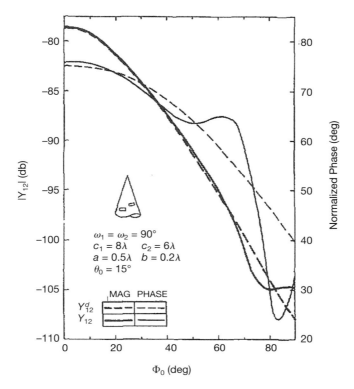

Figure 6.74. Mutual admittance between two circumferential slots on a cone versus the angular separation when separated by a small radial distance [Lee 1978].

taken from these references to illustrate the characteristics of single-element patterns and embedded-element patterns when the elements are located at the conical surface.

Single-Element Pattern. Figure 6.78 shows the geometry used in [Pridemore-Brown and Stewart 1972], in which a single circumferential slot is located at a distance a from the apex. The slot has a small height and goes all around the cone; thus, the radial electric field across the slot is assumed to be constant. Figure 6.79 shows the radiation pattern (axial cuts of $|E_\theta|^2$) for $k_0 a = 10$ and $k_0 a = 30$, respectively. The cone half-angle is $\theta_c = 10°$. The resulting lobe patterns have a regular, periodic structure. This can be explained by the interference between the power radiated directly from the slot and that scattered from the tip.

The final graph (Figure 6.80) shows the single-element pattern as a function of θ_c for $k_0 a = 20$. The major difference is that the peaks and minima of the radiation patterns are shifted forward, as expected.

Embedded-Element Patterns. The embedded-element patterns of a conical array, shown in Figure 6.81, are discussed next. The array considered here has 11 rings, each with 37 equispaced rectangular apertures. A triangular grid is obtained by adjusting every second ring, as seen in Figure 6.81. The rings are numbered, and the first ring (No. 1) is located closest to the tip at a distance of 79 λ from the apex. The spacing be-

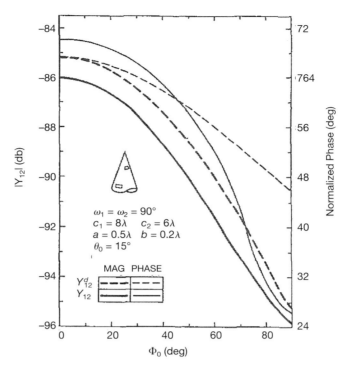

Figure 6.75. Mutual admittance between two circumferential slots on a cone versus the angular separation when separated by a large radial distance [Lee 1978].

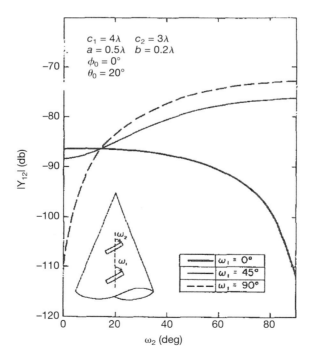

Figure 6.76. The mutual admittance between two arbitrarily oriented slots on a cone [Lee 1978].

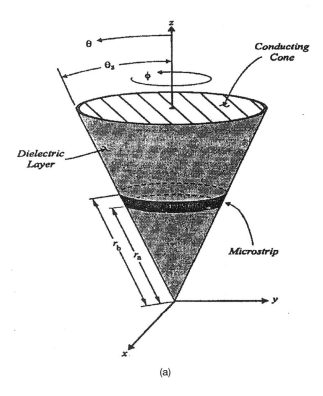

(a)

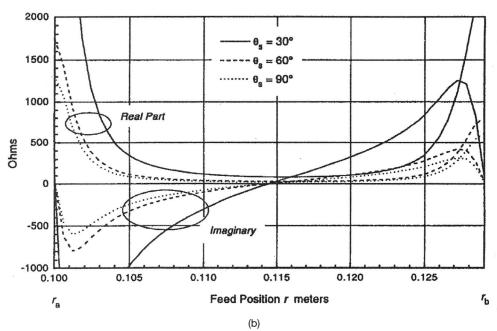

(b)

Figure 6.77. Wraparound microstrip antenna on a cone. (*a*) Geometry and coordinate system. (*b*) Input impedance with TM_{01} excitation [Meeks and Wahid 1996].

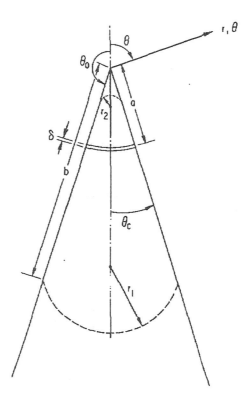

Figure 6.78. The cone geometry [Pridemore-Brown and Stewart 1972].

tween the rings is $h = 0.4\ \lambda$, and the circumferential spacing in ring 1 is $d = 1.17\ \lambda$ and in ring 11 it is $d = 1.229\ \lambda$. The apertures are filled with a dielectric material ($\varepsilon_r = 2.54$) and their dimensions are $2b = 0.5\ \lambda$ and $2w = 0.2\ \lambda$. The cone is sharply pointed with a cone half-angle of $2.5°$.

The first result (Figure 6.82) shows the axial element patterns (E plane) for elements in three different rings: ring no. 1, the center ring, and ring no. 11. The results are obtained by matching the fields at the single-mode apertures, where the external field is expanded in eigenfunctions, or phase modes, based on the periodicity of the structure. The results are shown for the forward hemisphere toward the tip of the cone. Included also are patterns for the equivalent planar array in an infinite ground plane. The spacing between the planar rows equals the ring spacing in the conical array, and the spacing between the apertures in each row equals the spacing of the radiators in the ring being considered. The results of the planar array are very similar to the conical array pattern, except in the region close to the apex. Near the tip of the cone, fast oscillations are seen, in particular in Figure 6.82 (*a*) for $-5° \le \theta \le 20°$. These oscillations are reduced when moving away from the apex. The pattern of the element in ring 11 is more affected by the presence of the array than by the tip.

Figure 6.83 shows the circumferential pattern ($\theta = 90°$), that is, the H plane for an element in the center ring. The pattern obtained using an infinite cylindrical array model is also shown. The results are found using the same approach as for the cone. The ring spac-

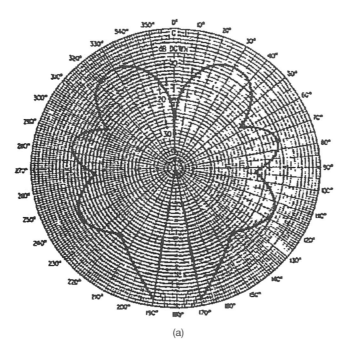

(a)

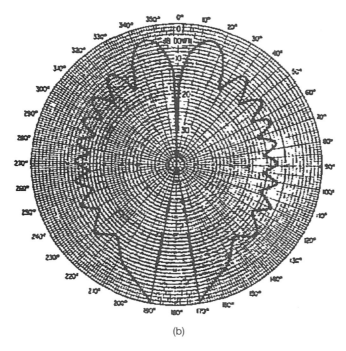

(b)

Figure 6.79. Calculated single element pattern due to a circumferential slot for the half-cone angle $\theta_c = 10°$ [Pridemore-Brown and Stewart 1972]. (a) $k_0 a = 10$. (b) $k_0 a =$

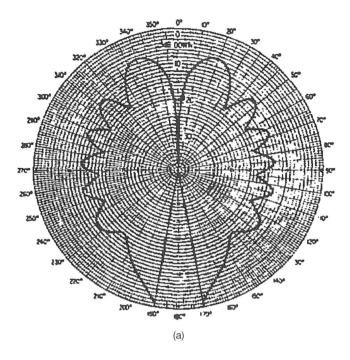

(a)

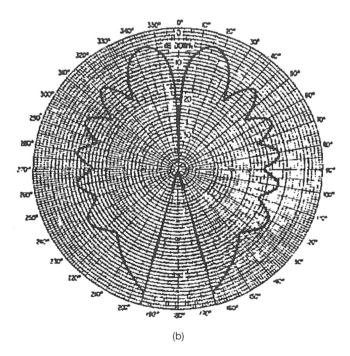

(b)

Figure 6.80. Calculated single element patterns due to a circumferential slot for $k_0 a =$ 20 [Pridemore-Brown and Stewart 1972]. (*a*) Cone angle $\theta_c = 10°$. (*b*) Half-cone angle $\theta_c = 15°$.

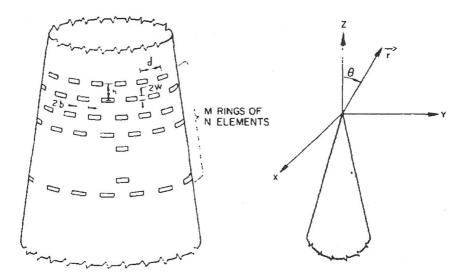

Figure 6.81. The conical array geometry [Balzano and Dowling 1974].

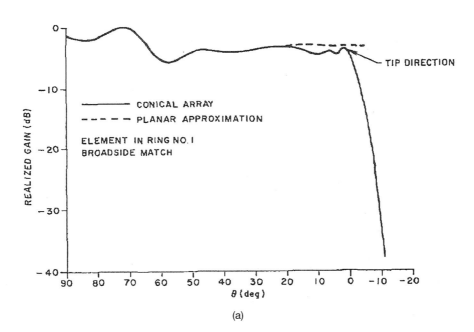

(a)

Figure 6.82. Embedded-element patterns (axial cut and *E* plane) for elements in different rings [Balzano and Dowling 1974]. (*a*) Element in ring No. 1. (*continued*)

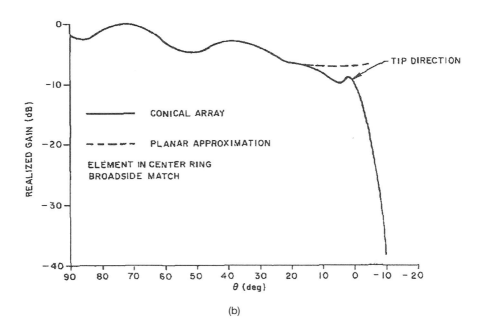

(b)

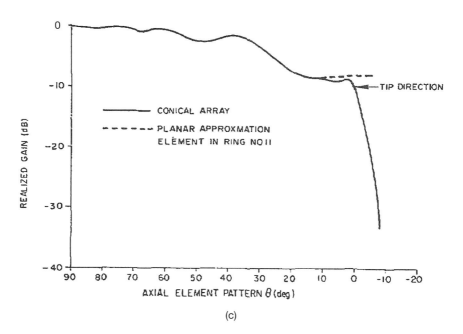

(c)

Figure 6.82. (*cont.*) (*b*) Element in the center ring. (*c*) Element in ring No. 11. Dashed curve is corresponding planar case.

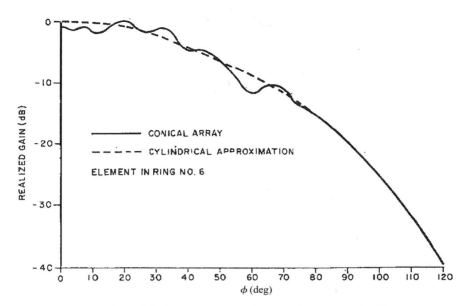

Figure 6.83. Circumferential element pattern (*H* plane) compared with the pattern of an equivalent cylinder array [Balzano and Dowling 1974].

ing of the cylindrical array is the same as in the conical array and the circumferential spacing is taken from ring 6. For $\phi > 60°$, the cylindrical array pattern agrees well with the conical array pattern, whereas at smaller angles the presence of the nonuniformily spaced elements in the other rings of the conical array is seen as oscillations, which is not obtained with the cylindrical solution.

6.7.3.2 Microstrip-Patch Antennas. There are very few references in the literature in which the radiation characteristics of microstrip-patch antennas on cones have been analyzed. The results found were obtained using a cavity model of the patch, as in [Descardeci and Giarola 1992, Shavit 1997]. Some measured results are also available [Smith et al. 2001]. Unfortunately, there are very few results in which the radiation characteristics are studied versus relevant parameters. One single example is found, however, in [Descardeci and Giarola 1992], in which the radiation pattern of a microstrip patch antenna is shown versus ε_r of the substrate. This example is shown here.

Figure 6.84 (*a*) shows the geometry of the microstrip-patch antenna located on a cone with the cone half-angle $\theta_0 = 33.88°$. The thickness of the coating is defined by $\Delta\theta = 0.57°$, the circumferential extension of the patch is given by $\phi_0 = 9.17°$, and the radial distances to the patch edges are $r_a = 0.10$ m and $r_b = 0.129$ m, respectively. Two values of the relative permittivity were chosen, namely, $\varepsilon_r = 2.32$ and 10.68. The results are found using a cavity model (note that tip diffraction is not included).

The *E*-plane radiation patterns are shown in Figure 6.84 (*b*). We can see that the patterns are wide and that more energy is guided along the dielectric layer when the dielectric constant is increased, causing higher radiation levels in the shadow region. The directivity is 8.0 dB when $\varepsilon_r = 2.32$ and becomes 7.12 dB with a larger permittivity ($\varepsilon_r = 10.68$).

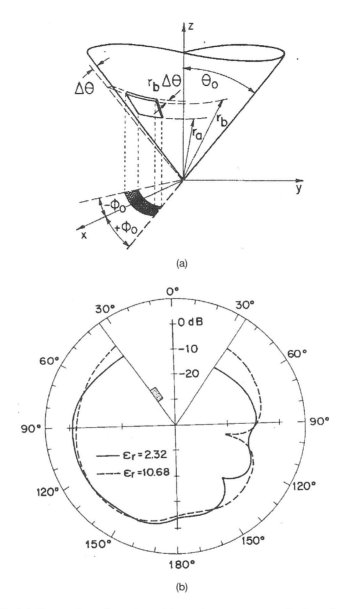

Figure 6.84. (*a*) Geometry of a microstrip antenna on a conical surface. (*b*) Single-element pattern cut for the antenna operating in the TM_{01} mode (*E* plane pattern) [Descardeci and Giarola 1992].

REFERENCES

Agrawal A. and Powell W. (1986), "A Printed Circuit Cylindrical Array Antenna," *IEEE Transactions on Antennas and Propagation,* Vol. AP-34, No. 11, pp. 1288–1293.

Amitay N., Galindo V. and Wu C. P. (1972), *Theory and Analysis of Phased Array Antennas,* New York, Wiley-Interscience.

Anderson J. B. and Hass F. (1970), "Edge Effects in Dielectric Covered Phased-Arrays," in *Proceedings of 1970 Phased-Array Antenna Symposium,* June 8–5.

Ashkenazy J., Shtrikman S., and Treves D. (1985), "Electric Surface Current Model for the Analysis of Microstrip Antennas on Cylindrical Bodies," *IEEE Transactions on Antennas and Propagation,* Vol. AP-33, pp. 295–300.

Ashkenazy J., Shtrikman S., and Treves D. (1988), "Conformal Microstrip Arrays on Cylinders," *Inst. Electrical Engineering,* Vol. 135, Part H, pp.132–134, April.

Bailin L. L. and Silver S. (1956), "Exterior Electromagnetic Boundary Value Problems for Spheres and Cones," *IRE Transactions on Antennas and Propagation,* Vol. AP-4, pp. 5–16, January.

Balzano Q. and Dowling T. B. (1974), "Mutual Coupling Analysis of Apertures on Cones," *IEEE Transactions on Antennas and Propagation,* Vol. AP-22, No. 1, pp. 92–97.

Bonnedal M. (1996), "Coupling in a Rectangular Waveguide Array—Theory and Algorithms," Tech. Rep. SE/REP/0221/A, SAAB Ericsson Space, 1996.

Borgiotti G. V. and Balzano Q. (1970), "Mutual Coupling Analysis of a Conformal Array of Elements on a Cylindrical Surface," *IEEE Transactions on Antennas and Propagation,* Vol. AP-18, No. 1, pp. 55–63.

Borgiotti G. V. and Balzano Q. (1972), "Analysis and Element Pattern Design of Periodic Arrays of Circular Apertures on Conducting Cylinders," *IEEE Transactions on Antennas and Propagation,* Vol. AP-20, No. 5, pp. 547–555.

Calander N. and Josefsson L. (2001), "A Look at Polarization Properties of Cylindrical Array Antennas," in *Proceedings of the 2nd European Workshop on Conformal Antennas,* Den Haag, The Netherlands, 24–25 April.

Carter P. S. (1943), "Antenna Arrays Around Cylinders," *Proceedings of IRE,* Vol. 31, pp. 671–693, December.

Chan K. K., Felsen L. B., Hessel A., and Shmoys J. (1977), "Creeping Waves on a Perfectly Conducting Cone," *IEEE Transactions on Antennas and Propagation,* Vol. AP-25, No. 5, pp. 661–670.

Dahele J. S., Mitchell R. J., Luk K. M., and Lee K. F. (1987), "Effect of Curvature on Characteristics of Rectangular Patch Antenna," *Electronics Letters,* Vol. 23, pp. 748–749, July.

Da Silva C. M., Lumini F., Da S. Lacava J. S., and Richards F. P. (1991), "Analysis of Cylindrical Arrays of Microstrip Rectangular Patches," *Electronics Letters,* Vol. 27, pp. 778–780, April.

Descardeci J. R. and Giarola A. J. (1992), "Microstrip Antenna on Conical Surface," *IEEE Transactions on Antennas and Propagation,* Vol. AP-40, No. 4, pp. 460–463.

Ertürk V. B. (2000), *Efficient Hybrid MoM/Green's Function Technique to Analyze Conformal Microstrip Antennas and Arrays.* PhD Thesis, Ohio State University.

Ertürk V. B. and Rojas R. G. (2000), "Efficient Computation of Suface Fields Excited on a Dielectric Coated Cylinder," *IEEE Transactions on Antennas and Propagation,* Vol. 48, No. 10, pp. 1507–1516.

Ertürk V. B. and Rojas R. G. (2002), "Paraxial Space-Domain Formulation for Surface Fields on a Large Dielectric Coated Circular Cylinder," *IEEE Transactions on Antennas and Propagation,* Vol. AP-50, pp. 1577–1587, November.

Ertürk V. B. and Rojas R. G. (2003), "Efficient Analysis of Input Impedance and Mutual Coupling of Microstrip Antennas Mounted on Large Coated Cylinders," *IEEE Transactions on Antennas and Propagation,* Vol. AP-51, pp. 739–749, April.

Ertürk V. B., Rojas R. G., and Lee K. W. (2004), "Analysis of Finite Arrays of Axially Directed Printed Dipoles on Electrically Large Circular Cylinders, *IEEE Transactions on Antennas and Propagation,* Vol. AP-52, No. 10, pp. 2586–2595.

Felsen L. B. (1957), "Alternative Field Representation in Regions Bounded by Spheres, Cones and Planes," *IRE Transactions on Antennas and Propagation,* Vol. AP-5, pp. 109–121, January.

Gladman B. R. (1970), "Mutual Coupling on Cylindrical Array Antennas," in *Proceedings of the Conformal Array Antenna Conference,* held at NELC, San Diego, 13–15 January.

Gniss H. (1998), "Digital Rx-only Conformal Array Demonstrator," in *Proceedings of International Radar Symposium, IRS 98,* Munich, Germany, pp. 1003–1012.

Gobau G. (1964), "Waveguides—General," *IEEE Spectrum,* Vol. 1, No. 10, pp. 81–84.

Golden K. E., Stewart G. E. and Pridemore-Brown D. C. (1974), "Approximation Techniques for the Mutual Admittance of Slot Antennas on Metallic Cones," *IEEE Transactions on Antennas and Propagation,* Vol. AP-22, No. 1, pp. 43–48.

Habashy T. M., Ali S. M. and Kong J. A. (1990), "Input Impedance and Radiation Pattern of Cylindrical-Rectangular and Wraparound Microstrip Antennas," *IEEE Transactions on Antennas and Propagation,* Vol. AP-38, pp. 722–731, May.

Hansen R. C. (ed.) (1981), *Conformal Antenna Array Design Handbook,* Dept. of the Navy, Air Systems Command, September 1, AD A110091.

Hansen R. C. (1998), *Phased Array Antennas,* Wiley.

Held G. and Hasserjian G. (1958), "An Experimental Analysis of the Surface and Tip Excitation of a Cone by a Slot," Dept. Elec. Eng., Univ. Washington, Tech. Rep. 25, March.

Herper J. C., Hessel A. and Tomasic B. (1985a), "Element Pattern of an Axial Dipole in a Cylindrical Phased Array, Part I: Theory," *IEEE Transactions on Antennas and Propagation,* Vol. AP-33, pp. 259–272, March.

Herper J. C., Hessel A. and Tomasic B. (1985b), "Element Pattern of an Axial Dipole in a Cylindrical Phased Array, Part I: Element Design and Experiments," *IEEE Transactions on Antennas and Propagation,* Vol. AP-33, pp. 273–278, March.

Herscovici N., Sipus Z. and Kildal P.-S. (1997), "The Cylindrical Omnidirectional Patch Antenna," in *Proceedings of IEEE AP-S Symposium,* Montreal, pp. 924–927.

Hessel A. and Sureau J. C. (1970), "Element Patterns of Arrays on Cylindrical Surface," in *Proceedings of the Conformal Array Antenna Conference,* held at NELC, San Diego, 13–15 January.

Hessel A. (1972), "Mutual Coupling Effects in Circular Arrays on Cylindrical Surfaces—Aperture Design Implications and Analysis," in *Phased Array Antennas,* A. A. Oliner and G. H. Knittel (eds.), pp. 273–291, Artech House.

Holter H. and Steyskal H. (2002a), "On the Size Requirement for Finite Phased-Array Models," *IEEE Transactions on Antennas and Propagation,* Vol. AP-50, No. 6, pp. 836–840, June.

Holter H. and Steyskal H. (2002b), "Some Experiences from FDTD Analysis of Infinite and Finite Multi-octave Phased Arrays," *IEEE Transactions on Antennas and Propagation,* Vol. AP-50, No. 12, pp. 1725–1731.

Holter H. (2005), "Examination of a New Numerical Code for Slot Element Arrays on Conical Metal Surfaces, Calculations and Measurements," in *Proceedings of the 4th European Workshop on Conformal Antennas,* pp. 43–46, Stockholm, Sweden, May.

Huang C.-Y. and Chang Y.-T. (1997), "Curvature Effects on the Mutual Coupling of Cylindrical-Rectangular Microstrip Antennas," *Electronics Letters,* Vol. 33, pp. 1108–1109, June.

Hussar P. E., Oliker V., Riggins H. L., Smith-Rowlan E. M., Klocko W. R., and Prussner, L. (2000), "An Implementation of the UTD on Facetized CAD Platform Models," *IEEE Transactions on Antennas and Propagation Magazine,* Vol. 42, No. 2, pp. 100–106.

James J. R. and Hall P. S. (1989), *Handbook of Microstrip Antennas.* London: Peter Peregrinus Ltd.

Jamnejad V., Huang J., Levitt B., Pham T., and Cesarone R. (2002), "Array Antennas for JPL/NASA Deep Space Network," in *IEEE Aerospace Conference Proceedings,* Vol. 2, 9–16 March, pp. 2.911–2.921.

Jayakumar I., Garg R., Sarap B. K., and Lal B. (1986), "A Conformal Cylindrical Microstrip Array for Producing Omnidirectional Radiation Pattern," *IEEE Transactions on Antennas and Propagation,* Vol. AP-34, pp. 1258–1261, October.

Jin J.-M., Berrie J. A., Kipp R., and Lee S.-W. (1997), "Calculation of Radiation Patterns of Microstrip Antennas on Cylindrical Bodies of Arbitrary Cross Section," *IEEE Transactions on Antennas and Propagation,* Vol. AP-45, pp. 126–132, January.

Kanno M., Hashimura T., Katada T., Sato M., Fukutani K., and Suzuki A. (1996), "Digital Beam Forming for Conformal Active Array Antenna," in *Proceedings of IEEE International Symposium on Phased Array Systems and Technology*, pp. 37–40, October.

Keller J. B. (1956), "Diffraction by a Convex Cylinder," *IRE Transactions Antennas and Propagation,* Vol. AP-24, pp. 312–321.

Knott P. and von Winterfeld C. (1999), "A Polarimetric Antenna Model for Small Conformal Arrays Including Mutual Coupling Effects" in *Proceedings of IEEE AP-S International Symposium,* Orlando FL, USA, pp. 2314–2317, July.

Knott P. and von Winterfeld C. (2001), "Antenna Model for Conformal Array Performance Prediction," in *Proceedings of Smart Electronics and MEMS, SPIE,* Newport Beach, CA, USA, March, pp. 207–213.

Knott P. (2002), *Antennenmodellierung mit Diagrammsynthese zur Systemanalyse von Konformen Gruppenantennen,* Ph.D. Thesis, Rheinisch-Westfälischen Technischen Hochschule, Bonn, Germany. In German.

Krowne C. M. (1983), "Cylindrical-Rectangular Microstrip Antenna," *IEEE Transactions on Antennas and Propagation,* Vol. AP-31, No. 1, pp. 194–199.

Lanne M. and Josefsson L. (1999a), "Dielektrikumtäckt konform antenn, mätrapport för kopplingsmätningar," Tech. Rep. UA/U-1999:023, Ericsson Microwave Systems AB, Mölndal, Sweden, October. In Swedish.

Lanne M. and Josefsson L. (1999b), "Dielektrikumtäckt konform antenn, mätrapport för strålningsdiagramsmätningar," Tech. Rep. UA/U-1999:028, Ericsson Microwave Systems AB, Mölndal, Sweden, November. In Swedish.

Lee S.- W. and Mittra R. (1977), "Mutual Admittance Between Slots on a Cylinder or Cone," Univ. Illinois at Urbana-Champaign, EM Lab. Rep. No. 77–24.

Lee S.-W. (1978), "Mutual Admittance of Slots on a Cone: Solution by Ray Technique," *IEEE Transactions on Antennas and Propagation,* Vol. AP-26, No. 6, pp. 768–773.

Lindgren S. and Josefsson L. (1998), "Konforma antenner, mätrapport," Tech. Rep. SR/R-1998:033, Ericsson Microwave Systems AB, Mölndal, Sweden, November. In Swedish.

Luk K. M., Lee K. F., and Dahele J. S. (1989), "Analysis of the Cylindrical Rectangular Patch Antenna," *IEEE Transactions on Antennas and Propagation,* Vol. AP-37, No. 2, pp. 143–147.

Löffler D., Wiesbeck W., and Johannisson B. (1999), "Conformal Aperture Coupled Microstrip Phased Array on a Cylindrical Surface," *IEEE Antennas and Propagation Symposium Digest,* Vol. 2, pp. 998–1001, June.

Mailloux R. J. (1982), "Phased Array Theory and Technology," *Proceedings of IEEE,* vol. 70, No. 3, pp. 1357–1366.

Mailloux R. J. (1994), *Phased Array Antenna Handbook,* Artech House.

Marcus C., Pettersson L., and Persson P. (2003), "Investigation of the Mutual Coupling and Radiation Pattern Due to Sources on Faceted Cylinders," in *Proceedings of Antenn 03,* pp. 213–218, Kalmar, Sweden, June.

Marcus C. (2004), Faceted Conformal Antenna Arrays Analysed with the Uniform Theory of Diffraction, *Licentiate Thesis No. 1082,* Dept. of Physics and Measurement Technology, Linköping, Sweden.

Marin M. A. and Pathak P. H. (1992), "An Asymptotic Closed-Form Representation for the Grounded Double-Layer Surface Green's Function," *IEEE Transactions on Antennas and Propagation,* Vol. AP-40, pp. 1357–1366, November.

Martín Polegre A. J., Roederer A. G., Crone G. A. E., and de Maagt P. J. I. (2001), "Applications of Conformal Array Antennas in Space Missions," in *Proceedings of the 2nd European Workshop on Conformal Antennas,* Den Haag, The Netherlands.

Meeks D. N. and Wahid P. F. (1996), "Input Impedance of a Wrap-around Microstrip Antenna on a Conical Surface," in *Proceedings of IEEE AP-S Symposium,* pp. 676–679.

McNamara D. A., Pistorius C. W. I., and Malherbe J. A. G. (1990), *Introduction to The Uniform Geometrical Theory of Diffraction,* Artech House.

Munger A. D., Provencher J. H., and Gladman B. R. (1971), "Mutual Coupling on a Cylindrical Array of Waveguide Elements," *IEEE Transactions on Antennas and Propagation,* Vol. AP-19, No. 1, pp. 131–134.

Munson R. E. (1974), "Conformal Microstrip Antennas and Microstrip Phased Arrays," *IEEE Transactions on Antennas and Propagation,* Vol. AP-22, pp. 74–78, January.

Nakatini A., Alexopoulus N. G., Uzunoglu N. K., and Uslenghi P. L. E. (1986), "Accurate Green's Function Computation for Printed Circuit Antennas on Cylindrical Antennas," *Electromagnetics,* Vol. 6, pp. 243–254, July–Sept.

Pathak P. H. and Wang N. (1981), "Ray Analysis of Mutual Coupling Between Antennas on a Convex Surface," *IEEE Transactions on Antennas and Propagation,* Vol. AP-29, No. 6, pp. 911–922.

Pathak P. H., Wang N., Burnside W. D., and Kouyoumjian R. G. (1981), "A Uniform GTD Solution for the Radiation from Sources on a Convex Surface," *IEEE Transactions on Antennas and Propagation,* Vol. AP-29, pp. 609–622, July.

Persson P. (1999), "The Mutual Coupling between Apertures on Convex Cylinders," *Proceedings of RVK 99,* Karlskrona, Sweden, June.

Persson P. and Josefsson L. (1999), "Calculating the Mutual Coupling between Antennas on a Convex Surface," Tech. Rep. TRITA-TET 99-4, Royal Institute of Technology, February.

Persson P. and Josefsson L. (2001), "Calculating The Mutual Coupling between Apertures on a Convex Circular Cylinder Using a Hybrid UTD–MoM Method," *IEEE Transactions on Antennas and Propagation,* Vol. AP-49, No. 4, pp. 672–677.

Persson P. and Rojas R. G. (2003), "High-frequency Approximation for Mutual Coupling Calculations between Apertures on a Perfect Electric Conductor Circular Cylinder Covered with a Dielectric Layer: Nonparaxial Region," *Radio Science,* Vol. 38, No. 4, 1079, doi:10.1029/2002RS002745.

Persson P. and Josefsson L. (2005), "Investigation of the Radiation Characteristics of Antennas on Singly Curved Surfaces," in *Proceedings of the 4th European Workshop on Conformal Antennas,* Stockholm, Sweden, pp. 47–50, May.

Provencher J. H. (1972), "A Survey of Circular Symmetric Arrays," in *Phased Array Antennas,* A. A. Oliner and G. H. Knittel (eds.), pp. 292–300, Artech House.

Pridemore-Brown D. C. (1972), "Diffraction Coefficients for a Slot-Excited Conical Antenna," *IEEE Transactions on Antennas and Propagation,* Vol. AP-20, No. 1, pp. 40–49.

Pridemore-Brown D. C. and Stewart G. E. (1972), "Radiation from Slot Antennas on Cones," *IEEE Transactions on Antennas and Propagation,* Vol. AP-20, No. 1, pp. 36–39.

Pridemore-Brown D. C. (1973), "The Transition Field on the Surface of a Slot-excited Conical Antenna," *IEEE Transactions on Antennas and Propagation,* Vol. AP-21, pp. 815–816, November.

Raffaelli S., Sipus Z., and Kildal P.-S. (2005), "Analysis and Measurements of Conformal Patch Array Antennas on Multilayer Circular Cylinder," *IEEE Transactions on Antennas and Propagation,* Vol. 53, No. 3, pp. 1105–1113.

Rai E., Nishimoto S., Katada T., and Watanabe H. (1996), "Historical Overview of Phased Array Antenna for Defense Application in Japan," in *Proceedings of IEEE International Symposium Phased Array Systems and Technology,* pp. 217–221, October.

Railton C. J., Paul D. L., and Craddock I. J. (2003), "Analysis of a 17 Element Conformal Array of Stacked Circular Patch Elements Using an Enhanced FDTD Approach," *IEE Proceedings on Microwave Antennas and Propagation,* Vol. 150, No. 3.

Saez de Adana F., Gutierrez O., Perez J., and Catedra M. F. (1998), "Computer Tool for the Analysis of Antennas on Board Complex Bodies Modelled by Flat or/and Curved Facets," in *Proceedings of IEEE International Symposium on Antennas and Propagation,* Vol. 2, 21–26 June, pp. 1082–1084.

Shavit (1997), "Circular Polarization Microstrip Antenna on a Conical Surface," *IEEE Transactions on Antennas and Propagation,* Vol. AP-45, pp. 1086–1092, July.

Sipus Z., Kildal P-S., Leijon R., and Johansson M. (1998), "An Algorithm for Calculating Green's Function of Planar, Cylindrical and Spherical Multilayer Substrates," *ACES Journal,* Vol. 13, no. 3, pp. 243–254.

Sipus Z., Rupcic S., Lanne M., and Josefsson L. (2001), "Moment Method Analysis of Circular-Cylindrical Array of Waveguide Elements Covered with a Radome," in *Proceedings of IEEE AP-S International Symposium,* Vol. 2, Boston, pp. 350–353, July.

Sipus Z. and Kildal P.-S. (2003), Analyzing Conformal Antennas on Cylindrical and Spherical Multilayer Structures using G1DMULT: An Overview," in *Proceedings of Antenn 03,* Nordic Antenna Symposium, Kalmar, Sweden, 13–15 May, pp. 201–205.

Sohtell E. V. (1987), *Microwave Antennas on Cylindrical Structures,* Ph.D. Thesis, Chalmers University of Technology, Gothenburg, Sweden.

Sohtell E. V. (1989), "Microstrip Antennas on a Cylindrical Surface," in *Handbook of Microstrip Antennas,* J. R. James and P. S. Hall (eds.), London: Peter Peregrinus.

Smith P. A., Ball G. J., Watts J. P., and Foster P. R. (2001), "Microstrip Patch Antenna on Planar and Conical Surfaces," in *Proceedings of the 2nd European Workshop on Conformal Antennas,* Den Haag, The Netherlands.

Steyskal H. (1977), "Analysis of Circular Waveguide Arrays on Cylinders," *IEEE Transactions on Antennas and Propagation,* Vol. AP-25, No. 5, pp. 610–616.

Sureau J. C. and Hessel A. (1969), "Element Pattern for Circular Arrays of Axial Slits on Large Conducting Cylinders," *IEEE Transactions on Antennas and Propagation,* Vol. AP-17, No. 6, pp. 799–803.

Sureau J. C. and Hessel A. (1971), "Element Pattern for Circular Arrays of Waveguide-Fed Axial Slits on Large Conducting Cylinders," *IEEE Transactions on Antennas and Propagation,* Vol. AP-19, No. 1, pp. 64–76.

Sureau J. C. and Hessel A. (1972), "Realized Gain Function for Cylindrical Array of Open-Ended Waveguides," in *Phased Array Antennas,* A. A. Oliner and G. H. Knittel (eds.), pp. 315–322, Artech House.

Tam W. Y., Lai A. K. Y., and Luk K. M. (1995), "Mutual Coupling Between Cylindrical-Rectangular Microstrip Antennas," *IEEE Transactions on Antennas and Propagation,* Vol. AP-43, pp. 897–899, August.

Thors B. and Rojas R. G. (2003), "Uniform Asymptotic Solution for the Radiation from a Magnetic Source on a Large Dielectric Coated Circular Cylinder: Non-paraxial Region," *Radio Science,* Vol. 38, No. 5.

Thors B., Josefsson L., and Norgren M. (2003), "Radiation Characteristics and Matching Properties for a Cylindrical Conformal Finite Array Antenna Coated with a Dielectric Layer," Technical Report: TRITA-TET 03-05, Royal Institute of Technology, Div. of Electromagnetic Theory, Sweden.

Visser H. J. and Gerini G. (1999), "Beam Switching and Steering in Cylindrical Array Antennas Using a Locally Planar Array Approach," in *Proceedings of First European Workshop on Conformal Antennas,* Karlsruhe, Germany, pp. 46–49, October.

Wait J. R. (1969), "Electromagnetic Radiation from Conical Structures," in *Antenna Theory,* Part 1, R. E. Collin and F. J. Zucker (eds.), New York: McGraw-Hill.

Wong K.-L. (1999), *Design of Nonplanar Microstrip Antennas and Transmission Lines,* New York: Wiley.

Wu D. I. (1995), "Omnidirectional Circularly-Polarized Microstrip Array for Telemetry Applications," *IEEE Antennas and Propagation Symposium Digest,* Vol. 2, pp. 998–1001, June.

Zmak K. and Sipus Z. (2005), "Analysis of Waveguide Arrays Mounted on Conical Structures," in *Proceedings of the 4th European Workshop on Conformal Antennas,* Stockholm, Sweden, pp. 59–62, May.

7

ANTENNAS ON DOUBLY CURVED SURFACES

With the extension to doubly curved surfaces, a number of new features are introduced. Among others, the polarization becomes an important parameter in the analysis. However, the analysis of doubly curved surfaces is still in its cradle and more results will come in the future as the research is progressing. The aim of this chapter is to go through a number of cases in order to present the state-of-the-art knowledge about mutual coupling between elements on doubly curved surfaces and also what we can expect from a radiation point of view. Both waveguide-fed apertures and microstrip patches are considered. For both types, measured data will be presented to verify the calculated results. As in Chapter 6, we will only consider single-element characteristics; array performance is discussed in Chapter 8.

7.1 INTRODUCTION

This chapter continues the investigation of antennas on curved surfaces (see Chapter 6) by considering doubly curved surfaces. Thus, some of the background material needed, such as the nomenclature used, can be found in Chapter 6 and will not be repeated here. The aim of this chapter is to provide an overview of the present knowledge in terms of mutual coupling and radiation for antenna elements mounted on doubly curved surfaces. This is done by considering the key features of a number of examples. In many of the examples, measured results verify the calculated results. However, many questions remain to be answered by future research.

Conformal Array Antenna Theory and Design. By Lars Josefsson and Patrik Persson
© 2006 Institute of Electrical and Electronics Engineers, Inc.

The extension to doubly curved surfaces is challenging and interesting. One goal is to provide the tools for designing antennas with full hemispherical coverage. A doubly curved surface can more accurately model, for example, realistic airborne or space-borne antennas than a singly curved surface can. However, in many applications a singly curved approach gives results accurate enough to be useful in the first stage of the design of a doubly curved antenna. Nevertheless, new features are introduced with doubly curved conformal antennas or, at least, these features become important in a new way and must be accounted for. One such feature is the co-polar and cross-polar aspects of the radiation pattern, since all elements on a doubly curved surface point in different directions. We will illustrate this with some examples. However, the array properties and the need to control the polarization in each element are left for Chapter 8.

Unfortunately, the literature references in this area are not many and much remains to be investigated. The analysis tools needed for doubly curved PEC surfaces are well developed, but there is no general tool for electrically large, coated surfaces. Furthermore, even though some doubly curved surfaces can be analyzed it is difficult to find information about the performance of elements mounted on these structures.

In several references, the radiation patterns for arrays on doubly curved surfaces are discussed and analyzed using simplified models. For example, the radiation pattern is often calculated by using the element pattern from a planar geometry and using a number of coordinate transformations and a summation to obtain the full array pattern. Some references are [Harris and Shanks 1962, Hoffman 1963, Sengupta et al. 1968, Hsiao and Rao 1976, Debionne 1986, Saito and Heichele 1987, Douchin and Lemorton 1997, Guy 1999, Girvan et al. 2001]. This approach will sometimes give satisfactory results for radiation analysis, but it is not accurate enough to include mutual coupling effects. However, a full analysis of aperture elements (radiation patterns and/or mutual coupling) is found in, for example, [Wills 1987; Jha et al. 1987, 1991, 1993, 1995; Macon et al. 2001, 2002; Josefsson and Persson 2001; Persson et al. 2001, 2003; Persson and Josefsson 2003a,b]. Microstrip patches are likewise discussed in, for example, [Ke and Kishk 1991; Lima et al. 1992; Chen and Wong 1993, 1994; Chen et al. 1995; Luk and Tam 1991; Tam and Luk 1991a,b, 1995; Burum and Sipus 2002a,b; Sipus et al. 2003a–c; Burum et al. 2004; Sipus and Burum 2005]. In addition, experimental antennas on doubly curved surfaces have been investigated. Some examples are found in [Haruyama et al. 1989, Rai et al. 1996, Lanne and Josefsson 2000, 2002]. However, few of these references treat in detail the antenna characteristics (dependence on design parameters, geometry, polarization performance, etc.). Furthermore, the results are seldom verified with reference solutions.

Most of the analysis presented in this chapter is based on PEC general paraboloids of revolution (GPOR). There are three reasons for choosing this type of surface. First, by changing the shape of the surface, that is, changing the shaping parameter a (see Table 5.3), doubly curved surfaces differing in character can be studied, as discussed in Chapter 5. If a is chosen to be large, we get the surface near the vertex region that it is similar to a sphere, and if a is chosen to be small a more pointed surface is obtained. Thus, many types of interesting surfaces can be studied with this basic geometry. Second, the ray tracing procedure becomes easier since standard mathematical functions can be used. Finally, the results can be verified against measured results obtained from an experimental paraboloid antenna built at Ericsson Microwave Systems AB, Mölndal, Sweden. In the analysis, we will continue to use waveguide-fed apertures. However, the rectangular apertures used in Chapter 6 are now replaced with circular waveguide-fed apertures, which are better suited for two-dimensional, wide angle scanning and allow polarization control. Both mutual coupling and radiation properties are discussed in Section 7.2.

It is today still very demanding, if at all possible, to make the necessary calculations for an electrically large, dielectric coated, general, doubly curved surface. This is an area in which there is a need for new research. However, for certain special geometries there are solutions available, in particular for the sphere. This geometry will be used for a discussion about the mutual coupling and radiation properties of microstrip-patch antennas on doubly curved surfaces presented in Section 7.3. A comparison against measurements will also be given here.

7.2 APERTURE ANTENNAS

7.2.1 Introduction

In this section, we will study aperture antennas located on a PEC doubly curved surface: the paraboloid of revolution. Waveguide-fed aperture antenna elements were discussed in Chapter 6, where a rectangular cross section was considered. Here, circular waveguide-fed apertures are used in order to study polarization effects. Most of the calculated results have been obtained with the hybrid UTD–MoM method described earlier; see Sections 4.5.2 and 6.2.2. Note, however, that a single-mode approximation is used, that is, only the dominant TE_{11} mode is accounted for in the simulations.

In addition to calculations, many of the results have been verified with measured data [Lanne and Josefsson 2000, 2002]. To our knowledge, this is one of very few cases in which UTD-based calculations for doubly curved surfaces have been experimentally verified.

The doubly curved experimental antenna shown in Figure 7.1 was built at Ericsson Microwave Systems AB in Mölndal, Sweden, in 2000. The experimental antenna is shaped as a paraboloid of revolution with $f/d \approx 0.22$, and the diameter of the surface is approximately 600 mm with a depth of approximately 175 mm; see Figure 7.1 (a). The surface is extended with a conical sheet of metal to gain extra space for the absorbers at the edge. The same principle was used with the cylindrical antenna (cf. Figure 6.12), with excellent results. For practical reasons the circular waveguide-fed apertures with diameter 14.40 mm are filled with Rexolite ($\varepsilon_r = 2.53$). The cut-off frequency for the dominant TE_{11} mode is 7.65 GHz. The nearest mode in cut-off order is TM_{01} with 9.99 GHz cut-off frequency. A single isolated element has a return loss (S_{11}) of about –8 dB.

The surface has 48 circular apertures; the layout is shown in Figure 7.1. Figure 7.1 (b) shows a two-dimensional plot together with an identification of each element, which will be used for reference later in this section. As seen, the elements do not cover the surface completely; instead, they are located at selected positions. The positions have been chosen to cover most element positions of interest in a doubly curved array. Thus, the information obtained is expected to provide sufficient knowledge that can be extrapolated to the fully populated surface. In order to study polarization effects, one of two orthogonal polarization can be selected. This is achieved by rotating the waveguides by 90°. The two possibilities are denoted as R and φ polarizations, respectively. For the vertex element with coordinates ±0, R polarization is defined as y polarization and φ polarization as x polarization.

7.2.2 Mutual Coupling

Most of the results presented here are for isolated mutual coupling, defined in Section 6.1. However, some examples of array mutual coupling are also included in Section 7.2.2.2.

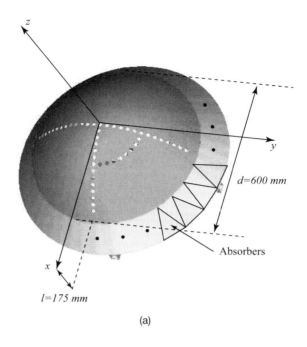

(a)

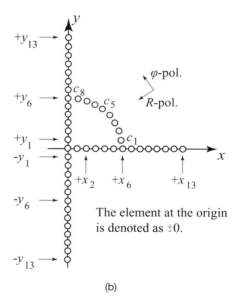

(b)

<u>Figure 7.1.</u> (*a*) The geometry of the experimental doubly curved antenna with the co-ordinate system used in the analysis. (*b*) The antenna from (*a*), together with the numbering of the elements. *(continued)*

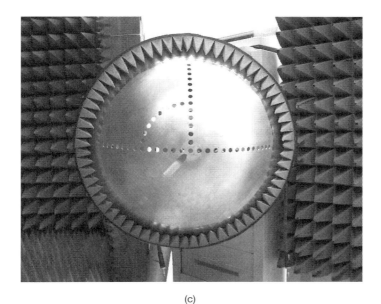

(c)

Figure 7.1. (*cont.*) (*c*) A picture of the antenna on location for mutual coupling measurements. See also color insert, Figure 3 for another view of the antenna. Courtesy of Ericsson Microwave Systems AB, Göteborg, Sweden.

7.2.2.1   *Isolated Mutual Coupling.*

Based on calculations and measurements for the experimental antenna, we will here discuss the salient features of the mutual coupling among elements on a doubly curved surface. Although the number of available elements in the antenna is limited, there are a lot of combinations to study, covering the main configurations of interest.

Polarization is an interesting parameter and for each case there are four different combinations of polarization, denoted as follows:

1. $[R, R]$ means that both the source and observation elements are R polarized.
2. $[\varphi, \varphi]$ means that both the source and observation elements are φ polarized.
3. $[R, \varphi]$ means that the source element is φ polarized and the observation element is R polarized.
4. $[\varphi, R]$ means that the source element is R polarized and the observation element is φ polarized.

For the results presented here, the frequency is 8.975 GHz. A single-mode approximation was used in the theoretical analysis. Observe that both the $[R, \varphi]$ and $[\varphi, R]$ cases are considered since they, in general, do not give the same results on doubly curved surfaces. In several of the figures presented here, the location of the source and observation elements are shown to the right of the graphs. In addition to the calculated result, measured data is included when available. It should be noted that not all observation elements were measured. Furthermore, in the simulations additional element positions were (in some cases) included in order to obtain a smoother curve. The phase delay corresponding to the geometrical separation is subtracted from all phase results. Note that in some graphs the results are plotted against the geodesic and a negative value of the geodesic length is used to indicate positions on the negative y axis [see Figure 7.1 (b)].

In the first example, we will look at the isolated coupling along the principal plane, the y axis in Figure 7.1 (b). This situation corresponds to the parabolic cylinder studied in Section 6.3.2.2. The first two figures (Figures 7.2 and 7.3) show the E-plane ([R, R] polarization) and H-plane ([φ, φ] polarization) couplings, respectively. As seen, the results do not exhibit any strange or surprising behaviors. There is a weaker coupling in the H plane, which decays faster than the E-plane coupling, as expected. If the surface had been more pointed, a slope change could have been observed when passing the vertex point ($s = 0$). Measured results are also included here and the agreement is good. Some of the differences between theoretical and measured results can be due to errors in the experimental setup, as will be discussed later. However, the differences at the very low levels are not unexpected; see, for example, Figure 7.3 where the level is below −90 dB.

It is interesting to study these results further. Figure 7.4 shows the calculated E-plane coupling ([R, R] polarization) from Figure 7.2 again, but compared with an approximate model. The comparison is made with results obtained from calculating the coupling with the same type of elements, but located on a parabolic cylinder with the same dimensions as the cross section of the paraboloid [the geometry is seen in Figure 7.4 (b)].

The results agree well and it seems to be possible to use a singly curved approximation for this case with some success. However, there is a deviation in amplitude that accelerates as the distance between the apertures increases. At large distances, the coupling amplitude of the parabolic cylinder solution is less than that of the paraboloid. An explanation can be that the parabolic cylinder, infinite in extent along the axial direction, behaves more as a "planar surface" in this region [cf. Figure 7.4 (b)]. Along this "planar surface," the energy is diffracted over a larger surface, resulting in the reduced amplitude (cf. the discussion at the end of Section 6.2.3.1).

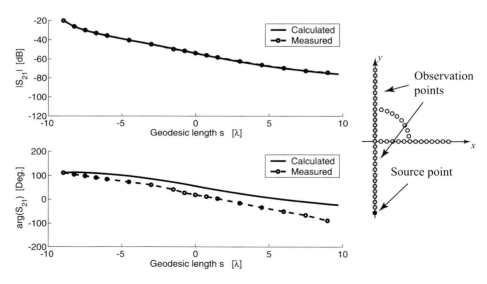

Figure 7.2. Isolated coupling along the y axis, E plane, or [R, R] polarization. The source point is element $-y_{13}$. Solid line: calculated; dashed line: measured.

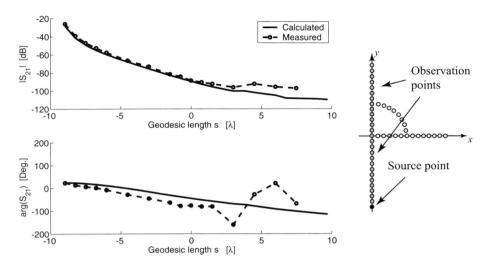

Figure 7.3. Isolated coupling along the y axis, H plane, or [φ, φ] polarization. The source point is element $-y_{13}$. Solid line: calculated; dashed line: measured.

The final combinations of interest along the principal plane are the cross polarization cases, that is, [R, φ] and [φ, R] couplings. Ideally, the coupling levels should be very low (cf. Section 6.2.3.2) and the calculated results are indeed small (about −100 dB). However, the measured results gave cross-polar levels of about −60 dB. This is higher than expected, and much above the leakage/noise floor of the measurements. A possible explanation for the high cross-polar levels in the measured results is the sensitivity to proper rotations of the radiating elements. To study this, additional simulations were made in which the elements were slightly rotated. Figure 7.5 shows the [R, φ] case with both the source and observation elements rotated 1°. This increased the calculated values by approximately 30 dB, resulting in a much better agreement. A similar result was obtained for the [φ, R] case. Thus, small rotation errors can explain the unexpected high measured cross-polarization levels. In the experimental set-up, dowel pins controlled the rotation of the elements; however, an accuracy of ±1° could not be guaranteed.

The next series of figures show the isolated mutual coupling along the circular arc shown in Figure 7.1 (b). For these examples, there is no z dependence since the positions of the apertures are varied with respect to the angle φ only. Figures 7.6–7.9 show the amplitude and phase for all four combinations of polarization. The amplitude decay is as expected since the geodesic length increases proportionally to the increase in φ. Furthermore, in this situation it is not possible to define E- and H-plane couplings since the polarization in the elements varies locally for each position. So, these four cases can be referred to as different co- and cross-polar type combinations for which the total amplitude and phase characteristics are very similar.

Once again, we can make a comparison with a singly curved solution for the circular arc case. This is done by using a circular cylinder with the same radius as the local radius of the circular arc in the xy plane. The results are presented in Figures 7.10 and 7.11 for two polarization cases, [R, R] polarization and [φ, φ] polarization. These two cases correspond to H-plane and E-plane coupling, respectively, for the circular cylinder. For each polarization, four different locations of the circle arc are considered, with the first arc

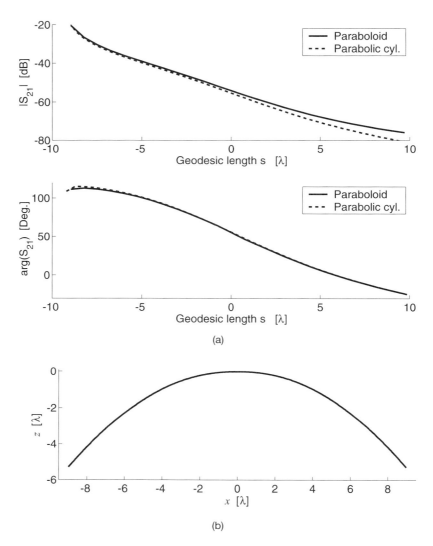

Figure 7.4. The same case as in Figure 7.2 ([R, R] polarization, solid line) compared with the results obtained using a parabolic cylinder (dashed line) with the same cross section as the paraboloid. (a) Amplitude and phase. (b) The cross section (cf. Figure 7.1).

starting at element $+x_6$. The next arc starts with element $+x_{13}$ and, finally, two fictitious arcs are used. These two fictitious arcs start with elements $+x_M$ and $+x_N$ located further away from vertex. The local radii in the xy plane for these four arcs are $r_{x6} = 4.3\ \lambda$, $r_{x13} = 8.3\ \lambda$, $r_{x_M} = 17.1\ \lambda$, and $r_{x_N} = 42.7\ \lambda$. The results show that the circular-cylinder approximation is in general much better for the $[\varphi, \varphi]$ polarization case than for the $[R, R]$ polarization case, as expected. Furthermore, the approximation becomes better and better (for both polarizations) when moving away from the vertex. Thus, using a singly curved surface (circular cylinder) as an approximation to the paraboloid solution is not as useful as it was when the elements along the principal plane were studied earlier. However, if a coni-

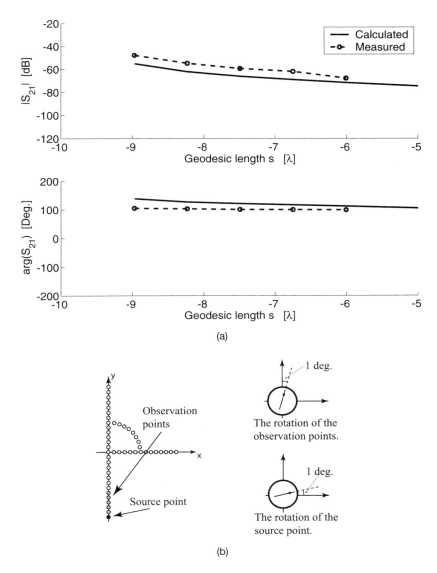

Figure 7.5. Isolated coupling along the y axis, $[R, \varphi]$ polarization. The source point is element $-y_{13}$. Calculated results with a small rotation of the elements. (a) Amplitude and phase. Solid line: calculated; dashed line: measured. (b) The geometry.

cal surface had been used we would most likely have obtained better agreement between the three-dimensional paraboloid solution and the approximate two-dimensional solution. Compare this to the conical array approximated by a cylinder solution in Section 6.7.

The results in the previous examples can be explained mainly by the surface field dependence on the geodesic length and the radius of curvature along the geodesic. In the next examples to be presented, the results are much more dependent on the polarization of the elements. Figures 7.12–7.15 show the isolated mutual coupling between an element

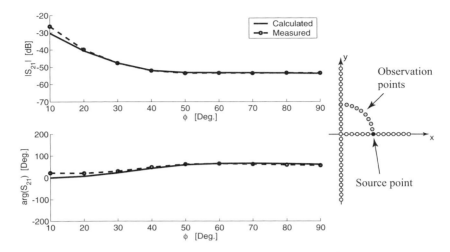

Figure 7.6. Isolated coupling along the circle arc, $[R, R]$ polarization. The source point is element $+x_6$. Solid line: calculated; dashed line: measured.

on the circular arc and elements along the positive y axis. Here, the results are explained mainly as polarization effects. To illustrate this, the geodesic length is included in Figure 7.12, where the $[R, R]$ polarization case is considered. The $[\varphi, \varphi]$ polarization case is shown in Figure 7.13, and the cross-polar results are shown in Figures 7.14 and 7.15. Note that these results are plotted against the parameter u; see Table 5.3, page 138.

To explain the polarization effects, the corresponding planar case is used as a model; see Figure 7.16. This figure shows a sketch of the magnetic field from the radiating element at the positions of the other elements. For the $[R, R]$ polarization case (Figure 7.12)

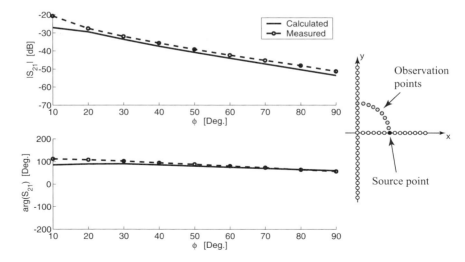

Figure 7.7. Isolated coupling along the circle arc, $[\varphi, \varphi]$ polarization. The source point is element $+x_6$. Solid line: calculated; dashed line: measured.

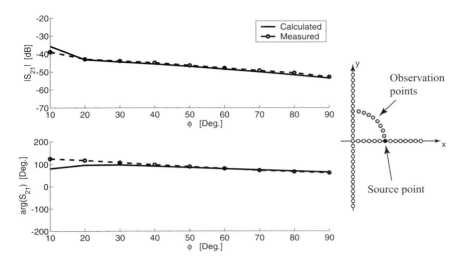

Figure 7.8. Isolated coupling along the circle arc, $[R, \varphi]$ polarization. The source point is element $+x_6$. Solid line: calculated; dashed line: measured.

the corresponding planar case is shown in Figure 7.16 (*a*). Here, the magnetic and electric fields are parallel to each other at two positions where, consequently, the mutual coupling is zero. This is the explanation for the double-hump behavior in Figure 7.12. Similarly, Figure 7.16 (*b*) shows the planar case for the $[\varphi, \varphi]$ polarization. For this case, shown in Figure 7.13, the polarization effect is not as dramatic along the positive *y* axis as it was for the $[R, R]$ case. However, continuing along the *y* axis in the negative direction it is possible to find a position with a local minimum. The same discussion can be used to explain the results in Figures 7.14 and 7.15.

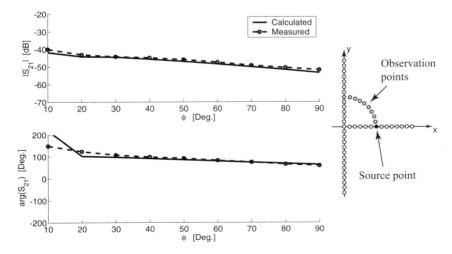

Figure 7.9. Isolated coupling along the circle arc, $[\varphi, R]$ polarization. The source point is element $+x_6$. Solid line: calculated; dashed line: measured.

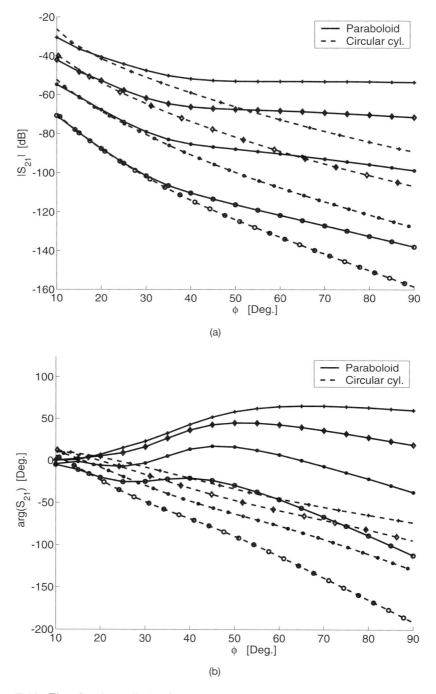

Figure 7.10. The circular cylinder (dashed lines) is used as an approximation for the isolated coupling along circle arcs located at different positions at the paraboloid (solid lines). The polarization is [R, R] and the different rows are starting with elements $+x_6$ (+), $+x_{13}$ (diamonds), and two fictitious rows located further away from the vertex (stars and circles), respectively. (a) Amplitude. (b) Phase.

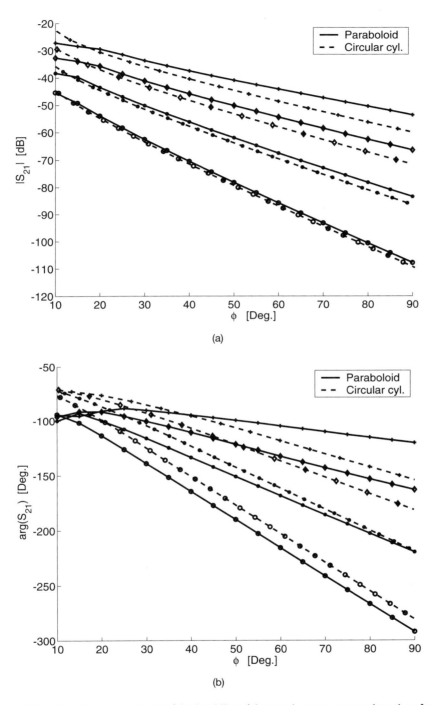

Figure 7.11. The circular cylinder (dashed lines) is used as an approximation for the isolated coupling along circle arcs located at different positions at the paraboloid (solid line). The polarization is $[\varphi, \varphi]$ and the different rows are starting with elements $+x_6$ (+), $+x_{13}$ (diamonds), and two fictitious rows located further away from vertex (stars and circles), respectively. (a) Amplitude. (b) Phase.

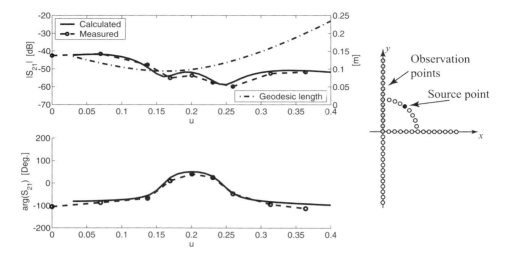

Figure 7.12. Isolated coupling, $[R, R]$ polarization. The source point is element c_5. Solid line: calculated; dashed line: measured, dashed-dotted line: geodesic length.

In addition to the full three-dimensional solution used here (or two-dimensional solution in special cases) it is possible to get a qualitative view of what to expect in terms of mutual coupling without performing a full analysis. Assuming the surface to be (locally) planar, the radiated field in the aperture plane varies as $\sin \phi$ [Balanis 1997, p. 608]; see the dashed lines in Figure 7.16. For most element locations, the two apertures are in the far fields from each other, so the coupling has a $\sin \phi_1 \sin \phi_2$ variation, to which we can

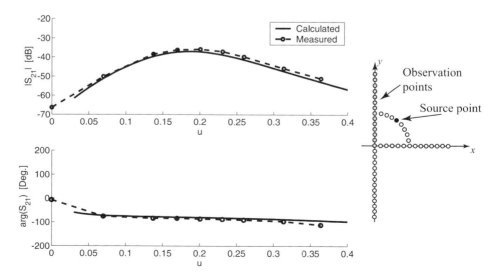

Figure 7.13. Isolated coupling, $[\varphi, \varphi]$ polarization. The source point is element c_5. Solid line: calculated; dashed line: measured.

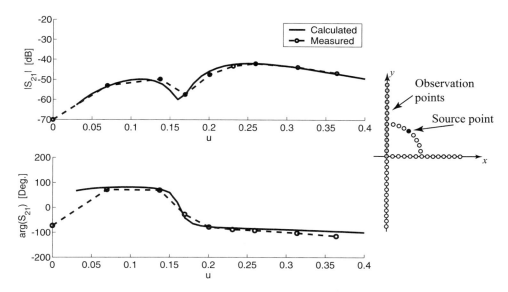

Figure 7.14. Isolated coupling, $[R, \varphi]$ polarization. The source point is element c_5. Solid line: calculated; dashed line: measured.

add the $1/s$ dependence. The result in Figure 7.12 is shown in Figure 7.17 for such an analysis. This approach gives satisfactory results on a qualitative basis and can be used for a first analysis of the problem.

As shown in this section, the agreement between measured and calculated results is generally very good. The minor disagreements (apart from some cross-polarization cases)

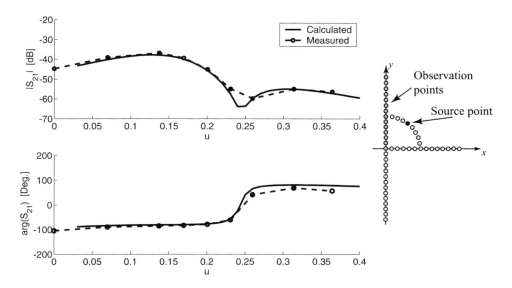

Figure 7.15. Isolated coupling, $[\varphi, R]$ polarization. The source point is element c_5. Solid line: calculated; dashed line: measured.

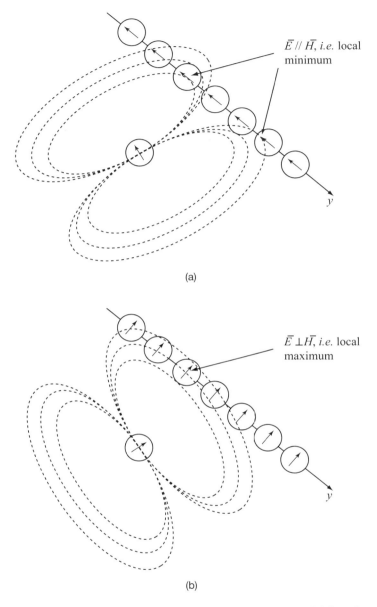

(a)

$\bar{E} /\!/ \bar{H}$, *i.e.* local minimum

(b)

$\bar{E} \perp \bar{H}$, *i.e.* local maximum

Figure 7.16. A sketch of the magnetic field for the planar case. (a) [R, R] polarization case; see also Figure 7.12. (b) [φ, φ] polarization case; see also Figure 7.13.

are probably due to one or more of the following factors: the lack of higher-order modes in the calculations and differences between the theoretical shape and the actual shape of the experimental antenna. Figure 7.18 shows the aperture locations and the surface shape of the theoretical model (the generating curve) compared with the experimental case. A third factor can be the geodesic constant used in the ray tracing procedure. This constant is critical, as discussed in Chapter 5, since all parameters in the UTD solution depend on

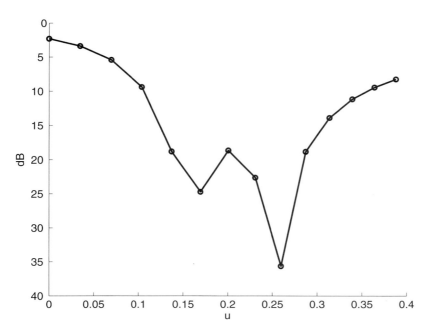

Figure 7.17. [*R, R*] polarization. Test of a geometrical approach for calculating the mutual coupling [Josefsson et al. 2002]; compare with Figure 7.12.

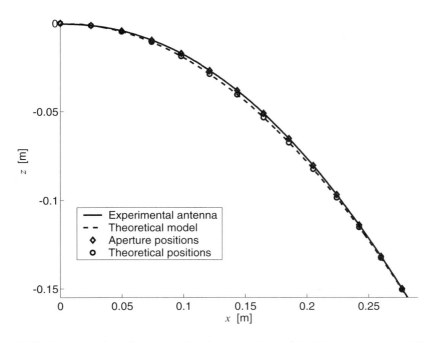

Figure 7.18. A comparison between the theoretical model of the antenna and the experimental antenna itself.

the geodesic constant. However, according to the discussion in Section 5.3.5, this is not a likely source of error. Higher-order modes were studied in Chapter 6 for the circular cylinder and it was found that they had a significant effect of the results, in particular for the array mutual coupling. For isolated coupling, however, they did not affect the results very much. So, for the examples shown in this section the main source of error is probably minor differences between the theoretical model and the experimental antenna and measurement errors.

7.2.2.2 *Array Mutual Coupling.*

Unfortunately, doubly curved arrays are not as symmetrical as singly curved surfaces. Thus, the analysis of doubly curved arrays is more time-consuming than singly curved arrays. Often, an element-by-element approach is needed and all combinations of element locations must be calculated. For example, about 500 apertures are required to fully cover the experimental antenna with elements (14 rings and 0.75 λ element spacings in a triangular lattice); see Section 8.5.3.1 and Figure 8.33.

Here, we will limit ourselves to the array mutual coupling along the principal y axis of the experimental antenna (Figure 7.1). All results presented are measured at the frequency $f = 8.975$ GHz. Several cases are included with different positions of the source along the y axis. During the measurements, the elements along the principal axis were terminated in matched loads (Figure 7.19), except the two between which the coupling was measured. The elements away from the y axis were all covered with metallic tape [Lanne and Josefsson 2002]. Both E- and H-plane array couplings are considered. For each polarization, three sets of measurements were made. The first set is when element $-y_{11}$ is fed and the array coupling is measured to the elements away from this element in the positive y axis direction. This set is named "Case A" for E-plane coupling and "Case B" for H-plane coupling. In the next set, element $-y_6$ is fed instead and this set is called "Case C" for E-plane coupling and "Case D" for H-plane coupling. Finally, the element at the vertex is fed and the array coupling is measured along the positive y axis. This set is called "Case E" for E-plane coupling and "Case F" for H-plane coupling. The results are shown versus

Figure 7.19. The inside of the experimental antenna when measuring the array mutual coupling along the y axis (cf. Figure 7.1). The probes, as well as the terminated elements in between are clearly visible. Courtesy of Ericsson Microwave Systems AB, Göteborg, Sweden.

the distance from the fed element. Thus, each result starts at the same position, as seen in Figures 7.20–7.23. Note that the phase delay corresponding to the geometrical separation is subtracted in all cases.

Figure 7.20 shows the array mutual coupling in the E plane and Figure 7.21 shows the results for the H plane; the behavior is similar for both polarizations. If element $-y_{11}$ is fed, the coupling does not decay as much as for the other cases since this part of the surface is fairly flat. The other results decay faster due to the greater curvature.

Finally, a comparison of the isolated and array mutual coupling along the y axis is shown for both polarizations. Note that different elements were used as the source: $-y_{11}$ was fed when studying the array mutual coupling but $-y_{13}$ was used when measuring the isolated coupling. However, these two elements are located close to each other where the surface has a similar shape. Thus, there should not be any substantial difference whether $-y_{13}$ or $-y_{11}$ is fed, and a comparison between these two cases is justified.

Figure 7.22 shows the results for the E-plane case. The results do not differ much from each other; in fact, the amplitudes of S_{21} are almost identical. This is not to be expected since similar results are not obtained in the planar and singly curved cases. We found (Section 6.2.3.3) that the array and isolated couplings experience the same type of decay for the singly curved surface, but the amplitude of the array coupling was less than the isolated coupling. The different type of behavior found here is probably explained as a combination of electrically small elements located on an electrically large surface with a curvature perpendicular to the geodesic path. Thus, the surface field is diffracted away along the path and the expected interaction among the elements is very small. Figure 7.23 shows the H-plane case with similar results.

7.2.3 Radiation Characteristics

In this section, we will investigate the typical radiation characteristics of aperture antenna elements on doubly curved surfaces. The results are obtained using the hybrid UTD–MoM approach. Thus, we include the effect of diffraction, calculate the radiated field in the shadow region of the source, and compare with measured data. In Section 8.6.6, we will compare array patterns using two element models: one based on the full UTD–MoM method and a reference model based on an idealized dipole.

We will here only discuss isolated-element patterns, leaving embedded-element patterns for future studies. A full analysis of a doubly curved array still needs some additional progress in order to be manageable. A significant complication is that the element positions are not in a regular grid; in addition, the polarization of the fields must be incorporated

The same PEC doubly curved surface is analyzed in this section as was studied previously in this chapter (see Figure 7.1). The first examples presented show the isolated-element patterns (normalized) for two element positions on the paraboloid. The patterns are measured and calculated in the yz plane (see Figure 7.1). All results are presented as a function of the θ angle; positive values correspond to the part of the yz plane where y is positive and negative values correspond to the part of the yz plane where y is negative. Furthermore, the theoretical results are calculated for the interval $|\theta| \leq 120°$, corresponding to the "front side" of the antenna. Outside this region, the inside of the antenna will be exposed toward the transmitting antenna during the measurements, giving undesired reflections and diffractions.

Since the elements can be rotated, both the E-plane and the H-plane patterns can be measured (and calculated) in the yz plane. Figure 7.24 shows the E-plane and the H-plane power patterns, respectively, for the element at the vertex. In this case, the shadow region

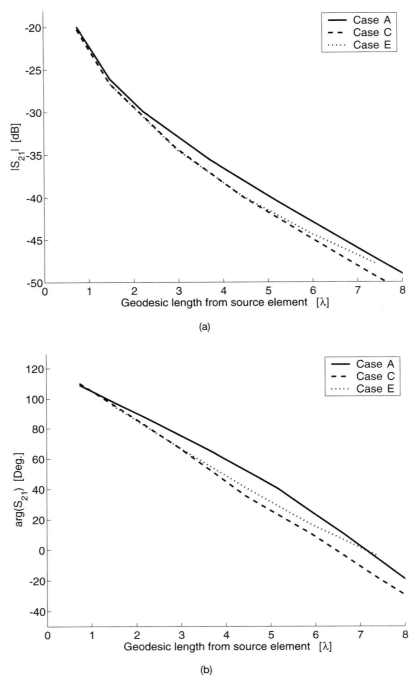

(a)

(b)

<u>Figure 7.20.</u> The array mutual coupling (*E* plane) along the *y* axis of the experimental antenna (*f* = 8.975 GHz). Three sets of data are shown. "Case A" shows the case when element $-y_{11}$ is fed, "Case C" shows the case when element $-y_6$ is fed, and "Case E" shows the case when the element at vertex is fed [Lanne and Josefsson 2002]. (*a*) Amplitude. (*b*) Phase.

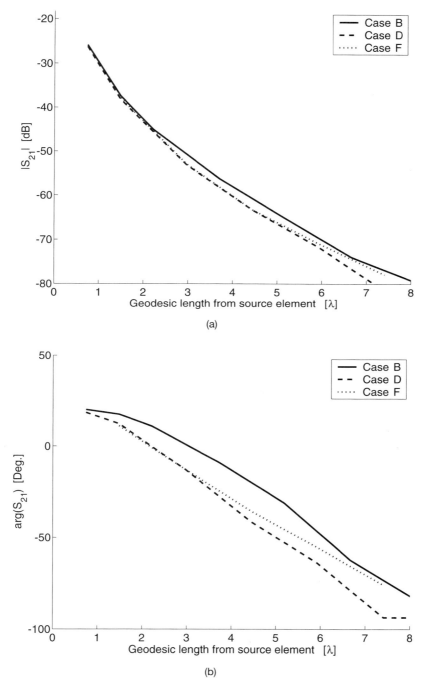

Figure 7.21. The array mutual coupling (*H* plane) along the *y* axis of the experimental antenna (*f* = 8.975 GHz). Three sets of data are shown. "Case B" shows the case when element $-y_{11}$ is fed, "Case D" shows the case when element $-y_6$ is fed, and "Case F" shows the case when the element at vertex is fed [Lanne and Josefsson 2002]. (*a*) Amplitude. (*b*) Phase.

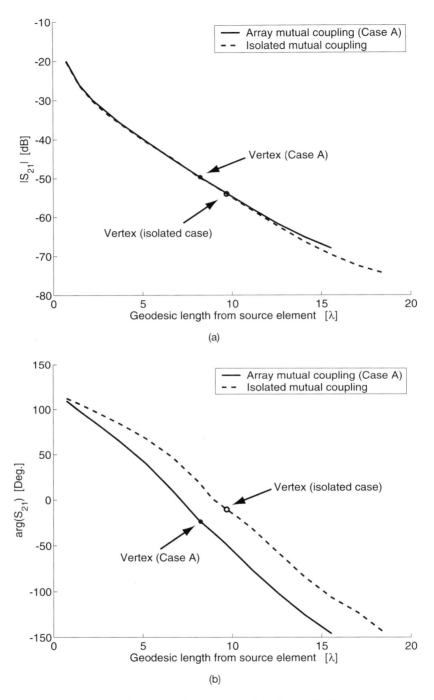

Figure 7.22. Comparison between the array and isolated E-plane mutual coupling along the y axis of the experimental antenna ($f = 8.975$ GHz). Note that the array mutual coupling shown is measured from element $-y_{11}$, whereas the isolated coupling is measured from element $-y_{13}$ [Lanne and Josefsson 2002]. (a) Amplitude. (b) Phase.

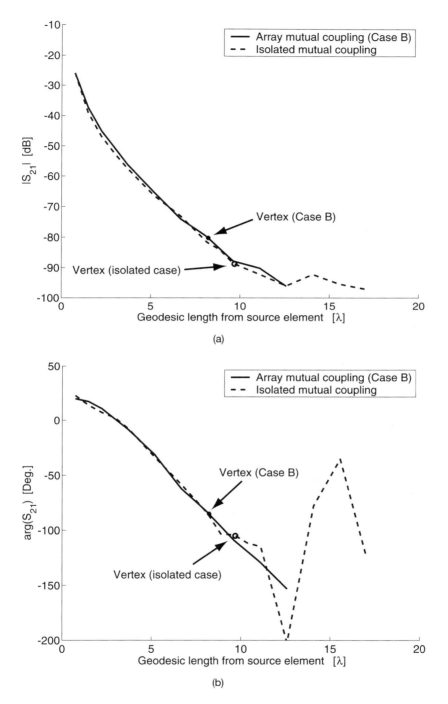

Figure 7.23. Comparison between the array and isolated *H*-plane mutual coupling along the *y* axis of the experimental antenna ($f = 8.975$ GHz). Note that the array mutual coupling shown is measured from element $-y_{11}$, whereas the isolated coupling is measured from element $-y_{13}$ [Lanne and Josefsson 2002]. (*a*) Amplitude. (*b*) Phase.

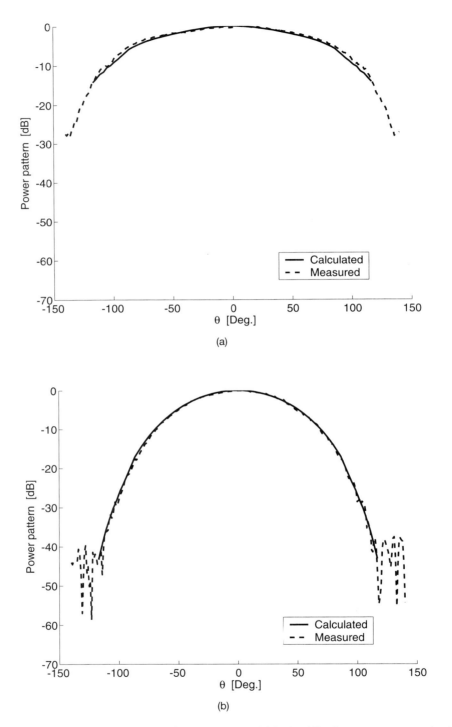

Figure 7.24. Calculated (solid line) and measured (dashed line) power patterns in the *yz* plane from a circular waveguide-fed aperture at vertex. (*a*) *E* plane. (*b*) *H* plane.

starts at $|\theta| > 90°$. The calculated results were obtained using a single-mode approximation; however, the agreement is very good.

As can be seen, the E-plane pattern is very wide. The 3 dB beamwidth is approximately 140° compared to approximately 130° for the same element mounted on a planar surface [Josefsson and Lanne 2003]. This is to be expected since the element is located at the vertex and the ground plane is swept backward, causing the lobe to widen. The H-plane diagram is not as wide as the E-plane diagram and the 3 dB beamwidth is approximately 80° for the doubly curved surface. According to Josefsson and Lanne [2003], the corresponding planar 3 dB beamwidth is approximately 60–80°. In the planar case, the H-plane pattern is wider than the E-plane pattern [Balanis 1997] for large element radii. Here, it is the opposite since the element radius is small compared to the wavelength.

The next example, Figure 7.25, shows the power pattern for the element at $+y_6$. The pattern is calculated in the yz plane with the element R-polarized [Figure 7.25 (a)] and φ-polarized [Figure 7.25 (b)]. Here, the shadow region starts at $\theta \approx -62°$. Once again, the E-plane pattern is much wider than the H-plane pattern and it is difficult to see that the main lobe has shifted in the E-plane pattern. The shift of the main lobe is easier to see in the H-plane pattern. Element $+y_6$ is located at $\theta = 29.5°$ and the peak of the main lobe in the H plane has moved almost the same amount.

7.3 MICROSTRIP PATCH ANTENNAS

7.3.1 Introduction

The analysis of a dielectric-coated, doubly curved surface is a difficult problem and there are few references available for this case, as mentioned in the Section 7.1. In this section we will present some results in an attempt to discuss the characteristics of microstrip-patch antennas on doubly curved surfaces, in this case the sphere. The geometry considered is shown in Figure 7.26. The radius of the grounded sphere is a and b is the radius of the coated sphere; thus, the coating thickness is $t = b - a$. The patches can be of different shapes; a wraparound patch is discussed in [Luk and Tam 1991], a circular disk is found in [Tam and Luk 1991a,b; Luk and Tam 1991; Ke and Kishk 1991; Tam et al. 1995], an annular ring is studied in [Luk and Tam 1991, Tam et al. 1995], and a (quasi-)rectangular patch is discussed by [Burum and Sipus 2002a,b; Burum et al. 2004; Sipus et al. 2003a–c, 2004; Sipus and Burum 2005].

The results shown here are obtained with a full-wave analysis in which the electric field integral equation is numerically solved by applying the method of moments. The coating can be multilayered and the problem is solved using a spectral-domain algorithm for calculating the Green's functions [Sipus et al. 1998]. Here, however, only a single layer is used. To facilitate the analysis, quasirectangular patches are placed on the coated spherical surface. Their dimensions are $W_\varphi = 2\varphi_p b$ and $W_\theta = 2\theta_p b$, and each patch is probe fed from a coaxial transmission line. Note that the shape of the quasi-rectangular patch follows the constant θ and φ lines as shown in Figure 7.26 (b). The two polarizations used here are θ polarization and φ polarization, corresponding to E-plane or H-plane coupling, depending on the orientations of the patches with respect to each other.

7.3.2 Mutual Coupling

Before discussing the mutual coupling among patch elements in an array, we will first look at the resonant frequency and the input impedance of a single element.

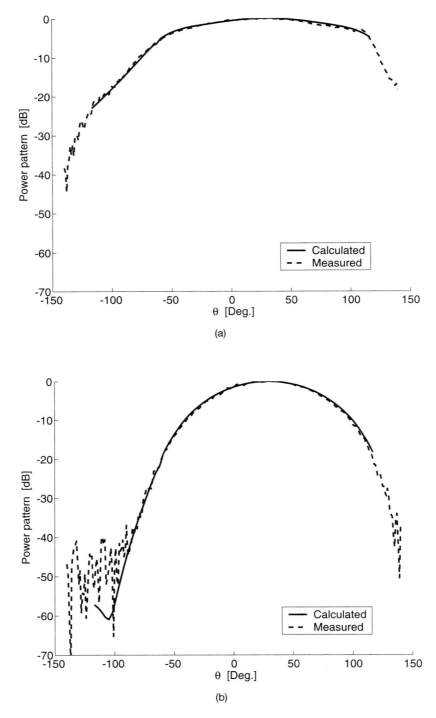

Figure 7.25. Calculated (solid line) and measured (dashed line) power patterns in the *yz* plane from a circular waveguide-fed aperture at position $+y_6$. (*a*) *E* plane. (*b*) *H* plane.

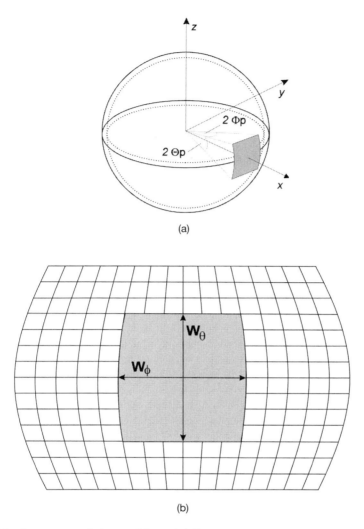

Figure 7.26. Geometry of the problem. (*a*) Rectangular patch on a spherical structure. (*b*) Azimuthal equatorial projection [Sipus et al. 2003a]. Copyright (2003) Wiley Periodicals, Inc. Reprinted by permission of John Wiley & Sons, Inc.

7.3.2.1 Single-Element Characteristics.

We will start by discussing the influence of the sphere radius on the input impedance of the quasirectangular patch. For comparison, the measured input impedance of a planar patch antenna with the same dimensions is also given. The size of the quasirectangular patch is 2.5 cm × 4.0 cm; it is printed on a single-layer dielectric substrate with dielectric constant $\varepsilon_r = 2.52$ and thickness $t = 1.58$ mm. Both θ- and ϕ-polarized patches are considered. The patch is assumed to be probe fed.

Figure 7.27 shows the normalized input impedance of a θ-polarized, quasirectangular patch; the ϕ-polarization case is shown in Figure 7.28. It can be seen that the radius of the sphere mainly influences the resonant frequency, and the change of resonant frequency is stronger for the ϕ-polarized patch. The resonant resistance is almost constant

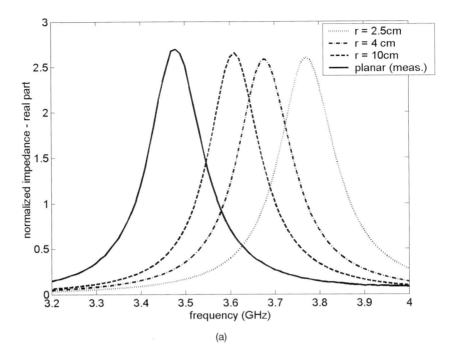

(a)

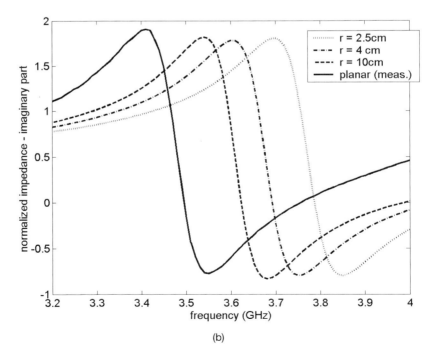

(b)

<u>Figure 7.27.</u> Normalized input impedance of a θ polarized quasirectangular patch printed on a spherical structure for different sphere radii. The planar case is also included. (*a*) Real part. (*b*) Imaginary part [Sipus et al. 2003a]. Copyright (2003) Wiley Periodicals, Inc. Reprinted by permission of John Wiley & Sons, Inc.

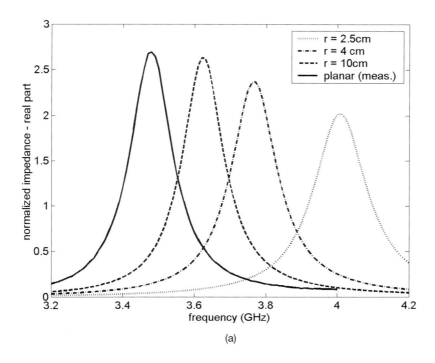

(a)

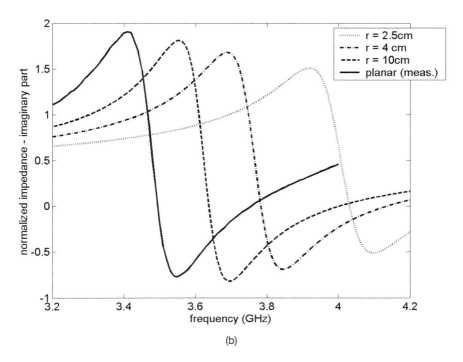

(b)

Figure 7.28. Normalized input impedance of a ϕ polarized quasirectangular patch printed on a spherical structure for different sphere radii. The planar case is also included. (*a*) Real part. (*b*) Imaginary part [Sipus et al. 2003a]. Copyright (2003) Wiley Periodicals, Inc. Reprinted by permission of John Wiley & Sons, Inc.

for the θ-polarized patch and it is slowly changing for the ϕ-polarized patch. Furthermore, the results move toward the planar case when the radius increases, as expected.

An option is to use a circular microstrip patch printed on a coated sphere. The resonant frequency for this type of patch has been discussed in [Tam et al. 1995] and the same type of behavior with regard to the radius was found. Thus, when comparing the results shown here for patches on doubly curved surfaces with those of the singly curved surface (Figure 6.61), it is found that the radius of the doubly curved surface has a much larger impact on the resonant frequency than does the singly curved surface. The main reason is probably that the resonance phenomenon for the sphere is more complicated due to the doubly curved structure. In the case of rectangular patches, the resonant mechanism is more complex than merely the interference of a forward- and a backward-propagating wave.

7.3.2.2 Isolated Mutual Coupling.
The results presented here are isolated coupling results, that is, there are only two elements present at the same time.

The first example shows the dependence on the sphere radius. Two quasirectangular patches of dimensions 10.57 cm × 6.55 cm are printed on a single-layer dielectric substrate with dielectric constant $\varepsilon_r = 2.5$ and thickness $t = 1.575$ mm. The position of the probe is 1.5 cm from the center of the patch. The working frequency of the equivalent planar patch is 1.41 GHz. However, for a spherical patch the working frequency is slightly changed in order to obtain a good match (the needed shift of resonant frequency depends on the sphere radius, as discussed in Section 7.3.2.1). The magnitude of the S_{21} parameter, both in the E and H planes, is shown in Figure 7.29 as a function of the spacing between the patch edges for different radii ($R = 0.5\ \lambda$, 1 λ, and 2 λ). For comparison, the measured mutual coupling of a planar patch antenna with the same dimensions is also given [Jedlicka et al. 1981]. It can be seen that the coupling is weaker for the spherical case in comparison with the planar case, especially in the H plane. For smaller radii, the mutual coupling starts to be oscillatory due to the interaction with surface waves encircling the sphere.

Next, a comparison between patches on a sphere and on a circular cylinder is discussed. In this example, quasisquare patches of dimensions 4.8 cm × 4.8 cm have been printed on a single-layer dielectric substrate with dielectric constant $\varepsilon_r = 2.38$ and thickness $t = 1.58$ mm. The working frequency is 2 GHz and the radius of the sphere and of the cylinder is 1 λ. The magnitude of the isolated mutual coupling is shown as a function of the spacing between patch centers. The coupling between spherical patches in the E plane is seen in Figure 7.30 (a) and the results for the H plane are shown in Figure 7.30 (b). These results are compared with the cylindrical counterpart. For the cylinder, both coupling in the axial direction ($y = 0$ in the figure legend) and in the φ direction ($z = 0$ in the figure legend) are considered. It can be seen that there is close agreement between the spherical and cylindrical cases only for coupling in the φ direction, as expected. Furthermore, the circular cylinder gives less coupling at large distances than the sphere. This feature was also discussed in Section 7.2, where a PEC paraboloid was compared with a PEC parabolic cylinder (cf. Figure 7.4).

In addition to the calculated results, measured data have been obtained from a laboratory model built from a copper half-sphere of radius $a = 18.7$ cm on which patches were mounted at different positions [Sipus et al. 2003c, Burum et al. 2004]. The antenna is shown in Figure 7.31. The dimensions of the patches are 5.1 cm × 5.1 cm, and small styrofoam cubes are used as distances between the patches and the grounded shell. Therefore, $\varepsilon_r = 1$ and the height of the styrofoam cubes define the thickness, $t = 5.2$ mm.

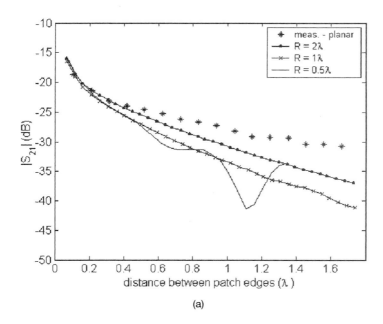

(a)

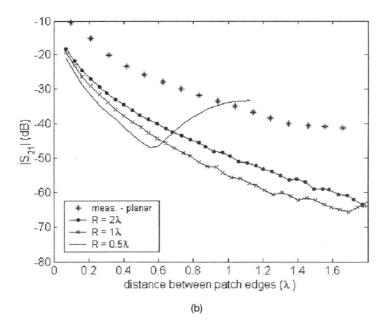

(b)

Figure 7.29. Magnitude of the isolated coupling of a two-patch array as a function of spacing between patch edges for different radii [Sipus et al. 2003b]. The planar case is also included by measured data (stars). (*a*) *E* plane. (*b*) *H* plane.

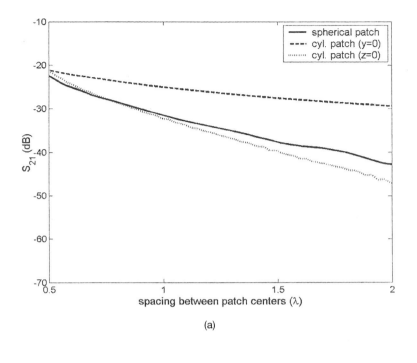

(a)

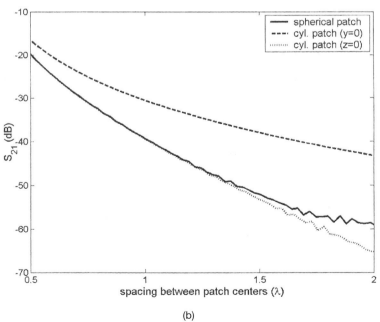

(b)

Figure 7.30. Magnitude of the S_{21} parameter shown as a function of spacing between patch centers [Burum and Sipus 2002b]. The spherical case (solid line) is compared with the cylindrical counterpart (dashed and dotted lines). (*a*) *E* plane. (*b*) *H* plane.

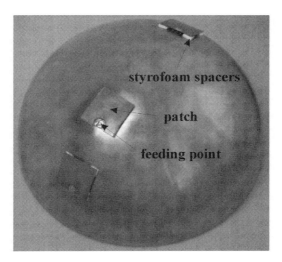

Figure 7.31. The experimental microstrip patch antenna [Burum et al. 2004]. Copyright (2003) Wiley Periodicals, Inc. Reprinted by permission of John Wiley & Sons, Inc.

The mutual coupling of two patches in the E plane is studied as a function of the frequency and the spacing between the elements and shown in Figure 7.32. Considering the differences and approximations in the experimental antenna and the simulations, the calculated and measured mutual couplings are in a good agreement.

7.3.3 Radiation Characteristics

In this section, we will give some examples of isolated-element patterns for quasirectangular patches (Figure 7.26). Element patterns for circular microstrip patches can be found in, for example, [Luk and Tam 1991, Tam and Luk 1991b].

The rectangular patch considered here is defined by the following parameters: a quasi-square patch of dimensions 4.8 cm × 4.8 cm is printed on a single-layer dielectric substrate with dielectric constant $\varepsilon_r = 2.38$ and thickness $t = 1.58$ mm. The working frequency is 2 GHz.

The first example shows the effect of the radius of the ground plane (grounded shell); see Figure 7.33. The following radii have been studied: $a = 0.25\ \lambda$, $0.5\ \lambda$, and $1\ \lambda$. For comparison, the radiation pattern of a planar patch antenna of the same dimensions is also given. It can be seen that with increasing radius the main lobe of the spherical patch antenna approaches the main lobe of the planar counterpart. As expected, the back radiation is smaller for spheres with larger radius.

It is also interesting to make a comparison of the radiation characteristics of a spherical patch and a circular–cylindrical patch. The radius of the sphere and cylinder is 0.5 λ; all other parameter values are the same as before. In Figure 7.34, the cylindrical patch is axially polarized, and in Figure 7.35 the cylindrical patch is φ polarized. It can be seen that the cylindrical patch can be used as an approximation of the doubly curved patch only in the azimuthal plane for both polarizations. In the elevation plane, the radiation pattern is significantly different, especially around the axis of the cylinder and in the back direction.

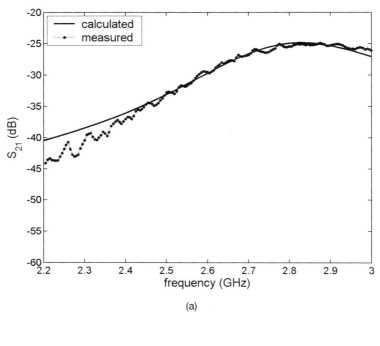

(a)

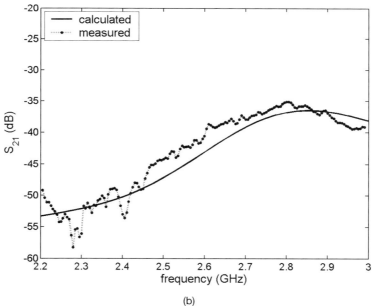

(b)

Figure 7.32. The comparison of calculated (solid line) and measured (dashed line) isolated mutual coupling coefficient of two patches in the *E* plane as a function of frequency and spacing between elements. (*a*) Spacing equals 100 mm. (*b*) Spacing equals 200 mm [Burum et al. 2004]. Copyright (2003) Wiley Periodicals, Inc. Reprinted by permission of John Wiley & Sons, Inc.

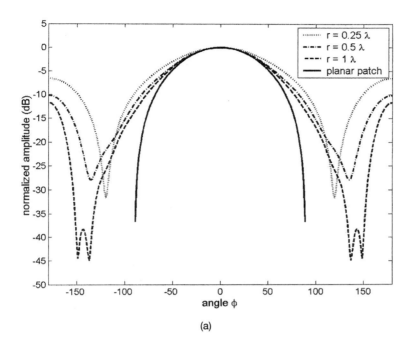

(a)

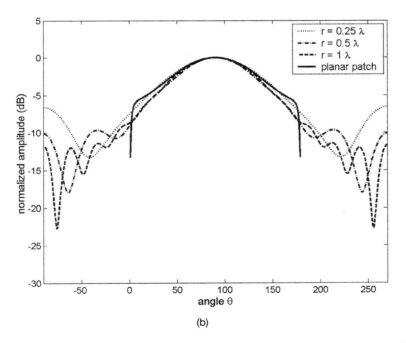

(b)

Figure 7.33. Radiation pattern of a spherical rectangular patch antenna for different sphere radii [Burum and Sipus 2002b]. The planar case is also included for comparison. (*a*) *H* plane. (*b*) *E* plane.

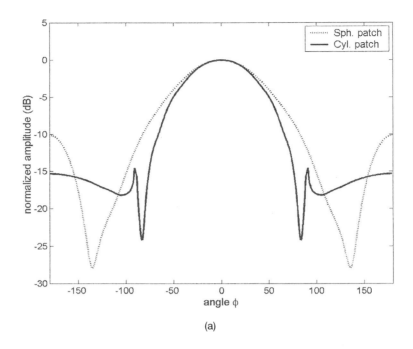

(a)

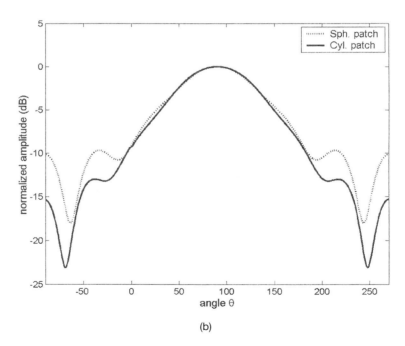

(b)

<u>Figure 7.34.</u> Comparison between radiation patterns of spherical patch antenna (dotted line) and ϕ-polarized cylindrical patch antenna (solid line) [Burum and Sipus 2002b]. (*a*) *H* plane. (*b*) *E* plane.

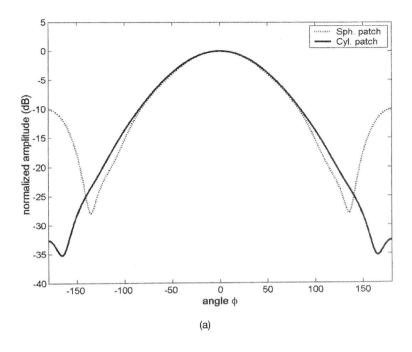

(a)

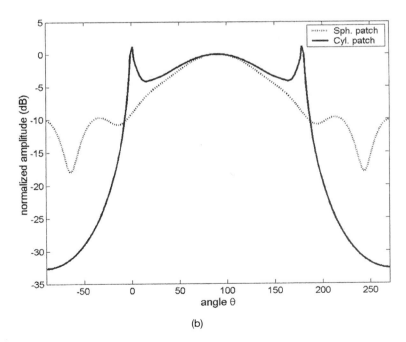

(b)

Figure 7.35. Comparison between radiation patterns of spherical patch antenna (dotted line) and axially-polarized cylindrical patch antenna (solid line) [Burum and Sipus 2002b]. (*a*) *H* plane. (*b*) *E* plane.

REFERENCES

Balanis C. A. (1997), *Antenna Theory, Analysis and Design,* 2nd ed., Wiley.

Burum N. and Sipus S. (2002a), "Radiation Pattern of Spherical Array of Rectangular Microstrip Patches," in *Proceedings of IEEE Antennas and Propagation Society International Symposium,* Vol. 1, 16–21 June, pp. 96–99.

Burum N. and Sipus S. (2002b), "Radiation Properties of Spherical and Cylindrical Rectangular Microstrip Patch Antennas," *Automatika,* Vol. 43, pp. 69–74.

Burum N., Sipus Z., and Bartolic J. (2004), "Mutual Coupling Between Spherical-Rectangular Microstrip Antennas," *Microwave and Optical Technology Letters,* Vol. 40, pp. 387–391, March.

Chen H. T. and Wong K. L. (1993), "Cross-polarization Characteristics of a Probe-fed Spherical-Circular Microstrip Patch Antenna," *Microwave Optical Technology Letters,* Vol. 5, pp. 705–710, September.

Chen H. T. and Wong K. L. (1994), "Analysis of Probe-Fed Spherical-Circular Microstrip Antennas Using Cavity-Model Theory," *Microwave Optical Technology Letters,* Vol. 7, pp. 309–312.

Chen H.-T., Chen H.-D., Chang T.-J., and Cheng Y.-T. (1995), "Analysis of the Spherical-Circular Microstrip Antenna With an Annular-Ring Parasitic Patch," in *Proceedings of IEEE International Symposium AP-S/URSI,* Vol. 4, June 18–23, pp. 1768–1771.

Debionne G. (1986), "Radiating Properties of Conformal Arrays," in *Proceedings of IEEE International Symposium AP-S/URSI,* pp. 937–940.

Douchin N. and Lemorton J. (1997), "Numerical Analysis of Active Conformal Arrays Including Crosspolarization Effects," in *Proceedings of IEE 10th International Conference on Antennas and Propagation,* April 14–17, pp. 1.200–1.205.

Girvan T. F. J., Fusco V. F., and Roberts A. (2001), "A Method for Calculating the Radiation Pattern of a Non-Planar Conformal Array," in *Proceedings of 6th IEEE High Frequency Postgraduate Student Colloquium,* 9–10 September, pp. 111–118.

Guy R. F. E. (1999), "Spherical Coverage from Planar, Conformal and Volumetric Arrays," in *Proceedings of IEE National Conference on Antennas and Propagation,* March 30–April 1, pp. 287–290.

Haruyama T., Kojima N., Chiba I., Oh-Hashi Y., Orime N., and Katagi T. (1989), "Conformal Array Antenna with Digital Beam Forming Network," in *Proceedings of IEEE International Symposium AP-S/URSI,* pp. 982–985, 1989.

Harris J. H. and Shanks H. E. (1962), "A Method for Synthesis of Optimum Directional Patterns from Nonplanar Apertures," *IRE Transactions Antennas and Propagation,* Vol. AP-10, pp. 228–236, May.

Hoffman M. (1963), "Convention for the Analysis of Spherical Arrays," *IEEE Transactions on Antennas and Propagation,* Vol. AP-11 , No. 4, pp. 390–393.

Hsiao J. K. and Rao J. B. L. (1976), "Computation of Far-Field Pattern and Polarization of Conformal Arrays," *IEEE Transactions on Antennas and Propagation,* Vol. AP-24, No. 2, p. 259.

Jedlicka R. P., Poe M. T., and Carver K. R. (1981), "Measured Mutual Coupling between Microstrip Antennas," *IEEE Transactions on Antennas and Propagation,* Vol. AP-29, No. 1, pp. 147–149.

Jha R. M., Sudhakar V., and Balakrishnan N. (1987), "Ray Analysis of Mutual Coupling between Antennas on a General Paraboloid of Revolution (GPOR)," *Electronics Letters,* Vol. 23, No. 11, pp. 583–584.

Jha R. M., Edwards D. J., and Bhakthavathsalam R. (1991), "Application of Novel Method for Surface-Ray Tracing over Hyperboloidal Scatterers," *Electronics Letters,* Vol. 27, No. 17, pp. 1501–1502.

Jha R. M., Bokhari S. A., and Wiesbeck W. (1993), "A Novel Ray Tracing on General Paraboloids of Revolution for UTD Applications," *IEEE Transactions on Antennas and Propagation,* Vol. AP-41, No. 7, pp. 934–939.

Jha R. M. and Wiesbeck W. (1995), "The Geodesic Constant Method: A Novel Approach to Analytical Surface-Ray Tracing on Convex Conducting Bodies," *IEEE Antennas and Propagation Magazine,* Vol. 37, No. 2, pp. 28–38.

Josefsson L. and Persson P. (2001), "An Analysis of Mutual Coupling on Doubly Curved Convex Surfaces," in *Proceedings of IEEE AP-S International Symposium,* Boston, July 8–13, pp. 342–345.

Josefsson L., Persson P., and Lanne M. (2002), "The Polarization Problem in Singly and Doubly Curved Conformal Array Antennas," in *Proceedings of IEEE International Symposium AP-S/URSI,* San Antonio TX, June 17–21.

Josefsson L. and Lanne M. (2003), Private communication.

Ke B. and Kishk A. (1991), "Analysis of Spherical Circular Microstrip Antennas," *IEE Proceedings part H,* vol. 138, pp. 542–548.

Lanne M. and Josefsson L. (2000), "Dubbelkrökt yta. Mätrapport för kopplingsmätningar," Tech. Rep. UA/U-2000:024, Ericsson Microwave Systems AB, Mölndal, Sweden, October. In Swedish.

Lanne M. and Josefsson L. (2002), "Dubbelkrökt yta. Rapport för mätning av aktiv koppling," Tech. Rep. EMW/UA/U-02:003, Ericsson Microwave Systems AB, Mölndal, Sweden, February. In Swedish.

Lima A., Descardeci J, and Giarola A. (1992), "Circular Microstrip Antenna on a Spherical Surface," *Micowave Optical Technology Letters,* Vol. 5, pp. 221–224, May.

Luk K.-M. and Tam W.-Y. (1991), "Patch Antennas on a Spherical Body," *IEE Proceedings part H,* Vol. 138, No. 1, pp. 103–108, February.

Macon C. A., Kempel L. C., and Schneider S. W. (2001), "Modeling Conformal Antennas on Prolate Spheroids Using the Finite Element-Boundary Integral Method," in *Proceedings of IEEE AP-S International Symposium,* pp. 358–361.

Macon C. A., Kempel L. C., and Schneider S. W. (2002), "Modeling Cavity-Backed Apertures Conformal to Prolate Spheroids Using the Finite Element-Boundary Integral Technique," in *Proceedings of IEEE AP-S International Symposium,* pp. 550–553.

Persson P., Josefsson L., and Lanne M. (2001), "Investigation of the Mutual Coupling between Apertures on a General Paraboloid of Revolution: Theory and Measurements," in *Proceedings of 2nd European Workshop on Conformal Antennas,* The Hague, The Netherlands, April 24–25.

Persson P. and Josefsson L. (2003a), "A Study of the Radiation Characteristics Due to Sources on Doubly Curved Surfaces," *IEEE AP-S International Symposium,* Columbus OH, USA, June 22–27.

Persson P. and Josefsson L. (2003b), "Radiation Characteristics of Doubly Curved Array Antennas," in *Proceedings of 3rd European Workshop on Conformal Antennas,* Bonn, Germany, October 22–23.

Persson P., Josefsson L. and Lanne M. (2003), "Investigation of the Mutual Coupling between Apertures on Doubly Curved Convex Surfaces: Theory and Measurements," *IEEE Transactions on Antennas and Propagation,* Vol. AP-51, No. 4, pp. 682–692, April.

Rai E., Nishimoto S., Katada T., and Watanabe H. (1996), "Historical Overview of Phased Array Antenna for Defense Application in Japan," in *Proceedings of IEEE International Symposium on Phased Array Systems and Technology,* pp. 217–221, October.

Saito J. and Heichele L. (1987), "Comparision of Element Arrangements of a Spherical Conformal Array," in *Proceedings of IEEE International Symposium AP-S/URSI,* pp. 125–128.

Sengupta D. L., Smith T. M., and Larson R. W. (1968), "Radiating Characteristics of a Spherical Array of Circularly Polarized Elements," *IEEE Transactions on Antennas and Propagation,* Vol. AP-16, No. 1, pp. 2–7, January.

Sipus Z., Kildal P.-S., Leijon R., and Johansson M. (1998), "An algorithm for Calculating Green's Functions for Planar, Circular Cylindrical and Spherical Multilayer Substrates," *Applied Computational Electromagnetics Society Journal,* Vol. 13, pp. 243–254.

Sipus Z., Burum N., and Bartolic J. (2003a), "Analysis of Rectangular Microstrip Patch Antennas on Spherical Structures," *Microwave and Optical Technology Letters,* Vol. 36, pp. 276–280, February.

Sipus Z., Burum N., and Zentner R. (2003b), "Analysis of Curved Microstrip Antennas," Final Report for Contract F61775-01-WE024 (EOARD), Faculty of Electrical Engineering and Computing, University of Zagreb, April.

Sipus Z., Burum N., and Bartolic J. (2003c), "Theoretical and Experimental Study of Spherical Rectangular Microstrip Patch Arrays," in *Proceedings of the 3rd European Workshop on Conformal Antennas,* pp. 69–72, Bonn, Germany, October.

Sipus Z., Burum N., and Skokic S. (2004), "Analysis of Spherical Arrays of Microstrip Antennas Using Moment Method in Spectral Domain and G1DMULT," in *Proceedings of Computational Electromagnetics—Methods and Applications (EMB04),* Gothenburg, Sweden, October 18–19, pp. 138–145.

Sipus Z. and Burum N. (2005), "Effect of Curvature on Characteristics of Spherical Patch Antennas," in *Proceedings of the 4th European Workshop on Conformal Antennas,* Stockholm, Sweden, pp. 51–54, May.

Tam W. Y. and Luk K. M. (1991a), "Resonances in Spherical-Circular Microstrip Structures," *IEEE Transactions on Microwave Theory Technology,* Vol. 39, No. 4, pp. 700–704.

Tam W.-Y., Luk K.-M. (1991b), "Far Field Analysis of Spherical-Circular Microstrip Antennas by Electric Surface Current Models," *IEE Proceedings part H,* Vol. 138, No. 1, pp. 98–102.

Tam W. Y., Lai A. K. Y., and Luk K. M. (1995), "Input Impedance of Spherical Microstrip Antenna" *IEE Proceedings on Microwaves, Antennas and Propagation,* Vol. 142, 1995, pp. 285–288.

Wills R. W. (1987), "Mutual Coupling and Far Field Radiation from Waveguide Antenna Elements on Conformal Surfaces," in *Proceedings of the International Conference RADAR 87,* London, October 19–21, pp. 515–519.

8

CONFORMAL ARRAY CHARACTERISTICS

This chapter deals with the characteristics of conformal array antennas in terms of radiation patterns, impedance, and polarization characteristics. We will make use of simple models, but also build upon the theories and analyses that were presented in the previous chapters. Furthermore, design issues, including mechanical aspects that can be critical for the practical realization of a conformal antenna system, will be considered. Several typical array configurations will be discussed, including circular, cylindrical, spherical, and conical arrays.

8.1 INTRODUCTION

In previous chapters, we have presented the theories and analysis methods that are applicable to conformal antennas. We have focused on the electromagnetic aspects of radiation and coupling from elementary sources on curved surfaces. Now we turn to the performance of conformal array antennas. Thus, we have to be more specific than before since the array shape, element lattice, and array excitation are among the factors that must be known in order to determine the characteristics of the array. The conformal shapes can be classified as slightly curved (almost planar); singly curved, including ring arrays and cylindrical arrays; and doubly curved. The slightly curved antennas behave more or less like planar antennas and exhibit the same limitations; the design principles are roughly the same. For the other types, it is hard to make general statements, since dimensions,

Conformal Array Antenna Theory and Design. By Lars Josefsson and Patrik Persson
© 2006 Institute of Electrical and Electronics Engineers, Inc.

shapes, element types, requirements, and so on can be so different. However, we will select and analyze typical examples, including case studies from the literature. In this way, we hope to provide a picture of the performance and characteristics of conformal antennas together with an understanding of important factors that govern their performance. Additional information is provided in the Chapters 9 and 10 on beam forming methods and pattern synthesis, respectively.

In order to illustrate the cylindrical case, we will several times use data derived from a comprehensive theoretical and experimental study of a cylindrical C-band array [Persson and Josefsson 2001]. This 600 mm diameter array has 54 azimuthally polarized waveguide aperture elements with 0.53 λ (at 4.3 GHz) spacings in the E plane. There are three rows with 18 elements each, extending over about 120° in the azimuth. Mutual coupling and other data for this array were presented in Chapter 6. Similarly, for the doubly curved case, we will use results from a study of an X-band test antenna with a paraboloidal shape, base diameter 600 mm, and circular waveguide aperture elements [Persson et al. 2003]. More details about this antenna are found in Chapter 7. Figure 8.1 shows the two experimental arrays.

The shape of a conformal antenna is, by definition, related to the mechanical structure with which it will be integrated. Hence, mechanical aspects of the design are very important, especially in the case of structures such as airplanes that are exposed to significant stress. The mechanical design also depends very much on the available technology for packaging and miniaturization of the electronic components, the methods for cooling, signal distribution, and so on. This is an extensive subject that is not possible to treat here in detail. However, we will in the next section present some viewpoints on the problems and possible solutions. In Section 8.3 radiation patterns will be discussed, followed by impedance and polarization properties in Sections 8.4 and 8.5. To conclude the chapter, a number of typical conformal arrays and their characteristics will be discussed, mainly based on literature information. Additional examples are found in Chapters 1, 3, 9, and 10.

8.2 MECHANICAL CONSIDERATIONS

8.2.1 Array Shapes

The strict definition of "conformal antennas" states that a conformal antenna must conform to a shape dictated by, among others, aerodynamical considerations. Thus, the shape is given and there is no room for optimizing the shape any further. However, compromises may, of course, be necessary in the practical case. Take, for example, a nose-mounted array for a fighter aircraft, the shape of which should ideally follow the rather pointed nose/radome shape of the aircraft. As a compromise, considering the limited room close to the pointed tip of the radome, rounded elliptical shapes have been proposed, with a pointed "radome cap" added to give the combination the desired streamlined form [Debionne 1986].

We include in the conformal antenna category also those designs whose shape is based on pattern and/or coverage considerations. The most common shape for wide-angular coverage requirements is the cylindrical shape, and we have included a lot of material on this type in various chapters throughout the book. A closely related shape is the cone, which, just like the cylinder, can be seen as a combination of several ring arrays placed above each other. In fact, the circular ring array is a basic element in all rotationally symmetric arrays. One could consider a building technique in which ring arrays of, in princi-

(a)

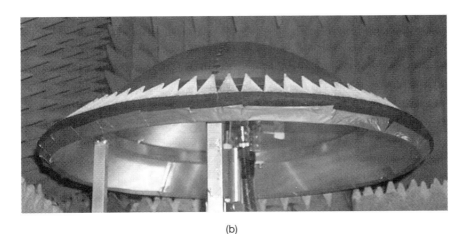

(b)

Figure 8.1. The C band cylindrical array (a), and the X-band paraboloidal array (b). (Courtesy of Ericsson Microwave Systems AB, Göteborg, Sweden.) See also color insert, Figures 2 and 3.

ple, identical designs are joined together; compare [Sindoris and Hayes 1981, Voskresensky and Ovchinnikova 2001].

In Chapter 3, comparisons of several rotationally symmetric antenna shapes for hemispherical coverage were presented. We analyzed performance in terms of gain in relation to cost. As a basis for cost we used the required total number of radiators. Conical shapes came out quite favorably in this comparison. It is interesting to note that chopping off the tip of the cone does not change its cost-effectiveness according to our criterion. Thereby, the problem with installing feed networks, radiating elements, and so on in the narrow tip region can be eliminated. However, a hole in the resulting projected aperture will in most cases lead to unacceptable high sidelobes, so this approach is best suited to, for example, ground-based surveillance and communication systems with 360° coverage (thus excluding the zenith direction).

8.2.2 Element Distribution on a Curved Surface

For grating-lobe suppression the element density should be sufficiently high (roughly half-wavelength spacings) and the elements distributed evenly over the surface. As in planar arrays, a triangular grid is usually preferred over a rectangular grid. However, regular periodic-element grids can only be defined for a few canonical shapes, such as the cylinder. Quasiperiodic grids, however, are possible. Near-uniform spacing of the elements on a sphere can be achieved to within ± 9% [Schrank 1970]. The quasi-periodic lattice can be further randomized, providing some reduction in sidelobe levels [Vallecchi and Gentili 2003].

Elements on a pointed cone (with a small cone half-angle, < 10–15°) can be distributed in a quasirectangular fashion, that is, in rings with an equal number of elements in each ring. This requires that the elements be near the base of the cone. A layout of elements on a blunt cone near the tip region is shown in Figure 8.2. Here, the four arcs of elements have from 7 to 13 elements each; still, the appearance is that of a uniform distribution. See also Figure 8.50.

8.2.3 Multifacet Solutions

Multiple facets can approximate a smoothly curved array quite well. As long as the facets are small, it can be expected that the electrical performance will be close to that of the continuous equivalent array; compare Section 6.4 [Marcus et al. 2003]. The piecewise planarity goes well with multilayer and MMIC technologies for RF front-end designs, and may require less costly tooling compared to curved structures. However, multifaceted antennas do not always meet the desired conformity requirements. A complication with a faceted antenna is the design of the joints between the facets, considering both the electromagnetic analysis and the mechanical design.

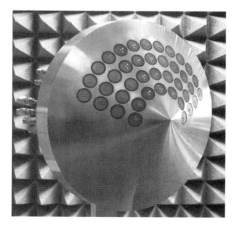

Figure 8.2. Dual-polarized patch elements on a cone. See also color insert, Figure 4. (Courtesy of FGAN Research Establishment for Applied Science, Wachtberg-Werthhoven, Germany.)

Multiple facets approximating the sphere (based on the regular icosahedron) were discussed in Chapter 3. In fact, there are many more possible polyhedra that approximate a sphere if nonregular ones are included. One example is the truncated icosahedron, an Archimedean solid [Weisstein 2004] that has 12 pentagonal and 20 hexagonal surfaces (Figure 8.3). The surfaces can be further subdivided into triangles, resulting in 180 triangular facets [Bondyopadhyay 2000]. These triangles can then again be further subdivided. See also [Fekete 1990].

A small faceted array with a general spherical shape was shown in Figure 1.8. It is based on the dodecahedron and contains 40 triangular facets. Figure 8.4 shows the electronics on the rear side of one of these facets. Each facet holds six dual-polarized radiating elements with transmit and receive functions.

8.2.4 Tile Architecture

In order to realize flush-mounted arrays, there is a need to design the RF front end (transmit/receive functions, control electronics, feeds, etc.) integrated together with the radia-

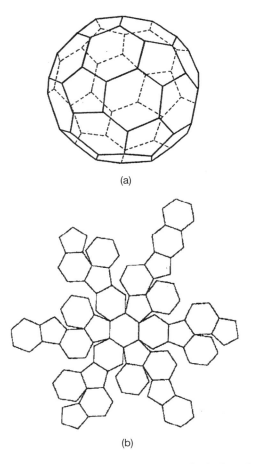

(a)

(b)

Figure 8.3. The truncated icosahedron (a) and its unfolded surface (b) [Bondyopadhyay 2000, 2001].

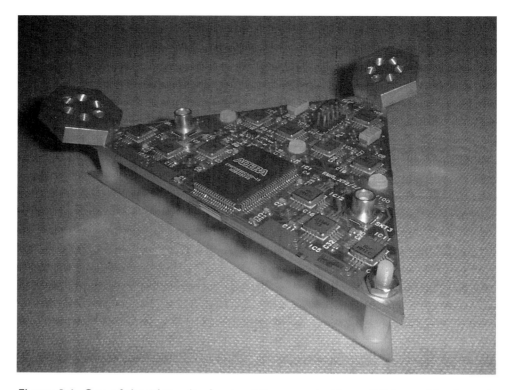

Figure 8.4. One of the triangular facets of the array in Figure 1.8 (rear side shown). Each facet has GaAs MMIC circuits for transmission and receiving functions. (Courtesy of Roke Manor Research Ltd., Roke Manor, Romsey, Hampshire, UK.)

tors in a thin package, usually referred to as *tile packaging;* see, for example [Sanzgiri et al. 1995]. This technology is well under way (Figure 8.5) and most easily realized at lower frequencies where the need (because of the larger dimensions) is also more pronounced. To make enough room for the electrical functions within an approximately $0.5\ \lambda \times 0.5\ \lambda$ cross section, a multilayer design with built-in interconnections between layers is required [Liu et al. 2000]. Tile modules have been designed for high frequencies, including the X- and Ku-bands [Midford et al. 1995, Schreiner et al. 2003]. A significant effort in this direction was the U.S. High Density Microwave Packaging Program [Cohen 1995]. Among the advantages of the tile technology are reduced cost and improved reliability, thanks to the highly integrated design and automated fabrication. Conformal broadband designs with dual polarizations and tile-type modules have been described [Mancuso 2002, Perpère and Hérault 2001, Barbier et al. 2001].

Improvements in optical components for conversion of RF signals to signals on optical fibers and vice versa can in the future result in even more compact and robust antenna modules. This will make remote beamforming possible, and provide increased immunity to electromagnetic interference. For receive-only functions, it could even be possible to use a 100% optical module interface, including an optical power supply fiber [Schaller and Beaulieu 1996].

The tile technology lends itself most easily to small, planar modules with one or rather few radiating elements and associated electronics on each tile; compare Figures 8.4 and

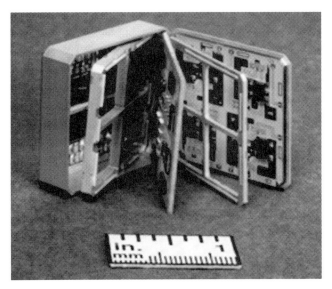

(a)

(b)

Figure 8.5. Four-element tiles. (a) X-band transmit/teceive module [Kemerley and Kiss 2000], (b) S- band receive module. (Courtesy of Ericsson Microwave Systems AB, Göteborg, Sweden.)

8.5. The planar tiles are then put together to form a faceted array that approximates the required shape. However, it is also possible to build truly curved arrays using appropriate materials and processing techniques; compare Figure 1.11. This is still a very advanced technology but some progress has been reported. Radiating elements and circuits produced directly on flexible substrates would allow shaping to the desired structure [Schneider et al. 2001]. Another approach is to first form the substrate layers to the desired shape.

Vacuum forming over a wooden mold to produce a spherical array was reported by [Chu-jo et al. 1990]. Photoetching the required circuits and radiators on a curved surface using a robot-controlled three-dimensional laser for activating photoresist patterns has been reported [Charrier et al. 2003].

8.2.5 Static and Dynamic Stress

Conformal arrays in the strict sense, that is, arrays integrated flush with, for example, the skin of an aircraft or other vehicle, entail a number of mechanical design challenges. The antenna will in this case replace a part of the original structure, so a fundamental question is whether the antenna is able to take the same mechanical loads as that structure in addition to fulfilling the electrical requirements [Lockyer et al. 1999]. Alternatively, a reinforced frame in the fuselage could take the original loads, leaving more freedom for the design of the antenna, which would then only need to withstand its own induced stresses. A general observation is that an opening in the skin of an advanced vehicle is much more expensive than the structural part taken away.

A load-bearing antenna integrated in the fuselage would have to withstand all kinds of mechanical stress, including rain erosion, bird impact, lightning strikes, and so on. A possible design has multiple dielectric layers, each providing the required stiffness, erosion protection, and so on, as well as the electrical functions [Sekora et al. 2003].

Aircraft and other vehicles suffer in operation both static and dynamic deformations that to some extent will be transferred to the structurally integrated conformal array antenna. The static change of the antenna shape is the easiest to analyze [Smallwood et al. 2003]. The effects of dynamic changes (vibrations) are more difficult to evaluate. They typically have implications for the signal characteristics, including antenna-related signal processing functions (adaptive nulling, direction finding, etc.). MTI and SAR are sensitive systems that depend on the spectral purity of the transmitted and received signals [Knott et al. 2003]. Thus, it may be required that the antenna be mounted with means for vibration isolation. Some compensation is possible with electronic phase corrections [Schippers et al. 2003].

8.2.6 Other Electromagnetic Considerations

Apart from analyzing radiation, mutual coupling between the radiating elements, matching the array under scanning conditions, and so on, there are other electromagnetic effects that deserve attention. Electromagnetic compatibility with other near or colocated RF systems is one such issue. As an example, leakage from the feeding systems and other components could propagate and interfere with sensitive parts of the electronics subsystem (analog and digital). One should, therefore, check what electrical modes can be excited in the antenna structure internally and externally. Resonances can affect the antenna function in unexpected ways if not accounted for in the design, as the following case illustrates.

The cylindrical array with aperture coupled patch elements shown in Figure 8.6 was found to support several mode resonances. The cylindrical structure is in fact a multilayer coaxial cavity composed of the metallic cylinder, dielectric layers, and the ground plane for the apertures that feed the patches. The most complex situations arise with dual-polarized patches. Some of the resonances that were examined are attributable to coaxial waveguide modes supported by the coaxial structure [Löffler et al. 2000, Löffler and Wiesbeck 2000]. There are also circumferential resonances in the substrate layers due to surface

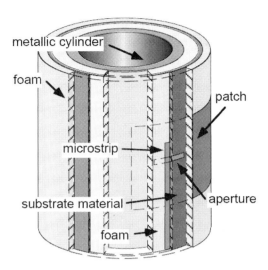

Figure 8.6. Aperture-coupled microstrip patch on a cylindrical structure [Löffler et al. 1999].

waves. The effects on the measured input reflection coefficient and the mutual coupling data can be quite severe. The conclusion is that the critical dimensions must be selected carefully so that undesired resonances do not appear, at least not inside the operational frequency band.

A different mechanical approach, also with aperture-coupled patches on a cylinder, is shown in Figure 8.7. Here, the part between the metal cylinder and the slot ground plane is divided into narrow channels, hence reducing most of the moding problem; compare Figure 3.25.

8.3 RADIATION PATTERNS

8.3.1 Introduction

In a broad sense, the radiation pattern characteristics of conformal arrays are not much different from the radiation characteristics of planar arrays. The same kinds of patterns— shaped beams, monopulse, low sidelobes, and so on—with similar performances can be realized with both types of arrays. However, there are many additional design issues to consider in the conformal case. We also need special methods for analyzing conformal arrays and predicting their performance. Advanced methods for doing this were discussed in detail in Chapters 4–7. In addition, there is a need for simple ways of estimating the performance that can be expected from different conformal antenna configurations. This can be relatively easy for planar arrays, for which there also exists a large knowledge base. Conformal arrays are more difficult to analyze; one complication is the lack of a scalar array factor.

In many early papers on conformal array analysis, rather crude simplifications were necessary. Sometimes cosine (dipole type) element patterns or isotropic patterns were assumed. This can be sufficient for an initial analysis of the general behavior of the radiation patterns. A better approximation is sometimes to compute the element pattern in a

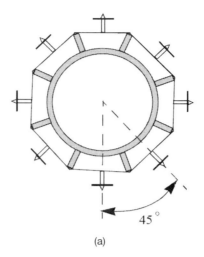

(a)

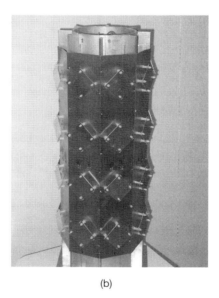

(b)

Figure 8.7. Cross section (a) of a cylindrical array demonstrator (b) with aperture-coupled patches [Raffaelli and Johansson 2003]. See also color insert, Figure 11. (Courtesy of Ericsson AB, Göteborg, Sweden.)

planar environment and then use this for all element positions in the conformal array. The best approach is, of course, to calculate the element patterns in the true curved environment and include the effects of mutual coupling.

However, simple models can provide helpful and easy to understand results in many cases. Some of the calculated results in the following are based on simple models. They serve well our purpose of illustrating fundamental characteristics of conformal array antennas. We will also discuss results from the literature that are typical for various conformal antenna types.

8.3.2 Grating Lobes

The periodicity of planar array aperture excitations means that their Fourier transforms (radiation patterns) also exhibit a corresponding periodicity (with grating lobes, etc.). This is different from conformal arrays. It is true that, for example, a cylindrical array can be periodic in its two dimensions, the axial and the circumferential directions, but there is no simple Fourier relation between excitation and radiation as with planar arrays. See also Section 2.2.

Another difference is the distinction between visible and invisible space. This distinction is very clear for planar arrays. Consider the planar array in Figure 8.8 (a) with the propagation vector $\bar{k} = (k_x, k_y, k_z)$. As long as we have

$$k_z^2 = k^2 - k_x^2 - k_y^2 > 0 \tag{8.1}$$

that is, k_x, k_y inside the circle in Figure 8.8 (b) (visible space), we have radiation of power away from the antenna. The region outside the circle represents imaginary k_z values, and, hence, no radiation, but stored (reactive) energy.

With curved arrays, there is no such clear distinction between visible and invisible space. In Chapter 2, we demonstrated for ring arrays the successive attenuation of phase modes with increasing mode order, which could be interpreted as a transition from visible to invisible space (Figure 2.12). A further difference relates to grating lobes. The periodic, planar array focused to generate a narrow beam produces narrow grating lobes that may radiate or remain in the invisible space. Also, a focused, curved array generates grating lobes. However, since the focusing is for the main beam, the grating lobes, radiating in other directions, are not as well focused and are, therefore, more dispersed and, hence, of less magnitude.

Let us take as a simple numerical example a circular (cylindrical) array with N elements evenly placed on the 360° circumference. Thus, the angular spacing is $\Delta\varphi = 360/N$ degrees. The elements are assumed to radiate only in the outward direction with an element pattern

$$\begin{aligned} EL(\phi) &= \cos\phi & |\phi| &< \pi/2 \\ EL(\phi) &= 0 & |\phi| &> \pi/2 \end{aligned} \tag{8.2}$$

We have from Chapter 2 the radiation pattern of the array:

$$E(\phi) = \sum_n V_n EL(\phi - n\Delta\varphi)e^{jkR\cos(\phi - n\Delta\varphi)} \tag{8.3} = (2.11)$$

where V_n is the element excitation.

In order to scan the main beam in the direction $\phi = \phi_0$, the phase to compensate for the curved shape of the array, cophasal excitation, is

$$V_n = |V_n|e^{-jkR\cos(\phi_0 - n\Delta\varphi)} \tag{8.4}$$

Thus,

$$E(\phi) = \sum_n |V_n|EL(\phi - n\Delta\varphi)e^{jkR[\cos(\phi - n\Delta\varphi) - \cos(\phi_0 - n\Delta\varphi)]} \tag{8.5} = (2.14)$$

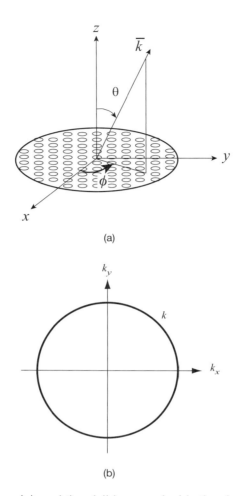

(a)

(b)

Figure 8.8. A planar array (a), and the visible space inside the circle with radius k (b).

If we include all elements in the summation, that is, also those on the "rear side," the rear elements will contribute only to the sidelobe region. It seems, therefore, logical to switch off those elements. Another approach is to use them just for shaping of the sidelobe region.

Figure 8.9 shows the pattern according to Equation (8.5) for $N = 16$, with equal amplitudes and the array phased for a 30° scan angle. The element spacing is half a wavelength. The small circles indicate the locations of the active elements.

If the elements on the rear side are turned off, we get the pattern in Figure 8.10. The two distinct sidelobes at about 120° off the main beam are parts of two grating lobes. They appear, in spite of the half-wavelength spacing, because the edge regions of the 180° active arc are scanned almost to endfire.

A denser element spacing eliminates the grating lobe problem, as shown in Figure 8.11, where we have taken twice as many elements; thus, $d/\lambda = 0.25$. We have now again 16 active elements, but on half the circumference.

Another way to reduce the grating lobes is to scan less, that is, to reduce the active sector. Figure 8.12 shows the result with only a 120° active sector. The element spacing is

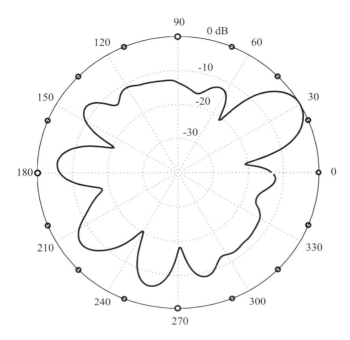

Figure 8.9. Computed pattern with 16 elements and 30° scan angle. All 16 elements are used; the element spacing is 0.5 λ (10 dB between circles).

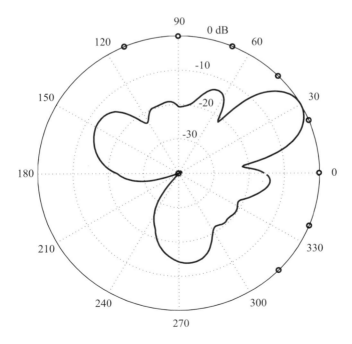

Figure 8.10. As for Figure 8.9, but with rear-side elements not excited.

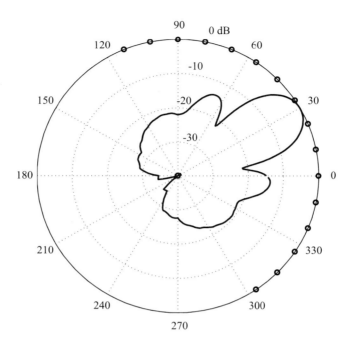

Figure 8.11. As for Figure 8.10, but with the element spacings reduced to 0.25 λ.

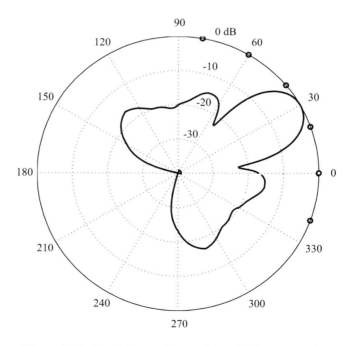

Figure 8.12. Radiation pattern with a 120° sector active.

back to 0.5 λ. This array has now only six active elements, but the beam width is still about the same as in the previous cases and the "grating lobes" have been reduced by about 5 dB compared to Figure 8.10.

Better sidelobe control is obtained with an amplitude-tapered excitation. The common weightings such as Dolph–Chebyshev [Dolph 1946] and Taylor [Taylor 1955] were developed for linear and planar antennas. Other, more general methods for low-sidelobe pattern synthesis are available for conformal arrays and will be discussed in Chapter 10. However, directly applying an amplitude taper according to, for example, Taylor to a cylindrical array can give useful results, as the following examples will demonstrate.

We show in Figure 8.13 three diagrams for a 16 element array with half-wavelength spacing on a 120° arc. The radius is 3.82 λ and the elements have cosine-shaped patterns. The three cases are an untapered case (a), tapered according to a sampled (-20dB, $\bar{n} = 3$) Taylor distribution (b), and a (-30dB, $\bar{n} = 5$) Taylor distribution (c). The patterns for corresponding linear arrays are also shown by dotted lines for comparison.

When we compare the circular arc array and the linear array with the same amplitude tapering, we find that the circular arc array has higher first sidelobes. The beamwidth is also slightly wider. Otherwise, the amplitude taper also successfully reduces the sidelobes in the circular case.

8.3.3 Scan-Invariant Pattern

With planar phased arrays, the radiation pattern changes with the direction of scan. Conformal arrays with rotational symmetry, however, can have scan-invariant patterns in at least one dimension. This is accomplished by moving the excitation around the array, for example, rotation in the azimuth for a cylindrical array. Doing this in steps of one element division each time (commutating) makes the radiation pattern move in corresponding angular increments. For a cophasal circular or cylindrical array, the increments are of about the same size as the azimuthal beam width (see Chapter 2). If better resolution is required, phase scanning can be used between the steps. Scanning by commutating can also be applied to spherical and conical shapes; in principle to all kinds of conformal antennas. However, it is only for rotationally symmetric arrays that we can get scan-invariant patterns. Commutating requires phase control of the radiating elements in order to compensate for the array curvature—the required compensation changes with the beam direction. In addition, elements have to be switched in and out when the active area is moved. Means for doing this will be discussed in Chapter 9.

8.3.4 Phase-Scanned Pattern

The beam can be moved by phase control of the element excitations if the active region is fixed, just like in planar arrays. However, with increasing scan angle the radiating elements on the far side of the active region point more and more away from the desired radiation direction. We can, therefore, expect the pattern shape to vary considerably with the beam direction. (In planar arrays the array factor shape does not (in principle) change with scanning when viewed in the sine space.)

Even with proper phase compensation, the radiating elements on the far side (near B in Figure 8.14) of the active region contribute mostly to sidelobes. The gain will, therefore, decrease and the sidelobe level increase as the scan angle increases. An example is shown in Figure 8.15, where the array geometry and mutual coupling data for the analysis were

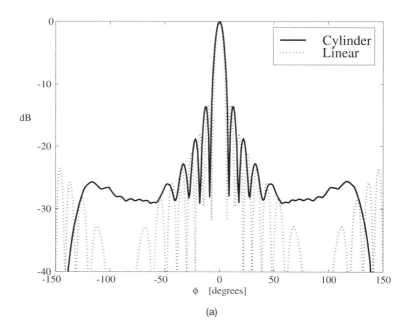

(a)

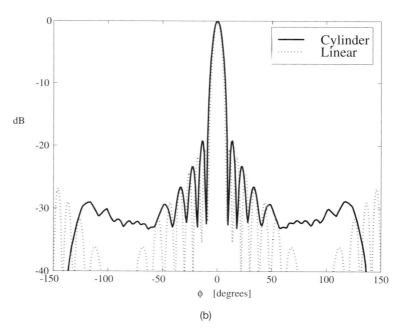

(b)

Figure 8.13. 16 elements with 0.5 λ spacings on a 120° arc. No tapering (a), −20 dB Taylor (b), and −30 dB Taylor (c). *(continued)*

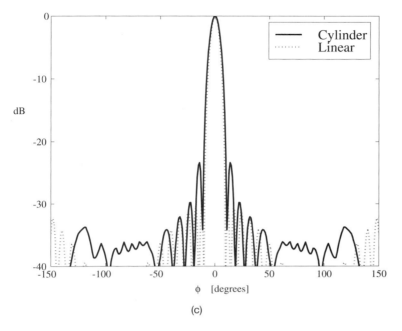

(c)

Figure 8.13. (*cont.*)

taken from the cylindrical array in Figure 8.1 (a). Phase steering with uniform amplitude excitation has been assumed. This example demonstrates an ability to scan to very large angles ($\pm$ 90°) but we also see beam broadening and increased sidelobes [Josefsson and Persson 1999b].

It would seem natural to turn off the elements that contribute so poorly, or at least to reduce their amplitudes in order to improve efficiency. This was tried; see Figure 8.16, where the element amplitudes were determined by an optimization procedure [Bucci et al. 1990]. This method and other pattern synthesis techniques for conformal arrays will be discussed in Chapter 10.

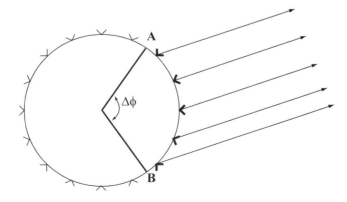

Figure 8.14. Phase steering with a fixed active region (between A and B).

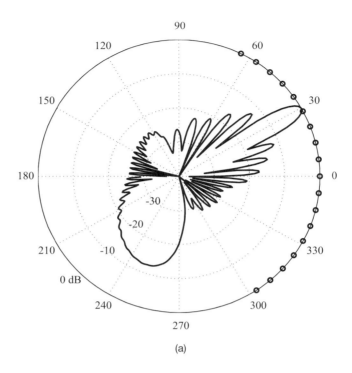

(a)

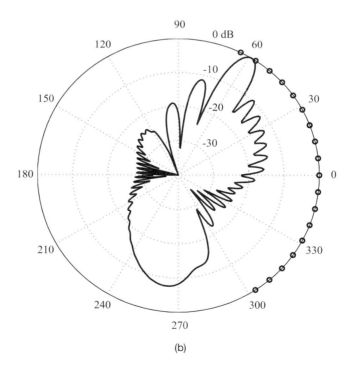

(b)

<u>Figure 8.15.</u> Phase scanning (fixed active region, uniform amplitude). The element locations are marked with small circles. (a) 30°, (b) 60°, and (c) 90° scan angles. *(continued)*

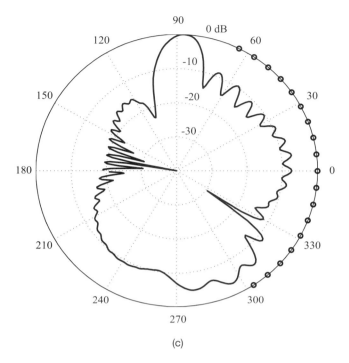

Figure 8.15. (*cont.*)

We notice that it was possible to restrict the sidelobes to about the –20 dB level; and the main beam is distinct, except at the extreme 90° scan angle, where the beam peak appears at about 80°. The calculated directivity with optimized amplitudes was found to be about 2 dB higher at large scan angles compared with the first case with equal amplitudes. The amplitude optimization results in a gradual shift of the most excited elements from the center of the active sector toward the edge (Figure 8.17). This example demonstrates that reasonably good pattern performance can be obtained with phase-steered conformal antennas. However, the price for this is a complex element control including both phase and amplitude (and sometimes polarization).

Similar observations have been made by [Humphrey et al. 1995] for a 19 element microstrip array on a near-spherical surface. In a study of a conical array [Morton and Pasala 2004], –30 dB sidelobes could be maintained when phase scanning to ±74° in the azimuth.

An analysis for a 148° circular arc array with assumed $\cos\varphi$ patterns for 31 elements with 0.5 λ spacing along the arc illustrates a similar shift in the amplitude weightings. The three patterns in Figure 8.18 were optimized for –35 dB sidelobes and 0°, 30°, and 60° scan angles respectively [Boeringer and Werner 2004].

8.3.5 A Simple Aperture Model for Microstrip Arrays

Following [Wait 1959, p. 42] and [Harrington 1961, p. 248] we can write the far field radiation from an aperture with a known field distribution in a perfectly conducting cylinder (see Section 4.3.2.2):

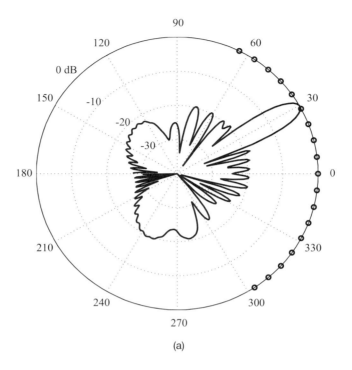

(a)

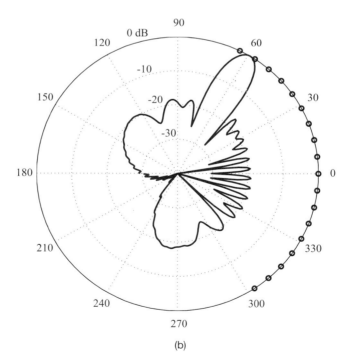

(b)

Figure 8.16. Phase scanning with optimized element amplitudes. Same array as in Figure 8.15 [Josefsson and Persson 1999a]. (a) 30°, (b) 60°, and (c) 90° scan angles. *(continued)*

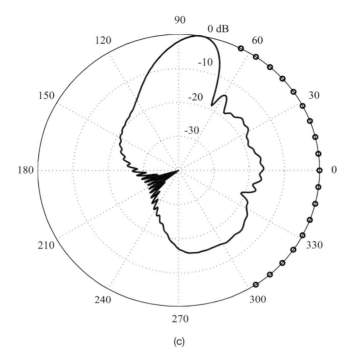

(c)

Figure 8.16. (*cont.*)

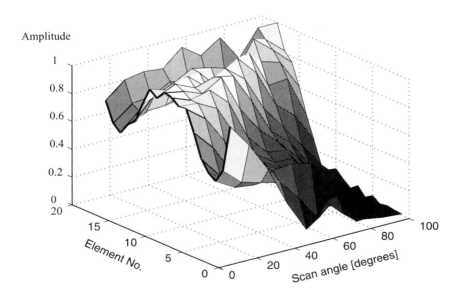

Figure 8.17. The variation of optimized amplitudes for a cylindrical array when the scan angle is changed.

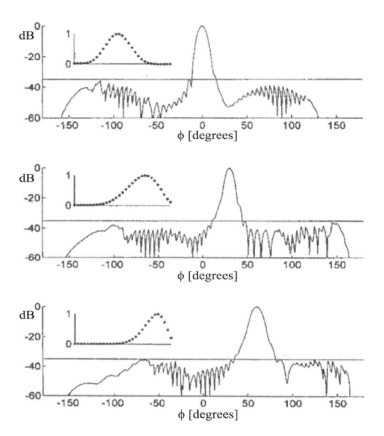

<u>Figure 8.18.</u> Patterns with amplitude weightings (see inserts) for a circular arc array when scanned at 0°, 30°, and 60° scan angles, respectively (top to bottom). After [Boeringer and Werner 2004].

$$\begin{pmatrix} E_\theta \\ E_\phi \end{pmatrix} = \frac{-je^{-jkr}}{\pi r} \sum_{-\infty}^{\infty} j^{n+1} e^{jn\phi} \begin{pmatrix} F_n(w) \\ G_n(w) \end{pmatrix} \tag{8.6}$$

where

$$F_n(w) = \frac{k\widetilde{E}_z}{jvH_n^{(2)}(Rv)} \tag{8.7}$$

$$G_n(w) = \frac{\widetilde{E}_\varphi}{H_n'^{(2)}(Rv)} + \frac{nw\widetilde{E}_z}{Rv^2 H_n'^{(2)}(Rv)} \tag{8.8}$$

and $\widetilde{E}_z$ and $\widetilde{E}_\varphi$ are the Fourier transforms of the aperture field. R is the cylinder radius and

$$\begin{aligned} w &= -k\cos\theta \\ v &= k\sin\theta \end{aligned} \tag{8.9}$$

$H_n^{(2)}$ is the Hankel function of second kind and order n, and $H_n'^{(2)}$ is its derivative with respect to the argument.

This model can be used for an approximate analysis of microstrip-patch antennas. Microstrip-patch antennas can be seen as two narrow-slot antennas carrying magnetic current elements [Derneryd 1976, Calander and Josefsson 2001]; see Figure 8.19.

The slot model for patch arrays was tried on a cylindrical array with 8×2 patches on a 1.6 λ radius cylinder; see Figure 8.20 (c). The patches are etched on a very thin substrate mounted on a 4 mm foam spacer. The center frequency is 5.2 GHz. The computed pattern, Figure 8.20 (a) agrees very well with the measured pattern (b). The excitation in this case was optimized for a beam pointing 22.5° in azimuth. The mutual coupling effects were neglected, but for this particular case, the simple model appears to work fine. More elaborate methods for microstrip patch calculations are discussed in Section 6.6.

8.4   ARRAY IMPEDANCE

8.4.1   Introduction

The active element impedance as seen on the feeding line of a radiating element in an active array depends not only on the element itself but also on the coupling from the other elements. Mutual couplings in various types of conformal arrays were discussed thoroughly in Chapters 6 and 7.

The reflected wave amplitude on the feed line of element number i can be written as

$$V_i^- = V_i^+ \cdot S_{ii} + \sum_{j \neq i} V_j^+ \cdot S_{ij} \tag{8.10}$$

where V_j^+ is the excitation of element j (as given by the feeding arrangement such as feed network, modules, phase shifters, etc). Hence, the active reflection coefficient for element i is

$$\Gamma_i(\theta_s) = V_i^-/V_i^+ = S_{ii} + \sum_{j \neq i} \frac{V_j^+}{V_i^+} S_{ij} \tag{8.11}$$

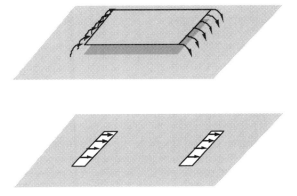

Figure 8.19. Equivalent slot model for a patch antenna.

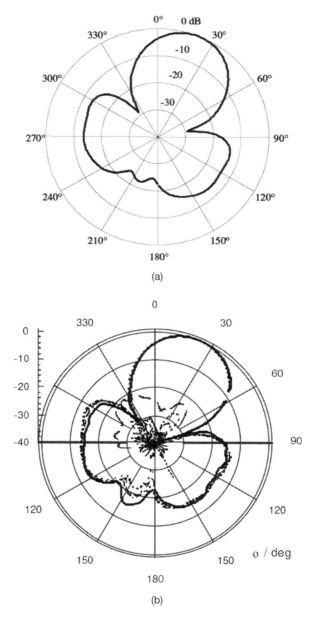

Figure 8.20. Computed pattern using the slot model (a). Measured (co- and cross polar) results at 5.0–5.4 GHz (b), and cylindrical array (c) [Löffler et al. 1999]. *(continued)*

Here, we indicate a dependence on the scan angle θ_s, since the element excitations, at least the phase values of V_j^+, depend on the chosen scan direction. More generally, the active reflection coefficient depends on the complex element excitations (both amplitude and phase). So far, there is no difference between linear/planar and conformal arrays.

Let us compare the curved array and the linear/planar array when they are excited with (a) equal amplitude and phase, and (b) proper phase for a focused main beam; see

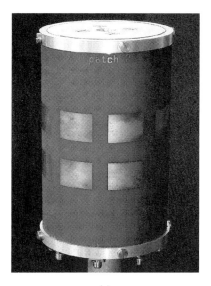

(c)

Figure 8.20. (*cont.*)

Table 8.1. We recognize that the situations (a) and (b) are in principle identical for the linear/planar array. However, for the curved array there is an important difference: the case with a focused beam results in an active reflection coefficient Γ that varies considerably from element to element, whereas for the linear/planar array the reflection coefficient is roughly the same for all elements (except for edge elements). Thus, for the curved case, it can be desirable to apply different matching circuits to different elements in the array, since the active impedances are not as uniform as in the linear/planar case.

We illustrate the above qualitative reasoning in Figures 8.21 and 8.22. The first example is the cophasal pattern in Figure 8.15 for 30° scan angle, which results in the computed active reflection coefficient variations of Figure 8.21. This graph is a polar representation of the complex reflection coefficient (i.e., a Smith chart) showing the values for each of the 18 elements along the 120° curved array. As seen, there are large impedance variations within the array.

Equal amplitude and phase excitation of the same array, on the other hand, gives a broad beam, Figure 8.22 (a), and the impedances in Figure 8.22 (b). All elements have about the same (rather low) reflection coefficient.

Table 8.1. Two typical excitations for curved and linear arrays

Excitation	Curved array	Linear/planar array
(a) Equal phase and amplitude	Broad beam $\Gamma \approx$ equal	Narrow beam $\Gamma \approx$ equal
(b) Focused beam (cophasal)	Narrow beam Γ varies	= above

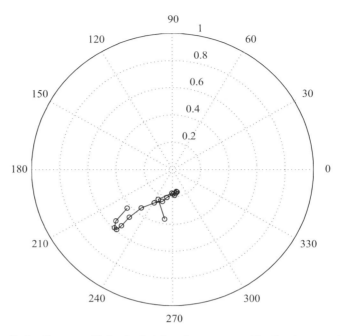

Figure 8.21. Reflection coefficient plot for the array with the beam steered 30° in Figure 8.15.

8.4.2 Phase-Mode Impedance

The element phase-mode impedance is the active element impedance in the array when the elements are excited with a particular phase mode. This impedance depends on the order of the phase mode. We can write the phase mode impedance Z^m for an N-element circular array excited with the mth phase mode [Longstaff et al. 1967]:

$$Z^m = \sum_{k=1}^{N} Z_{ik} e^{2\pi j(k-1)m/N} \qquad (8.12)$$

where Z_{ik} is the mutual impedance between elements i and k. The value of Z^m is the same for all elements, since we assume here that the array is circularly symmetric, hence it is independent of i.

We know from Chapter 2 that phase modes larger than about kR radiate poorly, even if they are excited by the feeding system of the array. What happens is that a nonradiating (reactive) near field is created, and the radiators become severely mismatched to the feeding lines. A simple numerical exercise illustrates this.

Let us consider a circular array of parallel dipole elements. The mutual impedances are easily computed using formulas in several textbooks, for example [Elliott 1981, p. 332; Kraus 1988, p. 424], and we can then evaluate the phase mode impedances from Equation (8.12). The results are shown in Figure 8.23 for an array with 16 elements and $kR = Nd/\lambda \approx 3$. The element spacings were in this case $\lambda/6$ and the dipole lengths $\lambda/2$. Modes up to about the third order can be expected to radiate from this array.

We find that the resistive part of the impedance goes to zero for $m > 3$, whereas the reflection coefficient approaches unity. Also shown in this graph is the Bessel function

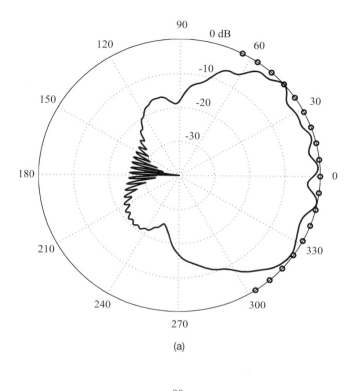

(a)

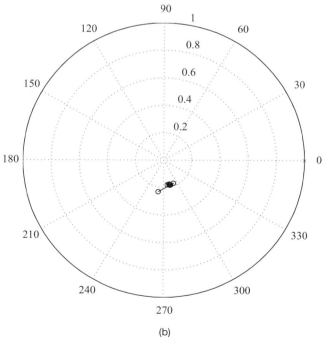

(b)

Figure 8.22. Curved array characteristics with equal amplitude and equal phase. Pattern (a) and reflection coefficient (b).

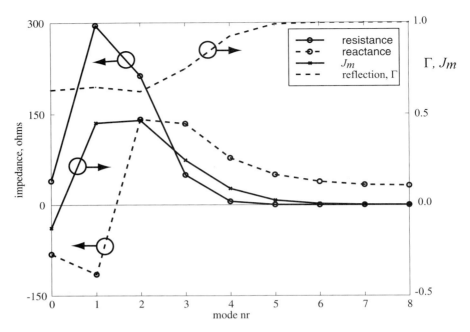

<u>Figure 8.23.</u> Circular dipole array phase-mode impedance performance. Solid line with circles: resistance. Dashed line with circles: reactance. Dashed line: reflection coefficient. Solid line with crosses: the Bessel function $J_m(kR)$.

$J_m(kR)$, which governs the transformation from excitation-phase modes to radiation-phase modes when the elements are isotropic; compare Eq. (2.25). As expected, J_m also approaches zero for $m > 3$. For the reflection coefficient, a feed-line impedance of 73 ohms was assumed.

Jones and Griffiths describe an analysis of a four-element circular array using quarter-wavelength monopoles on a ground plane. The element spacing was $\lambda/6$, corresponding to a kR value of 0.66. Thus, phase-mode orders beyond the $m = 0$ and $m = 1$ would be strongly attenuated. The active phase-mode resistance for the array was calculated to be about 90 ohms, 12 ohms, and about 2 ohms when the array was excited with the zeroth, first, and second phase mode, respectively [Jones and Griffiths 1988, 1989]. We got for the same case (assuming monopoles) 120 ohms, 12 ohms, and 0.6 ohms respectively.

For linear and planar arrays, the active impedance is sometimes presented as a function of scan angle, typically assuming a large or infinite array with uniform amplitude excitation. In this way, the useful scan range for a maximum acceptable reflection coefficient can be illustrated. The corresponding reference case for a circular array is the phase mode impedance, demonstrating the range of useful phase-mode orders. It should be remembered, however, that in practice a combination of phase-mode orders are used in order to synthesize a particular beam pattern [Josefsson and Persson 2005]; compare Section 10.4.

Each single phase mode produces an omnidirectional pattern in the azimuth, but with reduced radiation at high elevation angles when $m > 0$. This has found use in antifading antennas for long- and medium-wave broadcasting systems, where it is desired to suppress the reflected signal from the ionosphere (Section 2.4). Since these designs typically use a small-diameter array (in wavelengths), the radiation resistance can become quite small, as found from Equation (8.12). A low-order phase mode must, therefore, be chosen in order to avoid problems with matching and efficiency [Page 1948, Royer 1966].

For a cylindrical source, the field solution is conveniently written using a cylindrical-wave expansion. The mathematical formulation was given in Chapter 4, and we used a far-field expression in this chapter [Equations (8.6)–(8.9)]. Going back to the general expressions, we can easily obtain the near field in the two-dimensional case, that is, for an infinitely long PEC circular cylinder with a uniform source (along the cylinder, the z axis) [Sensiper 1957]. The source is assumed to be circumferentially polarized, that is, it is characterized by a z-directed magnetic current distribution defined by its Fourier coefficients C_m, the excitation phase-mode amplitudes. Thus, we obtain

$$E_\phi(r) = \sum_{-\infty}^{\infty} C_m \frac{H_m'^{(2)}(kr)}{H_m'^{(2)}(kR)} e^{jm\phi} \tag{8.13}$$

$$H_z(r) = \frac{1}{jZ_0} \sum_{-\infty}^{\infty} C_m \frac{H_m^{(2)}(kr)}{H_m'^{(2)}(kR)} e^{jm\phi} \tag{8.14}$$

Since the cylinder is considered to be infinitely long, these expressions represent the field magnitudes per unit length of the source. If we put r equal to the cylinder radius R, we get the near-field (source) expressions. For the radiated far field, we can use the asymptotic expressions for the Hankel functions as $r \to \infty$:

$$H_m^{(2)}(kr) \to \sqrt{\frac{2j}{\pi kr}}\, j^m e^{-jkr} \tag{8.15}$$

$$H_m'^{(2)}(kr) \to -\sqrt{\frac{2j}{\pi kr}}\, j^{m+1} e^{-jkr} \tag{8.16}$$

Thus, the far-field radiation function expansion [cf. Eq. (2.22)]

$$E_\phi(\phi) = \sum_{-\infty}^{\infty} A_m\, e^{jm\phi} \tag{2.22} = (8.17)$$

becomes in this case

$$E_\phi(\phi) = \sum_{-\infty}^{\infty} \sqrt{\frac{2j}{\pi kr}}\, e^{-jkr} \frac{-j^{m+1}}{H_m'^{(2)}(kR)} C_m\, e^{jm\phi} \tag{8.18}$$

where the amplitude and phase decays of the cylindrical wave are explicitly shown. This gives us a relation between the Fourier coefficients C_m and A_m (for the excitation and the far-field mode amplitudes, respectively; see Section 2.3). Note that this relation is not as simple as the one in Equation (2.26), which was for a case with omnidirectional source elements.

An expression for the radiated complex power density is obtained from the Poynting vector $\overline{S} = \frac{1}{2}\overline{E} \times \overline{H}^*$. Using Equations (8.13), (8.14) we obtain at the source ($r = R$) for each excitation phase mode m

$$(S)_m = \frac{1}{2}(E_\phi H_z^*)_m = \frac{j}{2Z_0} \frac{H_m^{(2)}(kR)}{H_m'^{(2)}(kR)} C_m^2 \tag{8.19}$$

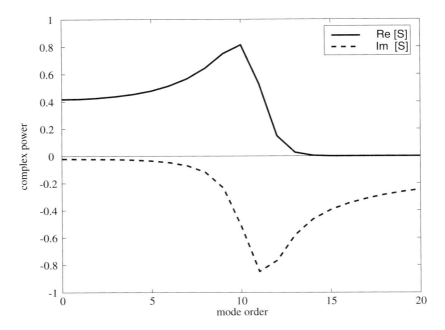

Figure 8.24. Real (solid line) and imaginary (dashed line) parts of the complex power (normalized); also representing the modal conductance and susceptance. Cylinder size $kR = 10$.

Here $Z_0 \approx 377$ ohms is the free-space impedance. The real and imaginary parts are shown in Figure 8.24 as a function of the mode number. The complex power density also represents the wave admittance, with a suitable normalization.[1] Hence, the conductance corresponds to the real part (radiated energy) of S, and the susceptance corresponds to the imaginary part (stored energy) of S [Rhodes 1964, Hill 1967].

A calculation for the same case ($kR = 10$) but for a slightly larger distance is shown in Figure 8.25 (a). Here the power density is computed at $kr = 11.5$, which is about $\lambda/4$ above the source, and we see how the reactive power has decreased significantly. At one wavelength from the source, the reactive power density is less than one-tenth of the value at the source.

In Figure 8.25 (b) we have $kr = kR = 100$, thus representing the complex power of the source field and the modal admittance for a much larger cylinder.

Note that in these cases the admittance is the modal admittance for the radial wave propagating away from the source. The characteristics of the radiating elements with their feeding lines must also be included [Sureau and Hessel 1971] in order to obtain the total active admittance of a particular cylindrical array (as we did for the dipole array example). However, the modal admittance gives significant information for the problem of matching the array, just like the "impedance crater" of infinite planar arrays [Wheeler 1966, Rhodes 1964]. High mode orders correspond to high Q values [Collin and Rothschild 1964]. For the example in Figure 8.24, the Q for $m > 13$ exceeds 100.

---

[1]The admittance is proportional to the total complex power radiated. For each phase mode, the complex power density is uniform in ϕ and z, and, hence, proportional to the complex power. The power density in Figures 8.24 and 8.25 is normalized to make the maximum of $|S_m| = 1$.

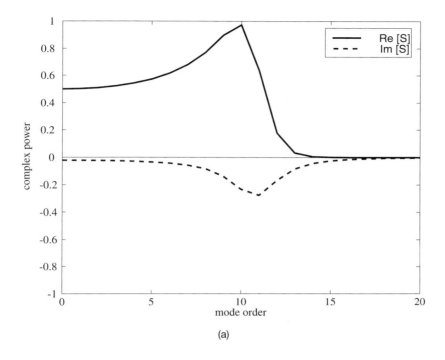

(a)

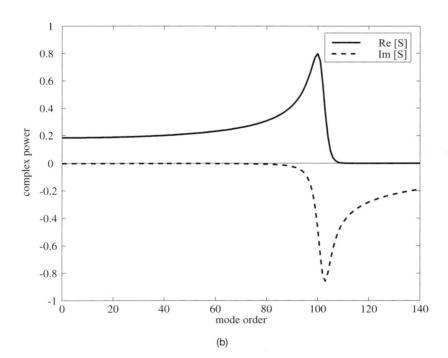

(b)

Figure 8.25. Real (solid line) and imaginary (dashed line) part of the complex power (normalized), also representing the modal conductance and susceptance. (a) Values at $kr = 11.5$ for a cylinder size $kR = 10$. (b) Cylinder size $kR = 100$.

The concept of phase modes finds use also in the analysis of spherical arrays [Witte et al. 2003].

8.5 POLARIZATION

For ordinary planar arrays, the radiation function can be divided into a (vector) element factor and a (scalar) array factor. The radiating elements carry the polarization information. Since the element part can generally not be factored out for conformal arrays, the analysis of the polarization becomes a complicated issue. It is also a complicated practical design issue, since polarization control at the element level often becomes necessary.

The fundamental problem is well illustrated in Figure 8.26 for a slot array on a conical structure. The slots in the figure are oriented for linear polarization in the axial direction, but they will clearly generate a depolarized pattern in other directions.

Even if the radiating elements themselves are free of cross polarization, the curvature of the antenna structure can give rise to cross polarization. If the curvature varies across the array, then the cross polarization contributions from different elements will also change. However, with independent control of two orthogonal radiating components at the element level it is possible to synthesize the radiation pattern for low cross polarization [Douchin and Lemorton 1997].

It has been shown [Gobert and Yang 1974] that the radiation function for an arc or ring array can be written in terms of θ and ϕ components whose amplitudes depend on scalar expressions involving a normalized array factor and its derivatives. The patterns of cylindrical, conical, and spherical arrays can, therefore, be constructed from a superposition of arc patterns. However, the method leads to complicated expressions and is furthermore limited to certain canonical-element types like Hertzian dipoles. Today, it appears more practical to employ direct numerical methods in the analysis.

8.5.1 Polarization Definitions

Polarization is defined by the direction of the radiated electric-field vector [IEEE 1993]. Usually, polarization characteristics are referenced to a particular coordinate system. The classical Ludwig definitions [Ludwig 1973] are shown in Figure 8.27.

Definition 2 (Figure 8.27) can be applied throughout all space, with the exception of the two singular directions along the Y axis. The Definition 2 is the most common

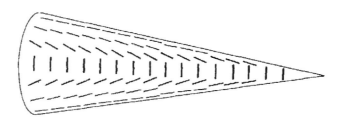

Figure 8.26. A conical array with slot elements oriented for correct linear polarization in the axial direction [Kummer 1974].

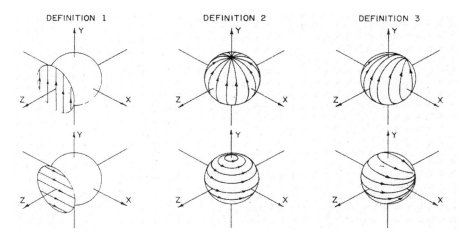

Figure 8.27. Polarization definitions according to [Ludwig 1973].

choice, and will be adopted here. The co- and cross-polar field lines correspond to the orthogonal θ and ϕ directions with Y (as in the Figure 8.27) as the polar axis. The corresponding field components can often be measured directly with a suitable antenna measurement setup.

For cylindrical antennas, the preferred orientation of the coordinate system is to make the vertical axis (polar or polarization reference axis) coincide with the cylinder axis (Y in Figure 8.27). Note, however, that in this book we name this (vertical/cylinder axis) the z axis. For more generally shaped conformal antennas, the choice of coordinates is not so self-evident.

8.5.2 Cylindrical Arrays

8.5.2.1 Dipole Elements.
Let us take a cylindrical or circular array with dipoles as radiating elements. If the dipoles are oriented vertically, that is, parallel to the cylinder axis, then they are all aligned in the same direction and the element factor can be brought outside the summation. The radiated electric field will be θ-directed and there will be no cross polarization (using Definition 2, polar axis = cylinder axis). If, on the other hand, the dipoles are horizontal they will point in different directions and the situation becomes much more complex.

In the coordinate system of Figure 8.28, a single horizontal dipole will have considerable E_θ radiation, that is, cross polarization off the symmetry planes even if the main (horizontal) polarization is ϕ-directed. The cross polarization (E_θ component) for this case is shown in Figure 8.29.

Take a ring array of 12 horizontal dipoles, of which seven dipoles are excited and focused in the azimuth plane. The resulting co- and cross-polarization patterns are shown in Figure 8.30. As we see, there is a considerable cross polarization outside the symmetry planes.

8.5.2.2 Aperture Elements.
Just as the cylindrical or ring array of vertical electric dipoles has no cross polarization, a cylindrical array of thin, horizontally polarized

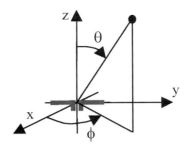

Figure 8.28. Horizontal dipole.

vertical slots has no cross polarization. This case can be seen as an array of vertical magnetic dipoles, analogous to the case with vertical electric dipoles.

It is often of interest to use dual polarizations, that is, polarization diversity in communication systems. Orthogonal components in the radiating elements, such as a combination of vertical and horizontal dipoles (or slots), do not, however, in general produce two orthogonal far-field components. This is apparent from the previous discussion about dipole arrays. We will take as an example a cylindrical array with square waveguide openings as radiating elements. The apertures can be excited with both vertical polarization and horizontal polarization simultaneously and independently. The aperture field distributions are assumed to be TE_{10} and TE_{01} respectively. The array is illustrated in Figure 8.31.

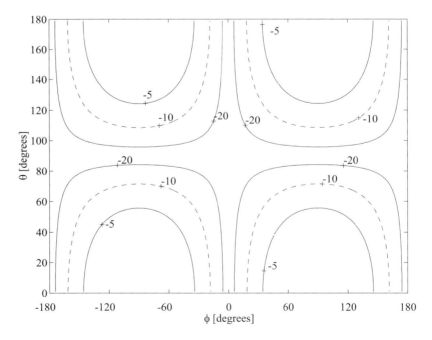

Figure 8.29. Cross polarization levels for the horizontal dipole when the polarization reference axis is vertical. Contour lines at –5, –10, and –20 dB levels.

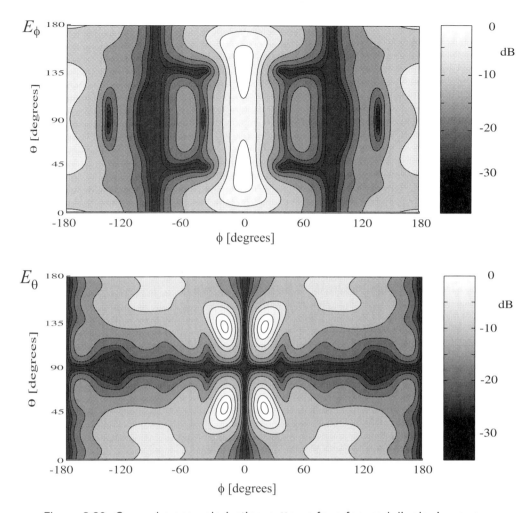

E_ϕ

E_θ

Figure 8.30. Co- and cross-polarization patterns for a focused dipole ring array.

This array was analyzed using the aperture model [Equations (8.6–8.8)]. The polarization for a beam steered 30° in elevation and 30° in the azimuth from the fixed 120° active sector is shown in Figure 8.32. The beam position is indicated by –3dB and –10 dB contour lines. Two contours are drawn, for vertical and horizontal element excitations, respectively, since the beam shape is slightly dependent on polarization. The element excitations (vertical and horizontal) could be expected to produce orthogonal linear polarizations in the far field. The source distribution for the horizontal polarization can be seen as a distribution of vertical magnetic (dipole) currents, and, hence, no cross polarization results. This corresponds to the horizontal lines in Figure 8.32. Vertical polarization, however, has a source of horizontal magnetic currents, resulting in considerable cross polarization, as indicated by the small ellipses. Only along the symmetry line where $\theta = 90°$ do we have pure vertical polarization and high isolation between the two components. In the center of the main beam, the isolation between the components in the far field was computed to be about 13 dB [Calander and Josefsson 2001].

Figure 8.31. Dual polarized circular cylinder array with square waveguide aperture radiators, spacings 0.5 λ. Six of the 18 columns are active, tapered to −20 dB sidelobes, and phase steered.

8.5.3 Polarization in Doubly Curved Arrays

The polarization issue is even more complicated for doubly curved arrays, compared to cylindrical arrays [Josefsson et al. 2002]. Since in general the radiating elements point in different directions, they must be individually polarized in order to generate a desired polarization in the main beam direction. Thus, for linear polarization the elements have to be

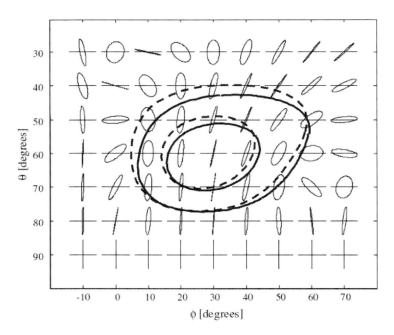

Figure 8.32. Polarization chart for a beam scanned 30° both in elevation and azimuth. The −3 dB and −10 dB contours of the main beam are shown. The solid and dashed contours represent horizontal and vertical polarization, respectively.

properly oriented in order to produce the desired (linear) polarization. Mechanical rotation can in practice be replaced by electronic means; see Section 8.5.4. With circular polarization, however, rotation is not necessary since the rotation does not change the polarization. If not correctly oriented, there will instead be a phase error, but this can be compensated for by the phase control that must be applied to each element in any case.

8.5.3.1 A Paraboloidal Array.
We take as an example an array of elements on a general paraboloid of revolution (GPOR). The radiating elements are assumed to have element patterns corresponding to those of short dipoles in free space (cosine patterns) and their polarization characteristics are also assumed to be those of short dipoles. Mutual coupling is neglected.

Since it is not possible to create a regular lattice of elements on a doubly curved surface, an approximation according to the following rule is used. A minimum spacing d along the surface is defined from grating-lobe considerations (typically $d \leq 0.5\lambda$). One element is placed at the vertex position. The first ring of elements is spaced d from the first element. Within such a ring, as many elements as possible are placed at regular angle (φ) intervals. However, the spacing along the ring is not less than d. The procedure is then continued with the next ring, and so on. The "first" elements of each ring are initially positioned along the same generating curve (at the same angle, thus on a parabolic arc). However, every second ring is then rotated one-half interelement spacing in order to spread the elements more evenly. An example of an element layout for a 14-ring array on a paraboloid as seen from the symmetry axis is shown in Figure 8.33.

In order to spread the "first elements" evenly along the parabolic arc (at constant φ) we need to know the total parabolic length from the vertex. This length normalized to the focal distance is

$$\frac{L}{f} = v\sqrt{1 + v^2} + \ln(v + \sqrt{1 + v^2}) \tag{8.20}$$

where f is the focal distance and $v = r/f$ is the (normalized) radial distance from the axis of the paraboloid. The element locations can be found iteratively from this equation. An example is shown in Figure 8.34.

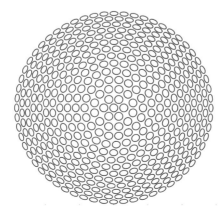

Figure 8.33. Element layout with about 500 radiators.

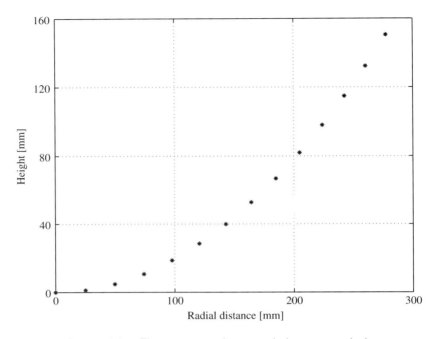

Figure 8.34. Elements evenly spaced along a parabola.

The active region is defined from the desired nominal beam direction $\hat{r}$ and the normal direction to the surface $\hat{n}$. The active region is obtained as the area encompassing all elements for which

$$\hat{r} \cdot \hat{n} \geq \cos \theta_{s\,max} \tag{8.21}$$

where $\theta_{s\,max}$ is a chosen maximum useful local scan angle, typically 45–60°. An example of an active region is shown in Figure 8.35.

In our model, the radiating elements are of the short-dipole type, that is, the element pattern is $E_0 \sin \theta$ with θ the angle from the dipole axis. The elements do not radiate "through" the parabolic surface, however. The dipoles are tangential to the curved surface and initially oriented in a vertical plane; see Figure 8.36.

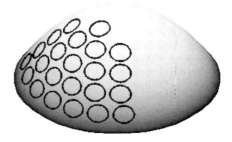

Figure 8.35. Active region for 60° scan angle (from zenith) and max 45° element scan angle.

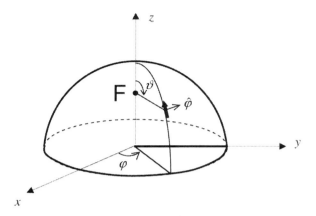

<u>Figure 8.36.</u> Dipole element at coordinates (ϑ, φ). F is the focal point of the parabolic surface.

In order to align the far-field contributions properly, each dipole is rotated until the far-field contribution in the main beam direction has the correct polarization. Additionally, the phase of each dipole is adjusted so that they add coherently to a plane phase front. However, it is generally not possible to obtain the same polarization throughout the pattern; thus, there will be some cross polarization. Figure 8.37(a) illustrates how the radiating elements are all aligned for the same (vertical) polarization in

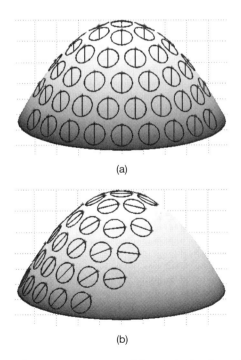

(a)

(b)

<u>Figure 8.37.</u> Radiation elements are rotated for polarization alignment in the main beam. The array is shown at 0° and 90° azimuth angles.

the main beam direction. When viewed from another angle (b), it is clear that a certain degree of cross polarization will radiate in the sidelobe region; compare Figure 8.26.

A first example in Figure 8.38 shows co- and cross polarization patterns for an array on a parabolic surface with h/D (height/base diameter) ratio of 0.1, thus relatively flat. The elements are dipoles arranged in six rings with 0.67 λ element spacing. The reference polarization definition is that of Ludwig Definition 2 (Figure 8.27) with horizontal polarization axis. The beam is scanned 45° in the H plane. The co-polar E_θ and cross-polar E_ϕ patterns in a cut perpendicular to the scan plane through the main beam are shown in Figure 8.38. (In the diagonal planes, the cross-polar level is about 10 dB larger.)

The next example is a larger array with 514 elements (Figure 8.33) scanned 30° from the zenith. The active region has 382 elements. The element spacing is 0.75 wavelengths and $h/D = 0.293$. This array corresponds to the experimental antenna in Figure 8.1(b), if fully populated. There is no amplitude taper. A three-dimensional pattern plot in (u, v) space of the co-polar field (theta component) is presented in Figure 8.39. The (u, v) coordinates are the normalized directional cosines (cf. Figure 8.8), hence

$$u = k_x/k_0 = \sin\theta\cos\phi$$
$$v = k_y/k_0 = \sin\theta\sin\phi$$

The co- and cross-polar components are shown in Figure 8.40 for a cut through the main beam. Note that essentially pure polarization is obtained in the main beam since the dipole elements were rotated to achieve this. However, the grating lobe, although in principle a replica of the main lobe, does not perform in the same way and demonstrates a significant cross-polarized component.

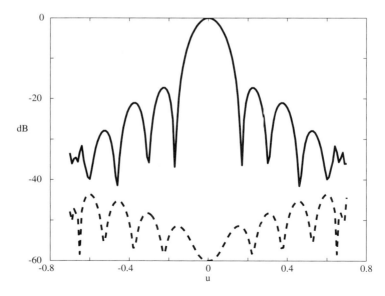

Figure 8.38. Radiation patterns (decibels vs. sine space) for a six ring paraboloidal array, $h/D = 0.1$. The solid line is the co-polar pattern and the dashed line the cross-polar pattern.

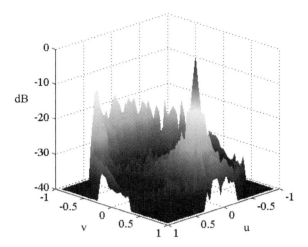

Figure 8.39. The E_θ component of the radiated field in (u, v) space.

In the next example, we take an array with just five rings of elements. The beam is directed towards the zenith and we have dipole elements as before; the Ludwig Definition 2 reference polar axis is horizontal. Figure 8.41 shows patterns with $h/D = 0.5$ and Figure 8.42 shows results with a reduced height, $h/D = 0.05$, thus there is less curvature.

By comparing the two cases, we see that the cross-polarized field dropped from around −20 dB to around −30 to −35 dB when we made the surface flatter. If it were perfectly flat,

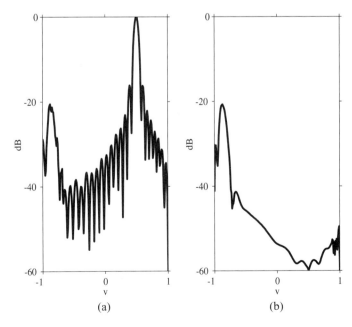

Figure 8.40. The E_θ (a) and E_ϕ (b) components in a cut through the pattern for $u = 0$ in Figure 8.39.

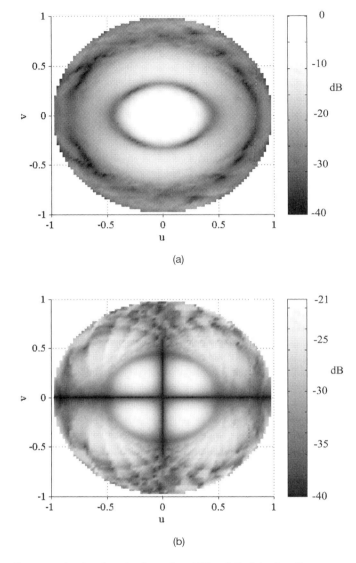

Figure 8.41. Patterns in the (u, v) plane for $h/D = 0.5$. (a): the E_θ component (co-polar), (b): the E_ϕ component (cross-polar).

there would be no cross polarization at all. ($h/D = 0.05$ corresponds to a total height of 3 cm on a diameter of 60 cm.)

Next, we show a case with the beam scanned 30° from the zenith. Other data are: six rings with crossed dipoles, the active region comprises 41 out of total 74 elements, and $h/D = 0.5$. The element spacing is $d/\lambda = 0.5$.

The cross polarization isolation illustrated in Figure 8.43 is calculated in the following way. Assume that the two fields are represented by their respective polarization unit vectors, $\hat{e}_1$ and $\hat{e}_2$. The isolation is obtained by the scalar product of the two polarization vectors; see *IEEE Standard Definition of Terms for Antennas* [IEEE 1993].

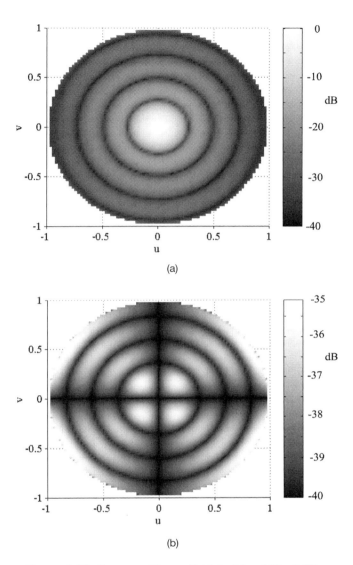

(a)

(b)

Figure 8.42. Same as Figure 8.41, but for $h/D = 0.05$.

The case in Figure 8.43 demonstrates a rather poor isolation in the far field between the two field components. This occurs in spite of the dipoles in each element being perfectly orthogonal to each other. It is a direct consequence of the characteristics of the dipole far-field radiation. In the limit for a flat surface, the isolation would be the same as for a single element with crossed dipoles; see Figure 8.44. Many elements have dipole characteristics. The polarization results shown here indicate that other element types with better polarization performance (better symmetry) would be desired.

An ideal element is the so-called Huygens source, which is a combination of electric and magnetic dipoles. With this element, cross-polarization free patterns can be synthesized in theory [Schmid 2001]. In practice, with real elements, we have also to consider the effects of mutual coupling and mismatch.

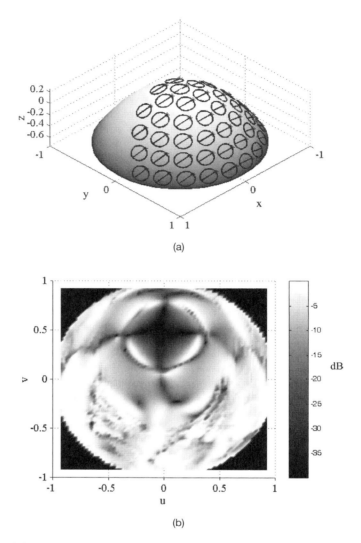

(a)

(b)

Figure 8.43. (a) The array configuration (first set of dipoles). (b) Shows the cross polarization expressed as isolation (in decibels) between the two fields that result from, respectively, the first set of dipoles and a second set of dipoles. The second set consists of dipoles that are orthogonal to the respective dipoles of the first set at each element location.

8.5.4   Polarization Control

In the previous examples, we assumed that the element polarization could be rotated in arbitrarily small steps. In practice, discrete steps with digital control are used. Changing between vertical and horizontal linear polarizations is here termed single-bit control; if the $\pm 45°$ planes are also included, we get two-bit control and so on (see Figure 8.45).

Figure 8.46 shows the cross-polarized component E_ϕ for a paraboloidal array, calculated with 1 and 3 bit polarization resolution in a cut through the main beam scanned at 30°.

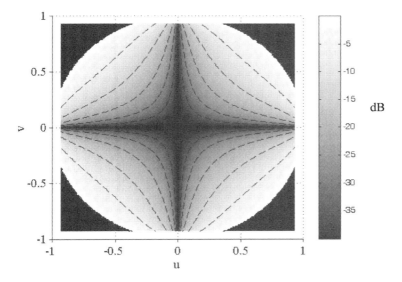

Figure 8.44. Cross polarization isolation for a dipole pair.

Three-bit resolution is not far from the ideal in this example and the cross polarization for this case is mainly due to the curvature alone. Other data are 5 rings and $h/D = 0.5$.

With two-port elements, the radiated polarization can be set by feeding the two ports with proper amplitudes and phases. There are many possible element designs: crossed dipoles, crossed slots, and crossed notch elements. Orthogonal feeds can be used in many radiators such as patches, square and circular waveguides, and so on. Let us assume that the two ports of a suitable element give vertical and horizontal polarizations, respectively. Feeding the two with equal amplitudes results in 45° slant polarization; any orientation can be obtained by a proper amplitude ratio. A variable power divider (VPD) is a suitable circuit for realizing this. A VPD can take the form shown in Figure 8.47—an equal power divider followed by two phase shifters and a 90° hybrid. The V and H polarizations (Figure 8.47) are obtained by setting the phase shifter difference to $\phi_1 - \phi_2 = +90°$ and $-90°$, respectively. Thus, each phase shifter needs only to be set from −45° to +45°. A larger range can be useful if the phase shifters also control the element phase settings.

In order to control all possible types of polarization, including circular and elliptical, a phase shifter in one of the output arms with a range of ±90° is needed. To preserve sym-

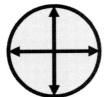 

Figure 8.45. Illustrating single-bit and two-bit polarization control.

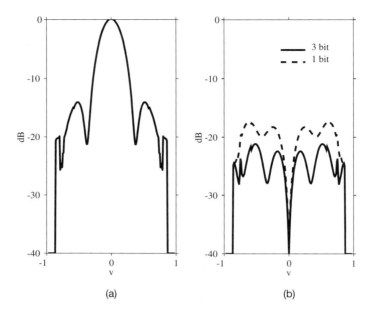

Figure 8.46. Co-polarization (a) and cross polarization (b) with different polarization control resolutions.

metry, one in each arm can be used. Thus, the polarization control function requires a lot of hardware. For receiving (RX), this can be implemented in an MMIC front end at marginal cost. However, for minimum loss in order to preserve noise performance, dual amplifier chains may be required. For transmitting (TX), dual transmit modules may also be desirable in order to minimize losses. With a suitable architecture, the VPDs can improve the isolation between the TX and RX functions.

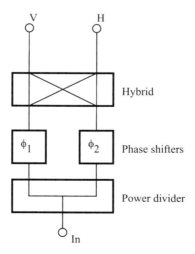

Figure 8.47. Variable power divider (VPD).

8.6 CHARACTERISTICS OF SELECTED CONFORMAL ARRAYS

8.6.1 Nearly Planar Arrays

Arrays with only a slight curvature perform in general like their planar counterparts. However, except for very small arrays, the nonplanarity requires at least some phase compensation. The curvature precludes the use of a common element factor in the analysis/design if low sidelobes are of interest. An example is given by the slightly curved circular arc array with 16 dipoles that was studied by Jiao and coworkers [1993]. This array extended over about 36° of the full circle. Mutual coupling was included by using a model by Herper and coworkers [1985] for the embedded-element patterns. Figure 8.48 shows results from a synthesis for –40 dB and –60 dB sidelobes with the main beam scanned 20° off the symmetric direction.

8.6.2 Circular Arrays

The circular array antenna is the basic building block of many conformal antenna arrays; we have discussed its characteristics previously. Some additional examples will be given in Chapters 9 and 10. Circular arrays have found use in, for example, omnidirectional communication systems, signal direction of arrival (DOA) measurement systems (see color plate 7), and navigation systems. Classical examples include the despun arrays in the Telstar and Meteosat satellites. Another classical example is the Wullenweber antenna [Gething 1966].

8.6.3 Cylindrical Arrays

Cylindrical arrays are among the most common of the conformal antenna species. Beam steering in elevation by phase control can be combined with azimuth steering by commutating; the latter causing no pattern degradation. Other beam steering principles are discussed in Chapter 9. Cylindrical arrays have found use in all fields including radar, communication, and sonar (see color plates 10 and 11).

Radiation pattern synthesis based on phase modes (cf. Chapter 10) is suitable for circular geometries like the cylindrical array. –40 dB sidelobes in the azimuth have been synthesized for a 3×60 waveguide aperture array on a 10 λ diameter cylinder [Martin and Pettersson 2003]. In this case, the array extended all around the cylinder.

8.6.4 Conical Arrays

Conical arrays can often use the same beam steering methods as cylindrical arrays. An example in which the conical shape matches the communication needs in a satellite application is shown in color plate 5.

Conical arrays can be of particular interest for applications in the nose of streamlined airborne vehicles, rockets, and missiles. A pointed conical shape offers wide-angle coverage, low radar cross section (RCS), and good aerodynamic performance. A small conical angle, however, poses problems with the installation of electronics in the tip region and the radiation performance in the forward direction is poor.

Arrays on less pointed cones or where the tip region is not used have a potential for hemispherical coverage (cf. Chapter 3), for instance, on ships, where the array could be mounted around a mast carrying other equipment. With the normal of the cone surface pointing about 30° above the horizon, there will also be some RCS advantages.

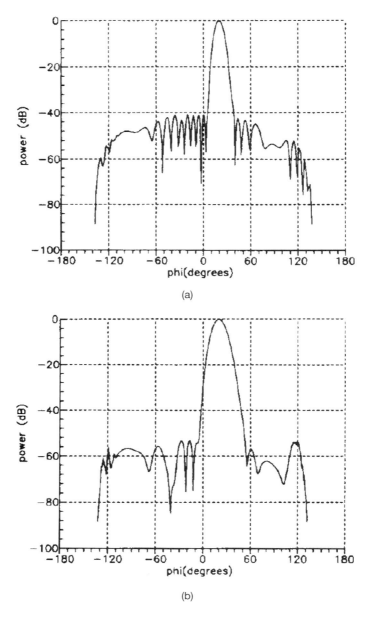

Figure 8.48. Theoretical patterns synthesized for –40 dB (a) and –60 dB (b) sidelobes. A 36° circular arc dipole array scanned 20° [Jiao et al. 1993].

A small conical array (0.5 λ length) with a total cone angle of only 11.5° was described by Thiele and Donn [1974]. With this small size it was possible to cover the forward ±60° region over a 2:1 bandwidth. The radiating elements were four small loop antennas mounted around the base of the cone and progressively phased in steps of $\pi/2$ around the base (phase mode $m = 1$; cf. Chapter 2). The small dimensions imply that the cone itself can be seen as the antenna, excited by the array elements.

A linear array of 65 slots along one generator of a metallic cone with (full) cone angle 30° has been described by Goodrich and coworkers [1959]. Phase steering of the beam was demonstrated for angles from 10° to 70° measured from the cone peak direction. The measured patterns agreed well with calculations; contributions from the slots included a geometric optical and a creeping wave term. Mutual coupling was not accounted for. It was shown that the tip diffraction could be neglected in this case; however, the nearest slot was as far as 16 wavelengths from the tip.

A similar scanning capability was demonstrated experimentally for an eight-element slot array along a generator on a 20° (full angle) cone [Villeneuve et al. 1974] (Figure 8.49). The problem with cross polarization was addressed in a related study [Villeneuve and Kummer 1981], in which 37 individually fed crossed slots on a 20.5° (full angle) cone were investigated. Linear slot arrays on elliptic cones have also been studied by Kaifas and coworkers [2003b].

A sector of a large conical array with dipole radiators was studied in the early 1970s [Boyns 1972, Munger et al. 1974]; see Figure 8.50. This array had 14 elements in each of the 50 columns; hence, 700 elements in total. The vertically polarized dipoles had a horizontal spacing that varied with the vertical position, from 0.42 to 0.53 λ. In order to excite the array, a space feed arrangement with horns illuminating the rear side of the array was used. Pick-up dipoles on the rear side connected via delay lines to the front face dipoles; compare the dome antenna (Figure 1.7). As expected, the peak measured gain was obtained at an elevation angle equal to the cone half angle, 25°. A cross polar component at the relative level of –22 dB in the radiation pattern was noted, caused by the dipole elements not being parallel to each other.

A much later study on a similar topic was presented in 2004 by Morton and Pasala [2004]. In their theoretical investigation, slot elements on a cone with a half-cone angle of 11° were considered. The slot elements were arranged for vertical polarization in nine 100° arcs with 13 slots in each, thus totaling 117 radiators. The theoretical excitation was optimized using a least mean squares method with sidelobe constraints. Since the slots are not parallel, due to the array curvature, some cross polarization is inevitable. At boresight scan, that is, normal to the array surface, the cross polarization was calculated to be 30 dB below the co-polarized main beam, but increasing with increased elevation scan.

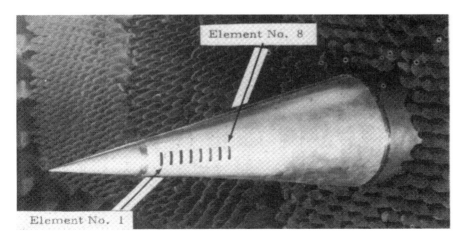

Figure 8.49. Linear slot array on a metallic cone [Villeneuve et al. 1974].

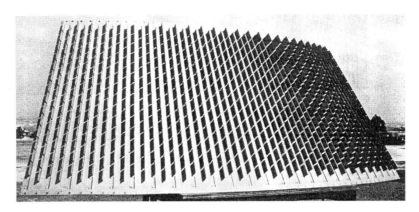

<u>Figure 8.50.</u> A sector of a conical array with 700 dipole elements [Munger et al. 1974].

8.6.5 Spherical Arrays

Spherical arrays represent another canonical case of great interest. One objective may be the realization of hemispherical coverage, although other shapes can also be used to obtain this; see Chapter 3. A near-uniform element distribution is possible. An attractive feature of spherical arrays is that all radiating elements see essentially the same environment, thus reducing the design effort. Pattern characteristics remain unchanged as the active region is moved over the spherical surface.

In an early study by Sengupta and coworkers [1968], radiation patterns from a spherical array were analyzed. To simplify the analysis, the elements had circular polarization with a cosine amplitude pattern. Uniform element excitation was assumed and mutual coupling was neglected. The element distribution was made as uniform as possible, starting from the 20 triangles of an icosahedron. On each triangle, the elements were evenly distributed and then projected onto the surface of a circumscribed sphere. On a complete sphere, there were 162 elements in total.

An active area was determined from the maximum-element look angle. For each beam direction, the element phase was adjusted to produce a plane wavefront and, hence, maximum directivity. As the beam direction was varied it was noticed that the sidelobe levels changed a few decibels as the active region moved over the discrete array surface. Figure 8.51 shows two typical patterns in two orthogonal cuts for a case in which the active region spanned 90° with 31 active elements on a sphere with a radius of 1.5 λ. The first sidelobe level was typically in the range of 17–20 dB below the main beam peak.

The patterns resemble the patterns from a uniformly illuminated plane circular aperture, with about an 18 dB first sidelobe ratio. With large element spacings, there are, of course, grating lobes, somewhat smeared out and not as high as the main beam. If the active region is enlarged, the directivity increases, unless the element spacing becomes too large, leading to increased losses in the grating lobes.

In [Saito et al. 1987], the grating lobe levels for spherical arrays were investigated. The problem was simplified to a scalar one, neglecting polarization and mutual coupling, and assuming cosine element patterns with uniform amplitude. As expected, a triangular element arrangement proved to be better than a rectangular one. For a hemisphere with a radius of 5 wavelengths and 316 elements in a "triangular" lattice (element spacing $d/\lambda =$

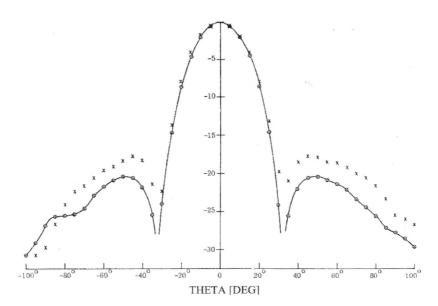

THETA [DEG]

Figure 8.51. Radiation pattern in two orthogonal cuts (solid line and crosses) for an array with 31 uniformly excited elements on a spherical surface with a 90° active region [Sengupta et al. 1968].

0.66), the typical grating lobe level was found to be –24 dB relative to the main beam; see Figure 8.52.

In a recent study, arrays of microstrip patches on a sphere were analyzed [Sipus et al. 2003]. The element impedance and radiation patterns were rigorously calculated using the method of moments for the multilayer spherical structure, but mutual coupling was not included. The element layout followed the icosahedron principle; an example is given in Figure 8.53. Resulting array patterns using 6, 16, and 31 elements with the array active area on the sphere corresponding to subtended angles 15, 30, and 45°, respectively, are shown in Figure 8.54.

A spherical array application for a satellite data link (see Figure 8.55), has been described by Stockton and Hockensmith [1977, 1979]. A hemispherical coverage with a steered moderate gain beam (7–15 dB) was required. In the design, a cluster of 12 circularly polarized patch elements out of a total 120 on a 2.7 λ radius sphere were activated at a time. This subarray was excited with equal phase and amplitude. Beam steering was accomplished simply by shifting the active part in small steps across the spherical array.

It might be of theoretical interest to note that the radiation pattern from a uniformly excited thin spherical shell with radius R and phased for maximum radiation in the direction $\theta = 0$ is simply a sinc function [Voles 1995]:

$$E(\theta) \propto 4\pi R^2 \, \frac{\sin(\alpha R)}{\alpha R} \tag{8.22}$$

where

$$\alpha = \frac{4\pi}{\lambda} \sin(\theta/2) \tag{8.23}$$

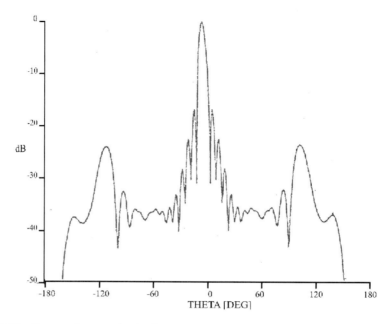

Figure 8.52. Pattern for a spherical array with 316 elements spaced 0.66 λ in a triangular lattice; from [Saito et al. 1987].

8.6.6 Paraboloidal Arrays

Paraboloidal arrays with a large length-to-base diameter ratio are similar to conical arrays, excepting the tip region. With a smaller ratio, on the other hand, the shape approaches that of a spherical surface. Thus, analysis models based on paraboloids can approximate many typical shapes of interest. However, not many investigations of paraboloidal

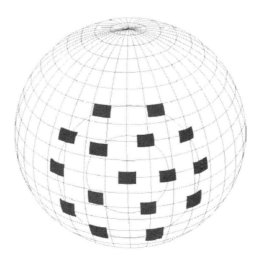

Figure 8.53. 16 patch elements on a sphere [Sipus et al. 2003]. Copyright 2003 Wiley Periodicals, Inc. Reproduced by permission of John Wiley & Sons, Inc.

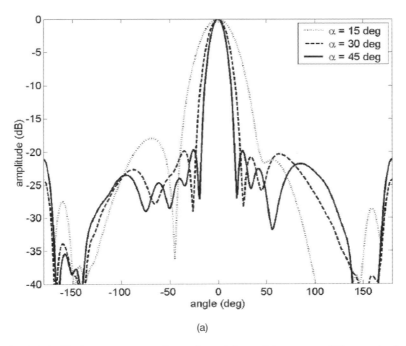

(a)

Figure 8.54 (a). *H*-plane patterns for microstrip patch arrays [Sipus et al. 2003]. Copyright 2003 Wiley Periodicals, Inc. Reproduced by permission of John Wiley & Sons, Inc.

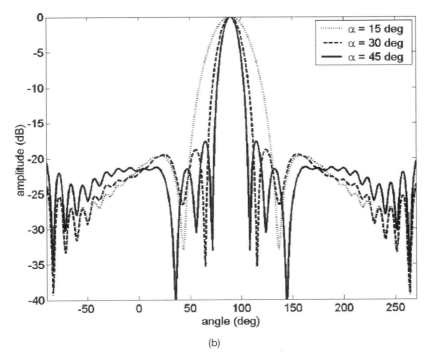

(b)

Figure 8.54 (b). *E*-plane patterns for microstrip patch arrays [Sipus et al. 2003]. Copyright 2003 Wiley Periodicals, Inc. Reproduced by permission of John Wiley & Sons, Inc.

Figure 8.55. The ESSA hemispherical array with patch elements (electronically switched spherical array) [Stockton and Hockensmith 1977, 1979; Mailloux et al. 1981].

arrays are reported in the literature. Some results for paraboloidal arrays, mainly regarding polarization characteristics, were discussed in Section 8.5.3.

A linear array of 21 narrow slots along the generator of a paraboloid was studied by Kaifas and coworkers [2003a]. The slots were considered as single-mode elements and mutual coupling was neglected. A Chebysjev pattern with –25 dB sidelobes was synthesized.

Persson and Josefsson [2003] analyzed radiation patterns for arrays of radiators on paraboloidal surfaces, also neglecting mutual coupling. Two radiating element models were compared. The first was a short dipole in free space with a cosine pattern, acting as a reference case. The other element model was more realistic; it assumed a radiating element in the form of a circular waveguide opening in the curved conducting surface. The aperture field in the opening was assumed to be the TE_{11} mode. The curvature of the parabolic surface was accounted for using UTD (see Chapter 7, Section 7.2). The effects of higher-order waveguide modes were not included in this analysis. However, computed isolated-element patterns agreed very well with measured patterns.

The patterns shown in Figure 8.56 were calculated for a 91-element array arranged in six rings near the vertex of a paraboloidal surface with $h/D = 0.293$. The beam is scanned 60° from the zenith, with 45 elements belonging to the active area. The assumed radiating element characteristics influence the array radiation pattern to a considerable degree, in particular the cross-polarized pattern (E_ϕ in this case). This component is generally very low in the symmetry plane, but outside this plane much higher values can be expected; compare Section 8.5.3.

8.6.7 Ellipsoidal Arrays

Ellipsoidal array antennas are similar to paraboloidal and spherical antennas, depending on the choice of parameters.

Wills [1987] presented a theoretical and experimental study of an array of waveguide elements on an ellipsoidal surface. The surface had a rather high and varying curvature over the array, comprising 29 elements. The nose radius was approximately 120 mm or about four wavelengths at 9.5 GHz. Patterns for five-element subarrays showed expected performance, but with high grating lobes due to large element spacings.

A numerical analysis of radiation patterns for ellipsoidal arrays was presented by Douchin and Lemorton [1997]; they considered large arrays with more than 1000 elements. By projecting the active area onto a plane perpendicular to the main beam direction, traditional taper functions (Taylor type) were applied. In their analysis, various ele-

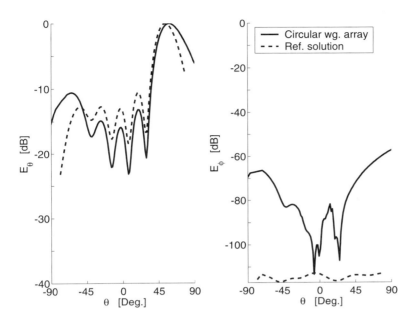

Figure 8.56. The E_θ and E_ϕ components in the symmetry plane with the beam scanned 60° from the zenith direction. Dashed line: idealized dipole element patterns. Solid line: TE_{11} mode circular aperture elements [Persson and Josefsson 2003].

ment pattern shapes could be selected, but mutual coupling was not included. It was shown that with dual polarization control in each element, the radiated cross polarization could be suppressed to very low levels around the direction of the main beam.

Similar results were reported by Schmid [2001] in a study of an ellipsoidal antenna designed for half-sphere scan coverage with 485 dual-port Huygens source elements. The use of idealized elements and neglect of mutual coupling and mismatch effects resulted in very low computed cross-polarization levels.

8.6.8 Other Shapes

A cylinder with a wing-shaped cross section with an embedded linear array of 116 patch elements has been used for a synthesis of low sidelobe patterns [Steyskal 2002]; see color plate 1. Representative array element patterns including mutual coupling and the local radius of curvature were used in the synthesis. Other studies relating to wing profiles have been based on cylinders with elliptical cross sections [Kanno et al. 1996, Bertuch 2003].

A lightweight aluminum cooking wok, approximately spherical in shape but with a flattened top, was used in an empirical investigation of an array of 19 microstrip patches at 3.2 GHz [Humphrey et al. 1995]. Conformal antennas need not be expensive!

REFERENCES

Barbier T., Mancuso Y., Favier I., Deborgies F., Reig B., and Monfraix P. (2001), "3D Microwave Modules for Conformal Phased Array Antennas," in *Proceedings of 2nd European Workshop on Conformal Antennas,* The Hague, The Netherlands, 24–25 April.

Bertuch T. (2003), "A Facetted Cylindrically Curved Array for a Conformal Radar Demonstrator," in *Proceedings of 3rd European Workshop On Conformal Antennas,* Bonn, Germany, 22–23 October.

Boeringer D. W. and Werner D. H. (2004), "Particle Swarm Optimization of a Modified Bernstein Polynomial for Conformal Array Excitation Synthesis," *IEEE AP-S International Symposium Digest,* pp. 2293–2296.

Bondyopadhyay P. K. (2000), "Geodesic Sphere Phased Arrays for LEO Satellite Communications," *IEEE AP-S International Symposium Digest,* pp. 206–209, July.

Bondyopadhyay P. K. (2001), "Geodesic Sphere Phased Array Antenna System," US Pat. 6,292,134 B1, 18 September.

Boyns J. E. (1972), "Cylindrical and Conical Array Investigations," in *Proceedings of Array Antenna Conference,* Naval Electronics Laboratory Center (NELC), San Diego, CA, USA, pp. 4-1–4-27, 22–24 February.

Bucci O., Franceschetti, G., Mazzarella G., and Panareillo G. (1990), "Intersection Approach to Array Pattern Synthesis," *IEE Proceedings Part H,* Vol. 137, pp. 349–357.

Calander N. and Josefsson L. (2001), "A Look at Polarization Properties of Cylindrical Array Antennas," in *Proceedings of 2nd European Workshop on Conformal Antennas,* The Hague, The Netherlands, 24–25 April.

Charrier M., Everett D., Fieret J., Karrer T., Rau S., and Valard J. (2003), "Manufacture of a Conformal, Multi-Layer RF Antenna Substrate Using Excimer Mask Imaging Technology and a 6-Axis Robot," http://www.exitech.co.uk, October 14.

Chujo W., Konishi Y., Ohtaki Y., and Yasukawa K. (1990), "Performance of a Spherical Array Antenna Fabricated by Vacuum Forming Technique," in *Proceedings of 20th European Microwave Conference,* Budapest, Hungary, 10–13 September, pp. 1511–1516.

Cohen E. D. (1995), "High Density Microwave Packaging Program," *IEEE MTT-S International Symposium Digest,* pp. 169–172.

Collin R. E. and Rothschild S. (1964), "Evaluation of Antenna Q," *IEEE Transactions on Antennas and Propagation,* Vol. AP-12, No. 1, pp. 23–27.

Debionne G. (1986), "Radiating Properties of Conformal Arrays," *IEEE AP-S International Symposium Digest,* pp. 937–940, June.

Derneryd A. (1976), "Linearly polarized microstrip antennas," *IEEE Transactions on Antennas and Propagation,* Vol. AP-24, No. 6, pp. 846–850.

Dolph C. L. (1946), "A Current Distribution for Broadside Arrays which Optimizes the Relationship between Beam Width and Side Lobe Level," *Proceedings of IRE,* Vol. 34, pp. 335–348, June.

Douchin N. and Lemorton J. (1997), "Numerical Analysis of Active Conformal Arrays Including Crosspolarization Effects," in *IEE Proceedings of 10th International Conference on Antennas and Propagation,* 14–17 April, pp. 1.200–1.205.

Elliott R. S. (1981), *Antenna Theory and Design,* Prentice-Hall Inc.

Fekete G. (1990), "Rendering and Managing Spherical Data with Sphere Quadtrees," in *Proceedings of First IEEE Conference on Visualization,* 23–26 October, pp. 176–186.

Gething P. J. D. (1966), "High-Frequency Direction Finding," *Proceedings of IEE,* Vol. 113, No. 1, pp. 49–61.

Gobert J. F. and Yang R. F. H. (1974), "A Theory of Antenna Array Conformal to Surfaces of Revolution," *IEEE Transactions on Antennas and Propagation,* Vol. AP-22, No. 1, pp. 87–91.

Goodrich R. F., Kleinman R. E., Maffett A. L., Schensted C. E., Siegel K. M., Chernin M. G., Shanks H. E., and Plummer R. E. (1959), "Radiation from Slots on Cones," *IRE Transactions on Antennas and Propagation,* Vol. 7, pp. 213–222, July.

Harrington R. F. (1961), *"Time Harmonic Electromagnetic Fields,"* Prentice-Hall.

Herper J. C., Hessel A., and Tomasic B. (1985), "Element Pattern of an Axial Dipole in a Cylindrical Array, Part I: Theory," *IEEE Transactions on Antennas and Propagation,* Vol. AP-33, No. 3, pp. 259–272.

Hill P. C. J. (1967), "Analytic Determination of Dipole Reactance by Radiation-Pattern Integration," *Proceedings of IEE,* Vol. 114, No. 7, pp. 853–858.

Humphrey P. J., Orton R. S., and Griffin J. M. (1995), "Planar and Conformal Patch Arrays," *Proceedings of IEE Workshop on Printed Antennas,* Lille, France, 25–26 September, pp. 113–120.

IEEE Standard 145-1993 (1993), *IEEE Standard Definition of Terms for Antennas,* IEEE.

Jiao Y-C., Wei W-Y., Huang L-W., and Wu H-S. (1993), "A New Low Side-Lobe Pattern Synthesis Technique for Conformal Arrays," *IEEE Transactions on Antennas and Propagation,* Vol. AP-41, No. 6, pp. 824–831.

Jones M. R. and Griffiths H. D. (1988), "Prediction of Circular Array Phase Mode Characteristics," *Electronics Letters,* Vol. 24, No. 13, pp. 811–812.

Jones M. R. and Griffiths H. D. (1989), "Broadband Pattern Synthesis from a Circular Array," *Proceedings of ICAP 1989,* pp. 55–59.

Josefsson L. and Persson P. (1999a),"Impedance and Radiation Characteristics of a Conformal Waveguide Array Antenna," in *Proceedings of 1st European Workshop on Conformal Antennas,* Karlsruhe, Germany, 1999.

Josefsson L. and Persson P. (1999b), "Scanning Characteristics of a Conformal Array Antenna," in *Proceedings of RVK 99,* Karlskrona, Sweden, June.

Josefsson L., Persson P., and Lanne M. (2002), "The Polarization Problem in Singly and Doubly Curved Conformal Array Antennas," in *Proceedings of IEEE International Symposium AP-S/URSI,* San Antonio, TX, 17–21 June.

Josefsson L. and Persson P. (2005),"Cylindrical Arrays: Pattern Synthesis, Impedance, and Bandwidth," in *Proceedings of Fourth European Workshop on Conformal Antennas,* Stockholm, Sweden, 23–24 May.

Kanno M., Hashimura T., Katada T., Sato M., Fukutani K., and Suzuki A. (1996), "Digital Beam Forming for Conformal Active Array Antenna," in *Proceedings of IEEE International Symposium Phased Array Systems and Technology,* 15–18 October, pp. 37–40.

Kaifas T. N., Samaras T., Vafiadis E., and Sahalos J. N. (2003a), "A UTD-OM Technique to Design Slot Arrays on a Perfectly Conducting Paraboloid," in *Proceedings of 3rd European Workshop on Conformal Antennas,* Bonn, Germany, 22–23 October.

Kaifas T. N., Samaras T., Vafiadis E., and Sahalos J. N. (2003b), "On the Design of Conformal Slot Arrays on Perfectly Conducting Elliptic Cones," in *Proceedings of 3rd European Workshop on Conformal Antennas,* Bonn, Germany, 22–23 October.

Kemerley R. T. and Kiss S. (2000), "Advanced Technology for Future Space-Based Antennas," in *Proceedings of IEEE International Symposium MTT-S.*

Knott P., Schippers H., Medynski D., Deloues T., van Lil E., and Gautier F. (2003), "Performances of Conformal and Planar Arrays," in *Proceedings of NATO Symposium on Smart and Adaptive Antennas in Military Communications,* RTO-MP-119, Chester, UK, 7–9 April.

Kraus J. D. (1988), *Antennas,* 2nd ed., McGraw-Hill.

Kummer W. H. (1974), "Preface, Special Issue on Conformal Arrays," *IEEE Transactions on Antennas and Propagation,* Vol. AP-22, pp. 1–3, January.

Liu B., Wang Y. A., Kang G., Ferendeci A. M., and Mah M. (2000), "Vertically Interconnected 3D Module for Conformal Phased Array Antennas," *Proceedings of Phased Array Systems and Technology, PAST'00.*

Lockyer A.J., Alt K. H., Coughlin D. P., Durham M. D., and Kudva J. N. (1999), "Design and Development of a Conformal Load-Bearing Smart-Skin Antenna: Overview of the AFRL Smart Skin Structures Technology Demonstration," *Proceedings of SPIE Conference,* Vol. 3674, Newport Beach, CA, USA., pp. 410–424, March.

Löffler D. and Wiesbeck W. (2000), "Dual Polarized Elements and Mutual Coupling in Conformal Arrays on Cylindrical Surfaces," *Proceedings of Antenn 2000,* Davos, Switzerland, 9–14 April.

Löffler D., von Hagen J., and Wiesbeck W. (2000), "Coupling Phenomena in Horizintal and Vertical Polarized Aperture Coupled Patch Antennas on Cylindrical Surfaces," in *Proceedings of ACES 2000, 16th Annual Review of Progress in Applied Computational Electromagnetics,* Monterey, CA, USA.

Löffler D., Wiesbeck W., and Johannisson B. (1999), "Conformal Aperture Coupled Microstrip Phased Array on a Cylindrical Surface," in *Proceedings of IEEE AP-S International Symposium,* Orlando FL, USA, pp. 882–885, July.

Longstaff I. D., Chow P. E. K., and Davies D. E. N. (1967), "Directional Properties of Circular Arrays," *Proceedings of IEE,* Vol. 114, No. 6, pp. 713–718.

Ludwig A. C. (1973), "The Definition of Cross Polarization," *IEEE Transactions on Antennas and Propagation,* Vol. AP-21, pp. 116–119. (See also Comments. Vol AP-21, pp. 907–908.)

Mailloux R. J., McIlvenna J. F., and Kernweis N. P. (1981), "Microstrip Antenna Technology," *IEEE Transactions on Antennas and Propagation,* Vol. AP-29, No. 1, pp. 25–37.

Mancuso Y. (2002), "T/R Modules Technological Trends for Conformal Phased Array Antennas," in *Proceedings of 12th International Symposium on Antennas, JINA 2002,* Nice, France, 12–14 November.

Marcus C., Persson P., and Pettersson L. (2003), "Investigation of the Mutual Coupling and Radiation Pattern Due to Sources on Faceted Cylinders," in *Proceedings of Antenn 03,* Kalmar, Sweden, 13–15 May, pp. 213–218.

Martin T. and Pettersson L. (2003), "Cylindrical FDTD with Phase Shift Boundaries for Simulation of Cylindrical Antenna Arrays," in *Proceedings of 3rd European Workshop on Conformal Antennas,* Bonn, 22–23 October.

Midford T. A., Wooldridge J. J., and Sturdivant R. L. (1995), "The Evolution of Packages for Monolithic Microwave and Millimeter-Wave Circuits," *IEEE Transactions on Antennas and Propagation,* Vol. AP-43, No. 9, pp. 983–991.

Morton T. E. and Pasala K. M. (2004), "Pattern Synthesis and Performance of Conformal Arrays," *Proceedings of 36th Southeastern Symposium on System Theory,* Atlanta GA, USA, 16 March, pp. 145–149.

Munger A. D., Vaughn G., Provencher J. H., and Gladman B. R. (1974), "Conical Array Studies," *IEEE Transactions on Antennas and Propagation,* Vol. AP-22, No. 1, pp. 35–43.

Page H. (1948), "Radiation Resistance of Ring Aerials," *Wireless Engineer,* pp. 102–109, April.

Perpère B. and Hérault J. (2001), "Conformal Broadband Phased Array Antenna," in *Proceedings of 2nd European Workshop on Conformal Antennas,* The Hague, The Netherlands, 24–25 April.

Persson P. and Josefsson L. (2001), "Calculating the Mutual Coupling between Apertures on a Convex Circular Cylinder Using a Hybrid UTD-MoM Method," *IEEE Transactions on Antennas and Propagation,* Vol. 49, No. 4, pp. 672–677.

Persson P. and Josefsson L. (2003), "Radiation Characteristics of Doubly Curved Array Antennas," in *Proceedings of 3rd European Workshop on Conformal Antennas,* Bonn, Germany, 22–23 October.

Persson P., Josefsson L., and Lanne M. (2003), "Investigation of the Mutual Coupling between Apertures on Doubly Curved Convex Surfaces: Theory and Measurements," *IEEE Transactions on Antennas and Propagation,* Vol. 51, No. 4, pp 682–692.

Raffaelli S. and Johansson M. (2003), "Conformal Array Antenna Demonstrator for WCDMA Applications," in *Proceedings of Antenn 03,* Kalmar, Sweden, 13–15 May, pp. 207–212.

Rhodes D. R. (1964), "On a Fundamental Principle in the Theory of Planar Antennas," *Proceedings of IEEE,* Vol. 52, pp. 1013–1021, September.

Royer G. M. (1966), "Directive Gain and Impedance of a Ring Array of Antennas," *IEEE Transactions on Antennas and Propagation,* Vol. AP-14, No. 5, pp. 566–573.

Saito J., Heichele L., Numasaki T., Orime N., and Katagi T. (1987), "Comparison of Element Arrangements of a Spherical Conformal Array," *IEEE AP-S International Symposium Digest,* pp. 125–128.

Sanzgiri S., Bostrom D., Pottenger W., and Lee R. Q. (1995), "A Hybrid Tile Approach for Ka Band Subarray Modules," *IEEE Transactions on Antennas and Propagation,* Vol. AP-43, No. 9, pp. 953–959.

Schaller M. and Beaulieu L. (1996), "Optoelectronics and Smart Skins: The Logical Symbiosis," *Proceedings of NATO Workshop on Smart EM Antenna Structures,* Bruxelles, 25–26 November, pp. 22.1–22.13.

Schippers H., Schaap W., Visser J. M. P. C. M., and Vos G. (2003), "Conformal Phased Array Antennas on Vibrating Structures," in *Proceedings of 3rd European Workshop on Conformal Antennas,* Bonn, Germany, 22–23 October.

Schmid O. (2001), "Pattern Properties of Singly and Doubly Curved Arrays," in *Proceedings of 2nd European Workshop on Conformal Antennas,* The Hague, The Netherlands, 24–25 April.

Schneider S. W., Bozada C., Dettmer R., and Tenbarge J. (2001), "Enabling Technologies for Future Structurally Integrated Conformal Apertures," *IEEE AP-S International Symposium Digest,* Boston.

Schreiner M., Leier H., Menzel W., and Feldle H.-P. (2003), "Architecture and Interconnect Technologies for a Novel Conformal Active Phased Array Radar Module," *IEEE MTT-S International Symposium Digest,* pp. 567–570.

Sekora R., Brand C., Nagy O., and von Storp W. (2003), "Conformal Antenna Array for X-Band Data Link Applications," in *Proceedings of 3rd European Workshop on Conformal Antennas,* Bonn, Germany, 22–23 October.

Sengupta D., Smith T., and Larson R. (1968), "Radiation Characteristics of a Spherical Array of Circularly Polarized Elements," *IEEE Transactions on Antennas and Propagation,* Vol. AP-16, No. 1, pp. 2–7.

Sensiper S. (1957), "Cylindrical Radio Waves," *IRE Transactions on Antennas and Propagation,* Vol. AP-5, No. 1, pp. 56–70.

Shrank H. E. (1970), "Basic Theoretical Aspects of Spherical Phased Arrays," in *Proceedings of Phased Array Symposium,* pp. 323–327.

Sindoris A. R. and Hayes S. T. (1981), "A Conical Array of Dielectric Filled Edge Slot Antennas," *IEEE AP-S International Symposium Digest,* pp. 467–470.

Sipus Z., Burum N., and Bartolic J., (2003), "Analysis of Rectangular Microstrip Patch Antennas on Spherical Structures," *Microwave and Optical Technology Letters,* Vol. 36, No. 4, pp. 276–280.

Smallwood B. P., Terzuoli A. J., and Canfield R. A. (2003), "Structurally Integrated Antennas for Remote Sensing," *Proceedings of IGARSS'03,* Vol. 7, 21–25 July, pp. 4252–4254.

Steyskal H. (2002), "Pattern Synthesis for a Conformal Wing Array," in *Proceedings of IEEE Aerospace Conference 2002,* Vol. 2, pp. 2-819–2-824.

Stockton R. J. and Hockensmith R. P. (1977), "Application of Spherical Arrays—A Simple Approach," *IEEE AP-S International Symposium Digest,* pp. 202–205.

Stockton R. J. and Hockensmith R. P. (1979), "Microprocessor Provides Multi-Mode Versatility for the ESSA Antenna System," *IEEE AP-S International Symposium Digest,* pp. 469–472.

Sureau J-C. and Hessel A. (1971), "Element Pattern for Circular Arrays of Waveguide-Fed Axial Slits on Large Conducting Cylinders," *IEEE Transactions on Antennas and Propagation,* Vol. AP-19, No. 1, pp. 64–74.

Taylor T. T. (1955), "Design of Line-Source Antennas for Narrow Beamwidth and Low Side Lobes," *IRE Transactions on Antennas and Propagation,* Vol. AP-3, pp. 16–28.

Thiele G. A. and Donn C. (1974), "Design of a Small Conformal Array," *IEEE Transactions on Antennas and Propagation,* Vol. AP-22, No. 1, pp 64–70.

Vallechi A. and Gentili G. B. (2003), "Broad Band Full Scan Coverage Polarization Agile Spherical Conformal Array Antennas: Pseudo-Uniform vs. Pseudo-Random Element Arrangements," in *Proceedings of IEEE International Symposium on Phased Array Systems and Technology 2003,* Boston, USA, 14–17 October, pp. 529–534.

Wheeler H. A. (1966), "The Grating-Lobe Series for the Impedance Variation in a Planar Phased-Array Antenna," *IEEE Transactions on Antennas and Propagation,* Vol. AP-14, No. 6, pp. 707–714.

Villeneuve A. T. and Kummer W. H. (1981), "Conical Phased Array Antenna Investigation," *IEEE AP-S International Symposium Digest,* pp.475–478.

Villeneuve A. T., Behnke M. C., and Kummer W. H. (1974), "Wide-Angle Scanning of Linear Arrays Located on Cones," *IEEE Transactions on Antennas and Propagation,* Vol. AP-22, No. 1, pp. 97–103.

Voles R. (1995), "Spherical Shell and Volume Arrays," *IEE Proceedings of Microwave Antennas and Propagation.,* Vol. 142, No. 6, December, pp. 498–500.

Voskresensky D. I. and Ovchinnikova E. V. (2001), "Wideband Conformal Phased Array with Spreaded Sector Scanning," in *Proceedings of 2nd European Workshop on Conformal Antennas,* The Hague, The Netherlands, 24–25 April.

Wait J. R. (1959), *Electromagnetic Radiation from Cylindrical Structures,* Pergamon Press.

Weisstein E. W. (2004), "Archimedean Solid," from *Mathworld*—A Wolfram Web Resource, http://mathworld.wolfram.com/ArchimedeanSolid.html.

Wills R. W. (1987), "Mutual Coupling and Far Field Radiation from Waveguide Antenna Elements on Conformal Surfaces," in *Proceedings of the International Conference on RADAR-87,* London, October 19–21, pp. 515–519.

Witte E. D., Griffiths H. D., and Brennan P. V. (2003), "Phase Mode Processing for Spherical Antenna Arrays," *Electronics Letters,* Vol. 39, No. 20, pp. 1430–1431.

9

BEAM FORMING

In this chapter, we will treat the feeding and scanning principles that apply to conformal array antennas, in particular cylindrical arrays. We will discuss solutions for commutating the active sector as well as beam steering and multiple beam generation for circular and cylindrical arrays. An additional problem in conformal antennas is the need for polarization control, a consequence of the lack of a common element factor in the radiation function. A universal approach to all these critical issues is provided in receiving arrays by digitizing the element signals and performing all corrections and control in computer software. This approach using digital beam forming may in the future also be applied to transmitting arrays.

9.1 INTRODUCTION

The feeding system for a conformal array must provide phase (or time delay) compensation for the array curvature. In many cases, the amplitude must also be steered and sometimes the polarization as well. Usually, only one part of the complete array is activated at a time, and this active sector must be moved according to the desired direction of observation. This last function is referred to as commutating. All these functions may lead to very complex networks with large losses. Several innovative arrangements have been proposed, however, aiming at reducing cost and complexity, and we will examine some of them. Most of the solutions apply mainly to circular and cylindrical arrays. Doubly curved arrays are even more difficult to feed. For large arrays, solutions may be found using optical and/or digital signal formats.

Conformal Array Antenna Theory and Design. By Lars Josefsson and Patrik Persson
© 2006 Institute of Electrical and Electronics Engineers, Inc.

9.2   A NOTE ON ORTHOGONAL BEAMS

Generating two or more beams from the same array antenna is often advantageous compared to scanning a single beam. The data rate is increased when there are several spatial channels, doppler performance in MTI surveillance radar systems is improved through longer dwell time on target, and more accurate bearing measurements are possible. Typically, each beam is associated with a physical beam port in the feeding system. For the ports to carry independent signals, they must be isolated from each other, or else a signal fed to one port will be radiated in more than one beam. On reception, poor isolation leads to signal losses.

It has been demonstrated [Allen 1961, White 1962, Stein 1962] that a necessary condition for the beam ports of a multibeam antenna to be isolated from each other is that the corresponding radiation patterns must be mutually orthogonal. The orthogonality requirement for two beams with radiation patterns $\overline{E}_1$ and $\overline{E}_2$ can be written quite generally as

$$\int_0^{2\pi} d\phi \int_0^{\pi} \overline{E}_1(\theta, \phi) \cdot \overline{E}_2^*(\theta, \phi) \sin \theta d\theta = 0 \qquad (9.1)$$

The orthogonality can be realized in several ways: (1) by using separate apertures, (2) by using orthogonal polarizations in the same aperture, and (3) by having angularly spaced beams from the same aperture.

A simple example of the last method is given by two patterns of the type $(\sin x)/x$. They will be orthogonal if they are separated in angle by a multiple of $x = \pi$ radians. The closest possible orthogonal spacing in this case results in -3.92 dB crossover depth between the main beams. This is the result obtained with a Butler-matrix-fed linear array antenna [Butler and Lowe 1961, Shelton 1961]. However, the fundamental requirement that the patterns be orthogonal and associated with a lossless feed system is given by the radiation characteristics, irrespective of the antenna physical shape as expressed by Equation (9.1). It is equally valid for curved arrays. Note, however, that in the case of digital beam forming, it is possible to obtain closely spaced receiving beams without the losses incurred in analog beam forming networks.

9.3   ANALOG FEED SYSTEMS

A feed system is a network connecting the antenna input to its radiators. The purpose is to transmit power to the elements (transmit mode) or to collect signals from the elements (receive mode). This must be done while maintaining the necessary phase and amplitude excitations for the required radiation performance. The feed network may also contain switches or other devices for scanning the beam, selecting between different antenna beam shapes, and switching between active sectors (commutating). Sometimes, active devices such as amplifiers are integrated into the feed.

We will briefly discuss some of the types of analog feed systems that have been proposed and studied. A substantial part of this development took place several years ago, as can be seen from the references cited, for example, the overviews [Provencher 1970, Hall and Vetterlein 1990], but there are also several recent developments. The main principles used, following [Uhlmann 1975a], are:

- Vector transfer matrix systems
- Switch matrix systems
- Butler matrix feed systems
- RF lens feed systems

There are also several combinations and variations of these principles. Digital feed systems (digital beam forming) will be discussed in Sections 9.3 and 9.4.

9.3.1   Vector Transfer Matrix Systems

A very general feeding arrangement would have all radiators fed via attenuators, phase shifters, and a power divider. By sufficiently attenuating those channels that do not belong to the intended active sector, the feed provides all functions: commutating, amplitude taper, and phase steering. This feed with amplitude and phase control is called a vector transfer feed. It is seldom used in this form, however, because of the complexity and large transmission-line losses, especially the losses in the attenuators.

9.3.2   Switch Matrix Systems

Basically, the number of attenuators and switches should not exceed the number of elements in the active sector. Thus, a "partial vector transfer feed" can be constructed, using output switches to choose between sectors. Giannini [1969] demonstrated one such system with 32 dipole elements of which eight where used at a time. Boyns and coworkers [1970] used a system for feeding and commutating an active 90° sector of a 128 element ring array. Their complete antenna hardware with feed and control elements is shown in Figure 9.1.

Gregorwich [1974] presented a cylindrical array for an electronically despun antenna for use on a spin-stabilized satellite in a geostationary orbit. For earth coverage, about an 18° beamwidth is required, which could be provided by a four-by-four array of crossed dipoles. In total, there were four rows of 64 elements each around the 190 cm diameter satellite operating over a 40% bandwidth at the S band. Since the satellite rotates, the antenna beam must rotate electronically in the opposite direction to keep the beam directed

Figure 9.1. A 128 element vector transfer-matrix-fed S-band array [Boyns et al. 1970].

on earth. For commutating the active sector of the array, a feed network with one power divider and several PIN diode switches was used. A simplified version (a smaller array with 18 elements) illustrating the principle is shown in Figure 9.2. Note that the combination of SPDT and SP3T switches is equivalent to one SP6T switch.

The feed network has equal amplitude outputs and there is no need for phase compensation since only four elements at a time are active (out of 64). Thus, this basic principle allows the beam to be scanned in steps corresponding to the element division along the circumference.

Reindel [1974] presented a simple beam scanning/commutating solution suitable for small cylindrical arrays. A matrix of interconnected 90° hybrids and fixed phase shifters around the array feeds the radiating elements. A second layer of hybrids connects to sum and difference beam ports. In this case, four elements are fed from each beam port with phases 0°, 90°, 90°, and 0°, respectively, thus approximately correcting for the array curvature. Multiple beams from a 22 dipole element circular array were successfully demonstrated over 2.4 to 3.4 GHz. The sidelobe levels were about –12 dB and monopulse difference pattern null depths were –15 dB or better.

A more complicated commutating scheme is used in the METEOSAT despun antenna array [Polegre et al. 2001]. This array has 32 element ports of which five are active at a time. By means of switches and variable power dividers, the power to the outer ports on each side of the active sector is gradually changed so that the power distribution is moved smoothly around the cylindrical array.

By adding phase shifters for fine steering to a switch-matrix feed such as the one in Figure 9.2, we get the system shown in Figure 9.3. SPNT stands for single-pole-N-throw switches. Each active sector is 360/N degrees wide. (In the example in Figure 9.2, we had $N = 6$ and each sector was 60°.)

In the simplest case, equal amplitudes are fed to the array elements, that is, no tapering is applied. Many times, however, it is desired to use a tapered amplitude distribution for sidelobe control. If the amplitude taper is built into the power divider, then the proper divider outputs must be connected to the corresponding element positions in the active sector. The solution is to insert a special matrix that reorders the signals appropriately as the active sector moves, a so-called transfer-switch matrix (Figure 9.4). With this arrange-

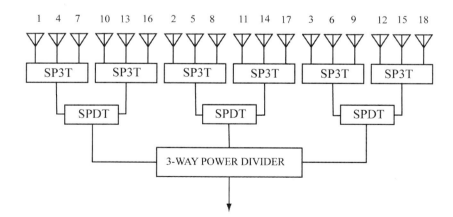

Figure 9.2. A switch matrix feed for activating 3 elements out of 18, after [Gregorwich 1974].

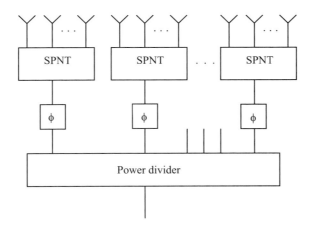

<u>Figure 9.3.</u> A switch matrix feed with phase shifters for fine steering of the beam.

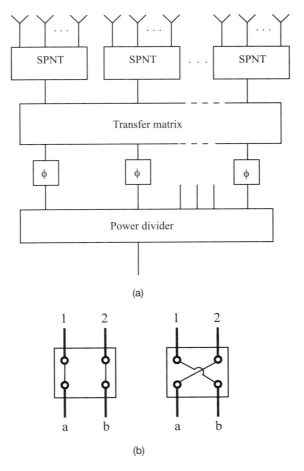

<u>Figure 9.4.</u> (a) A feed system using a transfer switch matrix. The transfer matrix allows routing of any input to any output. (b) The two states of a transfer switch.

ment, difference patterns for monopulse operation can also be generated, as was demonstrated by Giannini [1969] for a cylindrical array at about 1000 MHz. He had 32 dipole elements, of which eight where used at a time. Keeping and coworkers [1981] used the same principle for a 96 element cylindrical array at the *C* band.

The transfer-switch matrix consists of transfer switches with two inputs and two outputs; the two states are shown in Figure 9.4 (b). The complete matrix is quite complex, but the size can be reduced if the output amplitude is quantized to just a few amplitude values [Uhlmann 1975a].

The feeding and switching systems require careful phase trimming (and/or cable length adjustments). Another problem is the losses of the large network. For instance, a 9 dB loss was reported by Keeping and coworkers [1981] for a 96 element *C*-band cylindrical array (the active sector had 32 elements). To compensate for the losses, amplifiers can be inserted in the signal paths.

Lee and coworkers [2002] presented such a solution intended for an AWACS type nonrotating antenna at UHF. The active sector has 16 endfire radiators, in total 48 for the full array. A power divider with 16 outputs connects to a 16×16 transfer-switch matrix that maps the amplitude and phase distribution on to the active sector. The output SP3T switches connect to the 16 elements in the 120° active sector; see Figure 9.5.

Note in Figure 9.5 the transmit amplifiers inserted between the sector selection switches and the transfer switches. The receive signals are tapped directly at this point, and it is suggested that beam forming on receive be done digitally.

9.3.3   Butler Matrix Feed Systems

Several authors [Sheleg 1968, Brown 1971, Nelson and Britt 1972] discuss commutating by means of a Butler matrix [Butler and Lowe 1961]. A schematic of this approach is

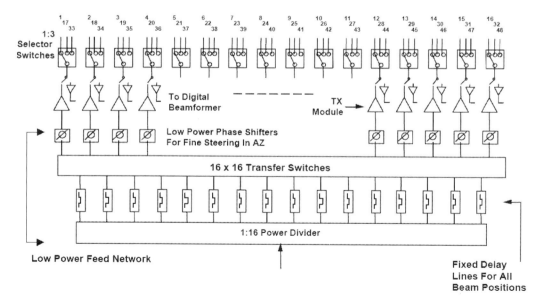

Figure 9.5. Beam forming network according to [Lee et al. 2002] with amplifiers in the transmit paths.

shown in Figure 9.6. The power divider and the fixed phase shifters control the desired excitation of a subset of the output ports (B), that is, the active sector. By appropriately setting the variable phase shifters to add a linear phase slope at the mode input ports (A) of the Butler matrix, the excitation is moved to a new subset of the output ports (B). See also [Holley et al. 1974], who used a Butler matrix for fine steering in combination with switches and a geodesic lens; compare Figure 9.15

The two sets of phase shifters in Figure 9.6 (fixed and variable) can in a realization of this principle be replaced by just one set of variable phase shifters. The same applies to the design in Figure 9.8.

A commutating and beam steering solution for space telemetry has been developed by the European Space Agency [Vourch et al. 1998, Polegre et al. 2001, Caille et al. 2002]. The conical array (see Figure 1.4 in Chapter 1) has 24 subarrays fed by 3×3 "Butler-like" matrices preceded by amplifiers and phase shifters. A simplified schematic is shown in Figure 9.7. A similar design using 4×4 matrices has also been developed [Polegre et al. 2003]. The appropriate active area is energized by proper phase inputs to the 3×3 (or 4×4) Butler matrices. The concept has the advantage that all the 24 power amplifiers are active simultaneously and operate at equal power. The scanning can be done in small steps, with minimum amplitude and phase jumps on the scanned beam. This design is called "semiactive" since the power amplifiers are located at an intermediate level of the beam-forming matrix.

A Butler matrix directly feeding a circular array creates linear phase variations among the radiating elements, that is, it creates phase-mode excitations (see Chapter 2). An $M \times M$ port Butler matrix can be used to excite M independent phase modes, each radiating an omnidirectional amplitude pattern, but with different phase variations with angle. These M patterns are mutually orthogonal and can be used in, for example, communication systems with several frequency channels, each channel using one of the omni mode patterns.

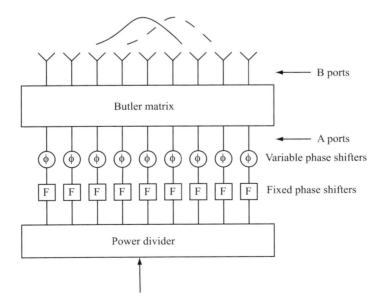

Figure 9.6. Commutating by means of a Butler matrix. The active output is shifted by means of the variable phase shifters.

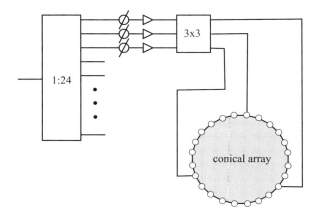

Figure 9.7. A "semiactive" phase-commutating feed system, after [Martín Polegre et al. 2001].

The isolation between the modes helps in maintaining a high isolation between channels, thus relaxing the requirements on frequency filters [Guy and Davies 1983].

Connecting the phase mode ports of the Butler matrix to a second network, as shown in Figure 9.8 [Sheleg 1968], makes it possible to create any (physical) beam pattern using phase-mode synthesis. The variable phase shifters can be used for scanning the beam since adding a phase shift with increments $\Delta\phi$ from one mode to the next will shift the excitation and also the radiation pattern by the same amount in angle. This fact, noted by Davies [1965], follows directly from the phase mode discussion in Chapter 2. Let us revisit Equation (2.22) describing the far field radiation expressed in phase modes with amplitudes A_m:

$$E(\phi) = \sum_{-\infty}^{\infty} A_m e^{jm\phi} \qquad\qquad (9.2) = (2.22)$$

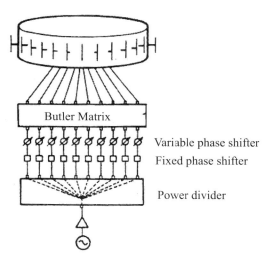

Figure 9.8. A Butler-matrix-fed array with variable phase shifters for scanning the beam [Sheleg 1968].

Now, the radiating phase mode amplitudes are proportional to their respective excitation phase mode amplitudes C_m. If we add phase increments that increase linearly with the phase-mode number, as could be obtained from the Butler matrix feed, we get $A_m^{new} = A_m^{old} \cdot e^{m\Delta\phi}$. Insertion into Equation (9.2) gives the radiation:

$$E^{new}(\phi) = \sum_{-\infty}^{\infty} A_m^{old} e^{jm(\phi+\Delta\phi)} = E(\phi + \Delta\phi) \qquad (9.3)$$

Continuous rotation of the pattern can be achieved by frequency scanning if the array ports are fed in series from a tapped delay line. At higher scan rates, some pattern degradation occurs, although less than in a comparable frequency scanned linear array [Arora and Patel 1983]. A simple microstrip cylindrical array with a series feed was described by Hall and Vetterlein [1992]. Scanning by amplitude control in a small dipole ring array has been demonstrated by Neff and Tillman [1960]. The TACAN air-navigational aid [Christopher 1974, Shestag 1974] also makes use of modulated element signals to provide a multilobed rotating pattern.

With a second Butler matrix replacing the fixed phase shifters and the power divider in Figure 9.8, multiple linear phase shifts along the phase-mode ports are added. In this way, multiple beams in the azimuth are created; a recent example is described in [Raffaelli and Johansson 2005]. The zero-order phase mode is not provided by the common Butler matrix with 90° hybrids, but this mode can be included by adding appropriate phase shifts on the output ports. With a matrix using 180° hybrids instead [Hering 1970], the zero-order mode is directly available. Rahim and coworkers [1981] have described a 16 × 16 Butler matrix that is split into an odd and an even phase-mode part, with the zero mode included.

Since the illuminated elements are confined to an active sector of about 90° to 120°, a reduced size Butler matrix can be used with output switches connecting to the appropriate sector [Skahill and White 1975]. Another way to save on matrix size according to [Sheleg 1973] makes use of the fact that the highest-order phase modes are seldomly needed, or will give distorted patterns (because of too large element spacing; cf. Chapter 2). In Sheleg's example, where two submatrices of size 16 × 16 are used instead of one full 32 × 32 matrix, the number of hybrids could be reduced by 20%.

The Butler matrix performs an operation analogous to the fast Fourier transform (FFT) [Nester 1968]. The FFT is often used for frequency filtering of broadband signals. Thus, the spatial and frequency processing are in principle quite similar. Raabe [1976] discusses a receiving sonar circular array with the successive operations

- FFT (frequency filtering of the element signals), obtaining frequency samples
- FMF (fast mode former), equivalent to the Butler matrix, for extracting phase modes
- Weight application to the mode outputs for beam shape control
- FBF (fast beam forming) for obtaining angular samples of the signals

The radiated phase modes differ in absolute phase by an amount that depends linearly on the azimuth angle. Thus, detecting the phase difference between modes on reception provides information about the direction toward the source. This has been exploited in so-called digital bearing discriminators for broadband ESM systems [Rehnmark 1980a,b]; see also Figure 1.6 in Chapter 1.

9.3.4 RF Lens Feed Systems

Many of the transmission line feed systems are quite complicated and suffer from large losses. Systems based on RF/optical principles are therefore attractive, especially for large arrays with many radiating elements. We will illustrate different kinds by examples and we will note that several RF/optical feed systems also incorporate various types of beam scanning functions.

RF refractive lenses have functions similar to their classical optical counterparts, using the refraction between dissimilar materials. Constrained lenses force the waves to follow given paths, as in a geodesic lens. The bootlace lens is one type of constrained lens in which the signals between input and output surfaces are routed on transmission lines. Sometimes, a conformal array feed makes use of combinations of lens types, or lenses and matrices [Provencher 1970].

An RF lens usually has a set of input/output probes that couple to the lens region. These probes are in fact small array antenna elements in an array environment, characterized by reflections and mutual coupling and the associated design problems. There can also be standing waves caused by reflections from the opposite side of the lens, in particular in circular lens designs. A further problem is the variation of the element phase center with frequency.

9.3.4.1 The R-2R Lens Feed. The R-2R lens feed (Figure 9.9) has feed ports on the circumference of a parallel-plate lens with radius R, illuminating the output ports on the opposite side of the lens. The latter are connected to the element ports on the $2R$ radius circular array with equal length cables. The number of element ports is twice the number of feeding ports. A study of the geometry [Boyns et al. 1968] reveals that the arrangement provides perfect focusing for all feed points, resulting in a plane-phase front. Moving the feed point an angle ϕ results in scanning the antenna beam at angle $\phi/2$.

By combining three to four adjacent feed ports (as shown in Figure 9.9) an illumination taper can be achieved, resulting in lowered sidelobes. The matrix-fed array shown in

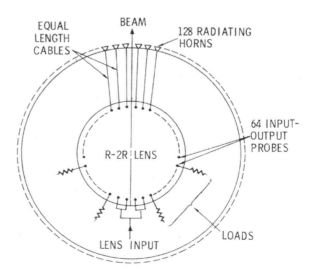

Figure 9.9. The R-2R lens feed system [Boyns et al. 1968].

Figure 9.1 was also tested with an R-2R lens feed with four ports combined. This resulted in a reduction of the sidelobes from -18 dB to -30 dB [Provencher 1970].

For scanning the beam, a set of switches on the lens ports has to be added. For multiple beam generation one needs to have access to several beam ports simultaneously. One solution could be to use half the lens for beam ports and connect the other half to a 90° arc array [Archer 1984].

The R-2R lens can be seen as a special case of the Rotman lens, which is mostly used for linear array feeds [Rotman and Turner 1963]. The Rotman lens can also be used for circular arcs up to about 90°; in fact, the curvature can be other than circular [Archer 1984], since the overall design, curvature of lens input and output lines, cable lengths, and so on can be optimized together with the array shape. Perfect focusing in the Rotman lens is obtained for three beam directions only.

9.3.4.2 The R-kR Lens Feed.
The R-kR lens feed system has the same number of ports on the lens as there are radiators on the circular array [Thies 1973, Archer 1984]. For 360° coverage, the lens ports have to be reused, both as feeding points and for connecting to the radiating elements. This can be accomplished with switches, circulators (Figure 9.10), or by using two lenses.

The radiators on radius R are connected by equal length cables to the ports of the circular lens with radius kR. An approximately planar phase front for rays within a sector of about 120° is obtained when k is about 1.9. This means that the lens is almost twice the size of the circular array, and it does not fit inside the array unless filled with a dielectric with permittivity 4 or higher. For large arrays (narrow beam widths), the lens also becomes very large, unless a very high-permittivity dielectric material is used in the lens.

The R-kR lens-fed circular array can be very broadband provided broadband radiators are used. Circulators or switches could, of course, limit the bandwidth. A critical design parameter is the phase center of the radiators, which must be located on the design radius

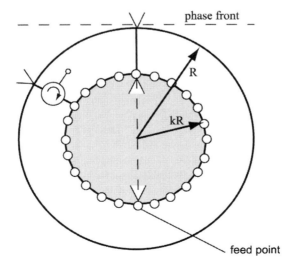

Figure 9.10. The R-kR lens, here with circulators.

R. Many element types have a phase center that changes position with frequency, thus limiting the focusing performance. However, designs with greater than octave bandwidths have been described [Archer 1984].

An elegant and compact R-kR lens-fed array using a microstrip lens and tapered-notch radiating elements in the same substrate with $\varepsilon_r = 10.2$ has been described [Sierra and Bernal 1989]; see Figure 9.11. 10 beams separated by 10° were obtained from one such array; hence, four arrays were needed for full azimuth coverage.

An R-kR network intended for feeding an array antenna for mobile satellite communication (MSAT) has been proposed by [Bodnar et al. 1989]. This vehicle-mount antenna has 360° coverage in the azimuth and covers 20°–60° in elevation. Thus, the focusing was optimized for an elevated beam. An inner ring and an outer ring of 12 microstrip elements, each with a pair of lenses combined with hybrids, provide full azimuth coverage. A compact multilayer design (see Figure 9.12) is predicted to cost less than a traditional phase-steered array for this application.

9.3.4.3 *Mode-Controlled Lenses.*

A radial transmission line forming a circular parallel-plate lens can act as a circular array feed. When excited by a set of probes near the center, the modes generated will direct the energy toward a part of the lens periphery. Thus, controlling the modes by phase shifters or a hybrid network connected to the input probes commutates the excitation. Pick-up probes can then connect to the radiating elements, via additional phase shifters if needed. Honey and Jones [1957] employed this mode-controlling technique in a coaxial structure, which was flared into a biconical direct radiating antenna. A 32 radiating port coaxial commutator was described by Irzinski [1981]. The parallel-plate approach was described as a commutating device by Bogner [1974]. Uhlmann [1975b] also used a parallel-plate structure with a central input probe surrounded by a number of pin diodes for mode control. Commutating was achieved by successively short-circuiting of diodes. On the output, switches and phase shifters limited the excitation to a 180° sector and focused the beam.

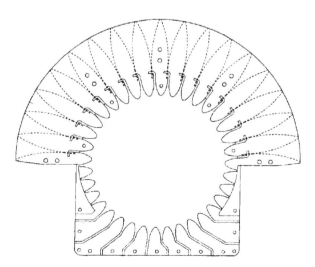

Figure 9.11. A broadband semicircular array with microstrip R-kR lens feed in printed circuit technology [Sierra and Bernal 1989].

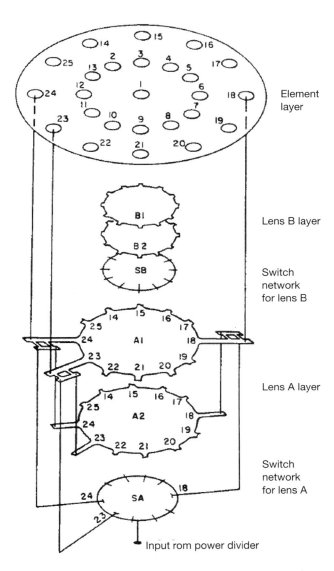

Figure 9.12. A four R-kR lens design for mobile communication [Bodnar et al. 1989].

9.3.4.4 *The Luneburg Lens.* The two-dimensional Luneburg circular lens has a refractive index that varies with the radius. With a feed on the perimeter, a collimated beam results, emerging from the opposite side of the lens (Figure 9.13). Switches, circulators, or feeds with 45° inclined linear polarizations [Cassel 1982] can be used for covering the full 360° sector. The three-dimensional Luneburg lens works similarly.

The refractive index *n* varies as

$$n(r) = \sqrt{2 - (r/a)^2} \tag{9.4}$$

where *a* is the radius of the lens and *r* is the distance from the center of the sphere. A typical design with the feed points on the lens surface has a variation of the dielectric con-

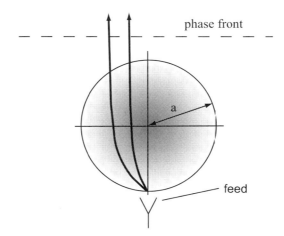

Figure 9.13. A Luneburg lens with radially varying refractive index.

stant $\varepsilon_r = n^2$ from two in the center to unity at the surface. In practice a number (<10) of concentric shells, each with fixed dielectric constant, are used.

A system with three-dimensional Luneburg transmit and receive lenses generating multiple beams was used in the Typhon experimental system [Fowler 1998].

Probes on the output side of a two-dimensional Luneburg lens can connect to a cylindrical array. A constrained Luneburg-type lens can also be constructed from a parallel-plate waveguide, which is deformed to two parallel "domes" between which the rays propagate with the desired electrical path. Thus, a geodesic "tin hat" equivalent to a Luneburg lens is obtained [Hollis and Long 1957, Peeler and Archer 1953].

9.3.4.5 *The Geodesic Lens.* A geodesic lens feed was used by Bodnar and coworkers [1980] for feeding and mechanically commutating a cylindrical Ku-band array. The design is shown in Figure 9.14. The rotating sector-shaped parallel plate has a phase-compensating geodesic lens segment.

Mechanical commutating of a cylindrical array has also been described [Wolfson and Cho 1979, Wolfson et al. 1980]. The rotating feed in their case is a stripline power divider with phase-compensating line lengths. The "organ pipe scanner" [Peeler et al. 1952, Peeler and Gabriel 1955] used a rotating feed for switching between several fixed waveguides connecting to feed horns in a lens antenna system, essentially a commutating device. The contactless coupling between array and rotor is a key element in this design, as well as in the previous ones. Although mechanically complicated, it replaces many active components in electronically switched systems; hence, lower loss can be anticipated. In the Wullenweber HF arrays of the 1960s, mechanical commutating devices with capacitive coupling between a rotor and a stator (goniometer) were used [Fullilove et al. 1959, Gething 1966].

A cylindrical geodesic lens can be used in a number of configurations. Holley and coworkers [1974] describe a feed system using a right-circular cylindrical geodesic lens with a minimum number of active devices. The active sector is commutated by selecting 8 out of 256 feed ports on the cylinder input. An 8 × 8 Butler matrix is used to shift the excitation between steps. On the output, diplexers and a second circular lens connect to a cylindrical array with 256 dipole line sources. Figure 9.15 shows this design with the signal flow in the transmit mode indicated.

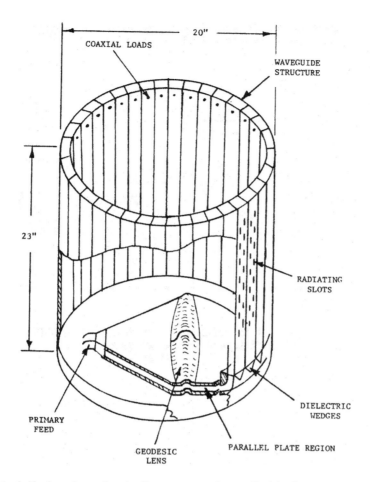

Figure 9.14. A Ku band mechanically commutating cylindrical array with a rotating geodesic lens feed [Bodnar et al. 1980].

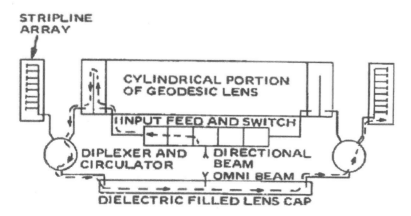

Figure 9.15. Geodesic lens feed system with signal flow in the transmission mode [Holley et al. 1974].

9.3.4.6 The Dome Antenna. The WASCAL or Dome antenna (Figure 9.16) [Bearse 1975, Schwartzman and Stangel 1975, Liebman et al. 1975] principle was briefly described in Chapter 1; see Figure 1.7. WASCAL stands for wide-angle scanning array lens. It allows scanning a pencil beam over the full hemisphere, in fact even a little more. The phase gradient of the lens can be optimized for a desired elevation coverage, as was studied by Steyskal and coworkers [1979] for a circular cylindrical case. Ellipsoidal-shaped lenses have also been suggested for increased gain at low elevation angles. For airborne applications, the dome antenna has been proposed for nose and tail mounting, giving very wide angle scanning capability. The dome concept has not found widespread use, however, due to relatively high cost, limited bandwidth, and somewhat poor sidelobe performance [Fowler 1998, Kinsey 2000]. A possible variation of the dome principle would be to insert active phase shifters in the constrained dome lens. The planar array would in this case work mainly as a commutating device.

A passive dome-type lens can, of course, also be realized using homogeneous dielectric materials. It can be less costly than the constrained dome antenna, but weight and fabrication difficulties limit its use to the X band and higher. Octave bandwidths have been quoted for real-time designs such as the one in Figure 9.17 [Valentino et al. 1980].

A wide-angle array-fed lens antenna (WAAFL) is described as a marriage between a two-dimensional dome antenna and the Rotman lens-fed array [Thomas 1979]. A curved feed array is used for the dome-type scan amplification lens; see Figure 9.18. Octave bandwidths have been reported and scanning works well over 180°.

9.4 DIGITAL BEAM FORMING

Digital beam forming (DBF) in array antenna systems represents the ultimate solution for flexibility, beam shaping, and multiple beam generation [Steyskal 1987]. DBF makes it possible to fully exploit the spatial dimension of the antenna, and it is sometimes combined with advanced processing in the temporal/frequency domain as well. Examples of signal processing antenna functions include adaptive interference suppres-

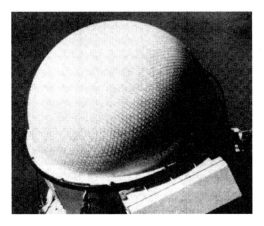

Figure 9.16. The dome array antenna with 3636 passive lens elements fed by a planar array with 805 elements and phase shifters [Fowler 1998].

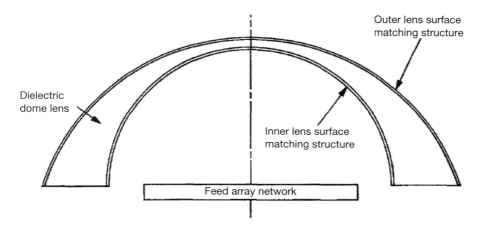

Figure 9.17. A dielectric dome lens antenna [Valentino et al. 1980].

sion, high-resolution direction finding, and multiple beam generation [Wirth 1978, Gröger et al. 1990]. STAP is an advanced space–time adaptive processing scheme for suppressing jamming signals and ground clutter in airborne surveillance radar systems [Klemm 1993].

It seems that everything you can do with analog beam forming can also be done digitally (Figure 9.19). The digital solution leads to rather extreme requirements on data transmission, storage, and signal processing, especially for large arrays. However, these problems are more easily solved today than in the past, thanks to the rapid growth of com-

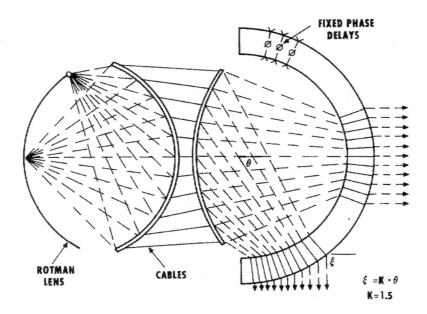

Figure 9.18. A broadband multiple array feed using a two-dimensional dome output lens [Thomas 1979].

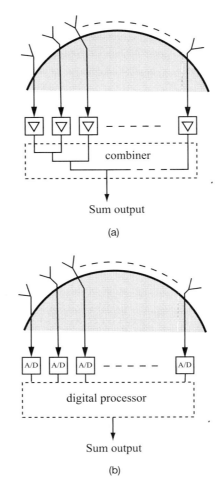

Figure 9.19. Analog beam forming (a) and digital beam forming (b) can in principle perform the same functions.

puter power, both software and hardware. In analog reception beam forming, the element signals are combined with weights determined by feed networks and/or phase and amplitude controlled receiver modules. The same operations can be performed in a digital computer on the element signals after analog-to-digital conversion. The digital approach allows the formation of simultaneous receive beams, even overlapping, without the feed losses that are incurred in an analog system. The element modules in the digital case have low noise amplifiers (LNA) preceding the analog-to-digital conversion. The signal-to-noise ratio is set by the LNAs and it is, therefore, not affected by transmission losses in further processing, resulting in "lossless beam forming."

Digital beam forming is also possible on transmission [Magill 2000, Jiaguo et al. 2003], but the advantages are less apparent. Once the beam is transmitted, there is no possibility to modify the beam shape or to do any other spatial signal processing. However, digital synthesis of the transmitted waveform on the element level combined with DBF on reception can provide interesting system capabilities in terms of, for example, LPI (low probability of intercept) radar with jamming resistance [Dorey et al. 1989]. A wide transmission beam illuminating the area of interest and multiple, narrow, digitally formed receive beams has also been suggested for LPI systems—"ubiquitous radar" [Skolnik 2003] and "OLPI radar" (Omnidirectional LPI) [Wirth 1995].

The capabilities of digital beam forming are well suited to the requirements posed by conformal array antennas. The need for amplitude and phase control, switching of the active sector (commutating), polarization control, compensating for element patterns in the beam steering algorithms, calibration, and so on are all aspects that can best be implemented digitally.

A DBF antenna system consists of combinations of several subsystems and components. Possible imperfections in these, which influence the overall system performance, include receiver channel imbalance, amplitude and phase errors, A/D converter offset errors, and frequency dependent errors. The imperfections can be more or less critical depending on the requirements and the type of processing used. Array calibration and special error correction schemes must typically be included in the antenna system design [Steyskal 1991].

Three experimental conformal digital beam forming X-band array antennas were described in a Japanese session at the 1996 Phased Array Symposium [Kanno et al. 1996a,b; Rai 1996]. One of these arrays had the shape of an elliptical cylinder matching an aircraft wing contour. The antenna used digital beamforming in two dimensions with 570 active elements. The synthesis of sector beams and low sidelobes (–40 dB) was demonstrated. The antenna is shown in Figure 1.2, page 3. The other two, also with about 600 active elements, were doubly curved, matching the shape of the nose of an aircraft and the fuselage, respectively. Haruyama and coworkers [1989] presented an experimental DBF array with 64 elements on a sphere. They were able to measure –44 dB sidelobes thanks to very precise weight control. The creation of multiple beams was also demonstrated.

Figure 9.20 shows one of several X-band arrays used for digital beam forming investigations [Gniss 1998]. The radiators are dual-polarized patch elements. 64 receive channels were used and beamforming quality including polarization characteristics was demonstrated.

Figure 9.20. Elliptical array with dual-polarized patch elements in cavities for DBF experiments. See also color insert, Figure 9. (Courtesy of FGAN Research Establishment for Applied Science, Wachtberg-Werthhoven, Germany.)

9.5 ADAPTIVE BEAM FORMING

9.5.1 Introduction

In adaptive array antenna systems [Gabriel 1992, Hudson 1981], special signal processing algorithms are applied to the digitized received signals. By deriving optimum weights for the element channels, based on the received signal characteristics, including noise and interference, the signal-to-noise-plus-interference ratio (SNIR) can be maximized. The result can be seen as creating nulls in the reception pattern in the directions of the interfering sources while preserving a high-gain beam toward the useful signal. A simple schematic for an adaptive beam forming system is shown in Figure 9.21.

We will here present a short outline of the mathematics used in adaptive nulling according to the sample matrix inversion (SMI) method [Applebaum 1976]. We will also illustrate the method by a simulation in which two jamming signals are to be suppressed using a DBF circular array. Finally, an example of superresolution techniques will be given. In Chapter 10, the adaptive method is applied to pattern synthesis problems.

Antenna pattern nulls can, of course, be generated by analog means as well. In the simplest case, this is done without adapting to the received signals using suitable phase-control elements in the feed circuit [Davies and Rizk 1977]. However, we will in the following focus on the more advanced adaptive principle, typically implemented in digital beam forming arrays.

9.5.2 The Sample Matrix Inversion Method

The signal vector received from an N-element array of general shape (planar, conformal, etc.), can be written as

$$\bar{x}(t) = [x_1(t), x_2(t), \ldots, x_N(t)]^T = V_S(t)\bar{x}_S^T \tag{9.5}$$

Here V_S is the signal strength and $\bar{x}_S$ the directional vector for the signal source direction; the superscript T stands for transpose. With complex weights $\bar{w} = [w_1, w_2, \ldots, w_N]^T$ we get the output $y(t)$ from the array:

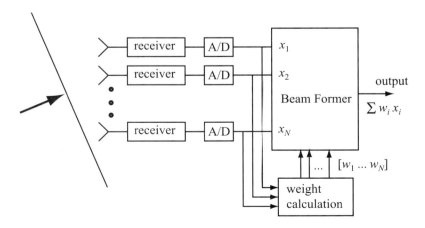

Figure 9.21. An adaptive beam forming antenna system with received signals [x_1, ..., x_N] forming the signal vector.

$$y(t) = \overline{w}^T \cdot \overline{x}(t) \tag{9.6}$$

The problem is to determine the weight vector such that an optimum (in some sense) output signal is obtained. One element $x_i(t)$ in the signal vector is

$$x_i(t) = V_S(t)EL_i(\phi_S)e^{j\overline{k}_S \cdot \overline{r}_i} \tag{9.7}$$

where $\overline{k}_S$ is the propagation vector in the signal direction. Equivalent expressions are obtained for the external noise sources. Note that the radiating elements with patterns $EL_i(\phi_S)$ and positions $\overline{r}_i$ can all be different and point in different directions. Hence, the theory applies equally well to planar and nonplanar arrays. The received signal power can be written as

$$P_S = |\overline{w}^T \overline{x}_S|^2 = \overline{w}^T \overline{x}_S^* \overline{x}_S^T \overline{w}^* \tag{9.8}$$

(the superscript * stands for complex conjugate). Summing over all noise sources we obtain

$$P_N = \overline{w}^T(\overline{x}_N^* \overline{x}_N^T + n_0 \overline{I})\overline{w}^* \tag{9.9}$$

where the internal noise n_0 is assumed to be of equal power in all channels, otherwise uncorrelated. $\overline{I}$ is the identity matrix. It can be shown that the signal-to-noise ratio, that is, the ratio between the quadratic forms, P_S/P_N, is maximized by the weight vector [Hudson 1981]:

$$\overline{w}^* = \mu \, \overline{R}^{-1}\overline{x}_D \tag{9.10}$$

where μ is an arbitrary proportionality constant, $\overline{x}_D$ is the signal vector for the desired signal, and $\overline{R}$ is the covariance matrix for the total noise:

$$\overline{R} = \overline{x}_N^* \overline{x}_N^T + n_0 \overline{I} \tag{9.11}$$

In operational adaptive antennas, the noise covariance matrix elements can be estimated by time averaging over several noise samples, thus

$$R_{ij} = \frac{1}{T}\int_0^T x_i^*(t)x_j(t)dt \tag{9.12}$$

9.5.3   An Adaptive Beam Forming Simulation Using a Circular Array

As an example, we will demonstrate the use of the adaptive nulling algorithm SMI applied to a simple case with a cylindrical array antenna. Compared with linear and planar array antennas, the adaptation includes also the "rear" side of the array. This part is usually without control in the planar or linear case. We choose a uniformly spaced 16 element circular array with 0.5 λ element spacing. The elements are assumed to have a cos α pattern, with no radiation for $|\alpha| > 90°$. Mutual coupling is neglected. The scenario is shown in Figure 9.22 with the desired signal incident from 30° direction and jammers at 0° and 270°.

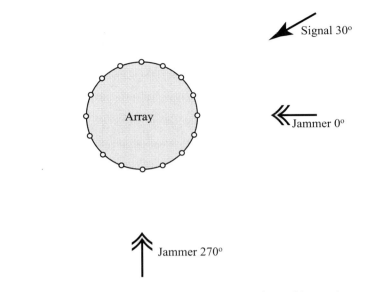

Figure 9.22. Test scenario with one signal competing with two jamming signals.

Whereas the desired signal vector for an M element (spacing d) uniform linear array (ULA) is simply

$$\bar{x}(u_d) = EL(\phi_d)[1, e^{ju_d}, e^{2ju_d}, \ldots, e^{(M-1)ju_d}]^T \qquad (9.13)$$

with $u_d = kd \sin \phi_d$, the corresponding signal vector for an N-element uniform circular array (UCA) is slightly more complicated:

$$\bar{x}(\phi_d) = [EL(\phi_d)e^{jkr \cos \phi_d}, EL(\phi_d - \Delta\varphi)e^{jkr \cos(\phi_d - \Delta\varphi)}, \ldots,$$
$$EL(\phi_d - (N-1)\Delta\varphi)e^{jkr \cos(\phi_d - (N-1)\Delta\varphi)}]^T \qquad (9.14)$$

For the above circular array, we simulate the following case:

- Noise level 0 dB
- Desired signal: 10 dB above noise, incident from an angle of $\phi_d = 30°$
- Interference signal 1: 40 dB level at $\phi_{i1} = 0°$
- Interference signal 2: 40 dB level at $\phi_{i2} = 270°$
- 100 samples are taken for estimating the covariance matrix.
- Noise and interfering signals are assumed random with a Gaussian distribution, independent from sample to sample.

Figure 9.23 shows the resulting pattern before (dashed line) and after adaptation (solid). The small circles indicate the positions of the 16 elements. The interfering signals have been suppressed by at least 25 dB. The pattern after adaptation remains roughly the same except for a slight broadening of the main beam.

Similar techniques can also be used for DOA (direction of arrival), that is, bearing

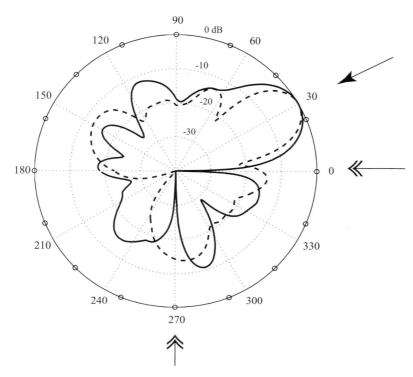

Figure 9.23. Cylindrical array pattern with (solid line) and without interference suppression (dashed line). Signal direction 30°, interfering signals at 0° and 270°.

measurements. With sufficient signal-to-noise ratio, closely spaced signals can be resolved, even when the angular spacing is much less than the antenna beamwidth ("superresolution"). As a simple example of digital superresolution techniques using a circular array, we will take the MUSIC algorithm (MUltiple SIgnal Classification) [Schmidt 1979]. The basis for this method is an eigen-decomposition of the covariance matrix into a signal subspace and an orthogonal noise subspace, $\overline{E}_n$. A testing vector $\overline{x}_s(\phi)$, representing a signal incident from an angle ϕ is scanned and projected onto the noise subspace. When the scan angle ϕ, is close to one of the signal directions, the projection becomes very small, due to the orthogonality between signal and noise subspaces. The inverse of this projection, the MUSIC spectrum, can be written as

$$P(\phi) = \frac{1}{\overline{x}_s^H(\phi)\,\overline{E}_n\,\overline{E}_n^H\,\overline{x}_s(\phi)} \tag{9.15}$$

where the superscript H indicates complex conjugate transpose. This spectrum exhibits sharp peaks at the locations of the signals as ϕ is varied. For the result shown in Figure 9.24, two signals at 25° and 35° with signal-to-noise ratios 20 dB were assumed. Other data are the same as for Figure 9.23.

In principle, a circular array can be used for DOA estimates in both the azimuth and elevation [Mathews and Zoltowski 1994], which is not possible with a linear array. Furthermore, the UCA covers 360° in the azimuth, whereas the ULA covers only about 120°.

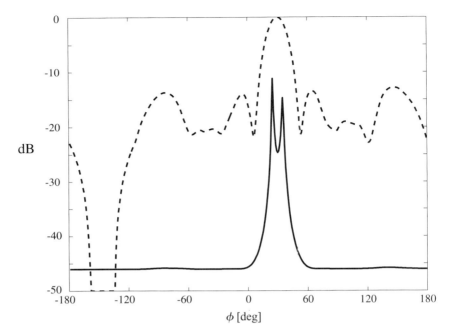

Figure 9.24. The MUSIC spectrum (solid line) with sources at 25° and 35°. The dashed line is the regular antenna pattern with the main beam pointing between the two sources.

However, it is desirable to be able to map the output signal vectors from a circular array onto a (virtual) linear array in order to be able to use existing fast ULA algorithms. This mapping principle has been analyzed by Hyberg and coworkers [2002]. It was assumed that the DOA was approximately known beforehand within a 30° sector of the circular array. [Recio et al. 2002] describe an interpolation scheme for transforming a spherical array onto a crossed ULA array for adaptive DOA processing, also taking mutual coupling effects into account. This requires that the mutual coupling be known accurately beforehand, which is not always the case.

There are many variations to the adaptive principles discussed in this section, and these are discussed extensively in the literature. The effects of errors, polarization, and mutual coupling can be included in the analysis. It is found that it is often necessary to make corrections by extensive calibrations [Steyskal 1991, Pettersson et al. 1997, Svantesson 2001].

9.6 REMARKS ON FEED SYSTEMS

Selecting and designing a feeding and scanning system for a conformal array antenna is not a simple matter. This chapter has provided some examples of designs that have been used in the past. The majority of them are intended for circular arrays. This reflects the standing of the technology today, since there are not yet so many doubly curved arrays in use.

One clever solution for a doubly curved array is the dome antenna. Although not used very much in practice, it illustrates the possibility of using optical means for feeding and commutating large arrays. Another approach for doubly curved, rotationally symmetric arrays could be to use stacked circular array solutions.

Cylindrical arrays can be used for two-dimensional scanning. It should be noted, however, that ordinary "row-and-column" steering does not work as with planar arrays. The focusing in the azimuth depends on the elevation beam steering. On the other hand, we have observed that a single ring array can be steered both in the azimuth and elevation.

The polarization issue is part of the beam steering problem, since element polarization often needs to be changed with the scan direction. The matter is further discussed in Chapters 8 and 10.

Many of the beam scanning solutions require feed matrices with a large number of switches, which typically results in networks with large transmission losses. The development of RF microelectromechanical systems (MEMS) switches may offer new possibilities, since these devices are very small, can be integrated with other circuits, and have low transmission losses. They are currently limited to low-power applications [Nguyen et al. 1998, Rebeiz and Muldavin 2001].

The final solution, at least for receiver arrays, may well be found in the use of digital beam forming, which is well suited to the requirements posed by conformal array antennas. Amplitude and phase control, commutating, polarization control, compensating for element patterns in beam steering algorithms, calibration, and so on can all be implemented digitally. Digital beam forming will probably also be used for the transmission function in the future. An alternative may be to employ wide floodlight transmit beams using a separate aperture and not requiring sophisticated scanning. Multiple beams, high resolution, and interference suppression will be realized in the digital domain. The digitized received signals can be routed from the array element modules to processing units on optical fibers.

REFERENCES

Allen J. L. (1961), "A Theoretical Limitation on the Formulation of Lossless Multiple Beams in Linear Arrays," *IRE Transactions on Antennas and Propagation,* Vol. AP-9, No. 4, pp. 350–352.

Applebaum S. P. (1976), "Adaptive Arrays," *IEEE Transactions on Antennas and Propagation,* Vol. AP-24, No. 5, pp. 584–598.

Applebaum S. P. and Chapman D. J. (1976), "Adaptive Arrays with Main Beam Constraints," *IEEE Transactions on Antennas and Propagation,* Vol. AP-24, No. 5, pp. 650–662.

Archer D. H. (1984), "Lens-Fed Multiple Beam Arrays," *Microwave Journal,* pp. 171–195, September.

Arora R. K. and Patel M. R. (1983), "Performance Degradation in Fast Frequency-Scanned Circular Arrays," *IEEE Transactions on Antennas and Propagation,* Vol. AP-31, No. 3, pp. 509–512.

Bearse S. V. (1975), "Planar Array Looks through Lens to Provide Hemispherical Coverage," *Microwaves,* July, pp.9–10.

Bodnar D. G., Barnes W. J., and Corey L. E. (1980), "A Lightweight HWL Radar Antenna," in *Proceedings of IEEE International Radar Conference,* pp. 259–262.

Bodnar D. G., Rainer B. K., and Rahmat-Samii Y. (1989), "A Novel Array Antenna for MSAT Applications," *IEEE Transactions on Vehicular Technology,* Vol. 38, No. 2, pp. 86–94.

Bogner B. F. (1974), "Circularly Symmetric RF Commutator for Cylindrical Phased Arrays," *IEEE Transactions on Antennas and Propagation,* Vol. AP-22, No. 1, pp. 78–81.

Boyns J. E., Munger A. D., Provencher J. H., Reindel J., and Small B. I. (1968), "A Lens Feed for a Ring Array," *IEEE Transactions on Antennas and Propagation,* Vol. AP-16, No. 2, pp. 264–268.

Boyns J. E., Gorham C. W., Munger A. D., Provencher J.H., Reindel J., and Small B. I. (1970), "Step-Scanned Circular-Array Antenna," *IEEE Transactions on Antennas and Propagation,* Vol. AP-18, No. 5, pp. 590–595.

Brown R. (1971), "The Uniqueness of the Butler Matrix as a Commutating Switch," *IEEE Transactions on Antennas and Propagation,* Vol. AP-19, No. 5, pp. 694–695.

Butler J. and Lowe R. (1961), "Beamforming Matrix Simplifies Design of Electronically Scanned Antennas," *Electronic Design,* Vol. 9, pp. 170–173, 12 April.

Caille G., Vourch E., Martin M. J., Mosig J. R., and Martin Polegre A. (2002), "Conformal Array Antenna for Observation Platforms in Low Earth Orbit," *IEEE Antennas and Propagation Magazine,* Vol. 44, No. 3, pp. 103–104.

Cassel K. E. (1982), "A Broad Band ESM/ECM Antenna with 360 Degrees Multibeam Coverage," in *Proceedings of Military Microwaves Conference,* London, pp. 512–518, October.

Christopher E. J. (1974), "Electronically Scanned TACAN Antenna," *IEEE Transactions on Antennas and Propagation,* Vol. AP-22, No. 1, pp. 12–16.

Davies D. E. N. (1965), "A Transformation Between the Phasing Techniques Required for Linear and Circular Aerial Arrays," *Proceedings of IEE, Part H,* Vol. 112, No. 11, pp. 2041–2045.

Davies D. E. N. and Rizk M. (1977), "Electronic Steering of Multiple Nulls for Circular Arrays," *Electronics Letters,* Vol. 13, No. 22, pp. 669–670.

Dorey J., Garnier G., and Auvray G. (1989), "RIAS, Synthetic Impulse and Antenna Radar," in *Proceedings of International Conference on Radar,* Paris, pp. 556–562, 24–28 April.

Fowler C. A. (1998), "Old Radar Types Never Die; They Just Phased Array or 55 Years of Trying to Avoid Mechanical Scan," *IEEE AES Systems Magazine,* pp. 24A-24L, September.

Fullilove N. N., Scott W. G., and Tomlinson J. R. (1959), "The Hourglass Scanner. A New Rapid Scan, Large Aperture Antenna," *IRE International Convention Record,* Vol. 7, pp. 190–200, March 1959.

Gabriel W. (1992), "Adaptive Processing Array Systems," *Proceedings of IEEE,* Vol. 80, No. 1, pp. 152–162.

Gething P. J. D. (1966), "High-Frequency Direction Finding," *Proceedings of IEE,* Vol. 113, No. 1, pp. 49–61.

Giannini R. J. (1969), "An Electronically Scanned Cylindrical Array Based on a Switching-and-Phasing Technique," in *Proceedings of IEEE International Symposium AP-S,* pp. 199–207, December.

Gniss H. (1998), "Digital Rx-only Conformal Array Demonstrator," in *Proceedings of Internat. Radar Symposium IRS 98,* 15–17 September, pp. 1003–1012.

Gregorwich W. S. (1974), "An Electronically Despun Array Flush-Mounted on a Cylindrical Spacecraft," *IEEE Transactions on Antennas and Propagation,* Vol. AP-22, No. 1, pp. 71–74.

Gröger I., Sander W., and Wirth W-D. (1990), "Experimental Phased Array Radar ELRA with extended Flexibility," *IEEE AES Magazine,* pp. 26–30, November.

Guy J. R. F. and Davies D. E. N. (1983), "Novel Method of Multiplexing Radiocommunication Antennas Using Circular-Array Configuration," *Proceedings of IEE, Part H,* Vol. 130, No. 6, pp. 410–414.

Hall P. S. and Vetterlein S. J. (1990), "Review of Radio Frequency Beamforming Techniques for Scanned and Multiple Beam Antennas," *Proceedings of IEE Part H,* Vol. 137, No. 5, pp. 293–303.

Hall P. S. and Vetterlein S. J. (1992), "Integrated Multiple Beam Microstrip Arrays," *Microwave Journal,* pp. 103–114, January.

Haruyama T., Kojima N., Chiba I., Oh-Hashi Y., Orime N., and Katagi T. (1989), "Conformal Array Antenna with Digital Beam Forming Network," in *Proceedings of IEEE International Symposium AP-S,* pp. 982–985.

Hering K. H. (1970), "The Design of Hybrid Multiple Beam Forming Networks," in *Proceedings of 1970 Phased-Array Antenna Symposium,* pp. 240–242, 2–5 June.

Holley A. E., DuFort E. C., and Dell-Imagine R. A. (1974), "An Electronically Scanned Beacon Antenna," *IEEE Transactions on Antennas and Propagation,* Vol. AP-22, No. 1, pp. 3–12.

Hollis J. S. and Long M. W. (1957), "A Luneberg Lens Scanning System," *IRE Transactions on Antennas and Propagation,* Vol. AP-5, No. 1, pp. 21–25, January.

Honey R. C. and Jones E. M. T. (1957), "A Versatile Multi-Port Biconical Antenna," *IRE Convention Record,* pp. 129–141.

Hudson J. E. (1981), "*Adaptive Array Principles,*" Peter Peregrinus Ltd., London.

Hyberg P., Jansson M., and Ottersten B. (2002), "Sector Array Mapping: Transformation Matrix Design for Minimum MSE," in *Proceedings of 36th Asilomar Conference on Signals, Systems and Computers,* 3–6 November, pp. 1288–1292.

Irzinski E. P. (1981), "A Coaxial Waveguide Commutator Feed for a Scanning Circular Phased Array Antenna," *IEEE Transactions on Microwawve Theory and Technology,* Vol. 29, No. 3, pp. 266–270.

Jiaguo L., Manqin W., Xueming J., and Zhenxing F. (2003), "Active Phased Array Antenna Based on DDS," in *Proceedings of Phased Array Systems and Technology 2003,* 14–17 October, pp. 511–516.

Kanno M., T. Hashimura, T. Katada, M. Sato, K. Fukutani and A. Suzuki (1996a), "Digital Beam Forming for Conformal Active Array Antenna," in *Proceedings of IEEE International Symposium on Phased Array Systems and Technology,* 15–18 October, pp. 37–40.

Kanno M., T. Hashimura, T. Katada, Watanabe T., Iio S., Soga H., Nakada T., and Miyano N. (1996b), "An Active Conformal Array Antenna with very Small Thickness," in *Proceedings of IEEE International Symposium on Phased Array Systems and Technology,* 15–18 October, pp. 155–158.

Keeping K. J., Rogers D. S., and Sureau J.-C. (1981), "A Scanning Switch Matrix for a Cylindrical Array," in *Proceedings of IEEE International Symposium on Microwave Theory and Techniques,* pp. 419–421, June.

Kinsey R. R. (2000), "An Objective Comparison of the Dome Antenna and a Conventional Four-Face Planar Array," in *Proceedings of 2000 Antenna Applications Symposium,* Allerton Park, Illinois, USA, 20–22 September, pp. 360–372.

Klemm R. (1993), "Adaptive Air- and Space-Borne MTI under Jamming Conditions," in *Proceedings of IEEE National Radar Conference,* pp. 167–172.

Lee J. J., Chu R. S., Livingston S., and Koenig R. (2002), "Ultra Wide Band Cylindrical Array and 360–Degree Beam Scan System," in *Proceedings of IEEE International Symposium AP-S,* San Antonio, 16–21 June, pp. 212–215.

Liebman P. M., Schwartzman L., and Hylas A. E. (1975), "Dome Radar—A New Phased Array System," in *Proceedings of IEEE International Radar Conference,* Washington, D.C., 21–23 April, pp. 349–353.

Magill E. (2000), "Digital Beamforming Phased Array Transmit Antenna," in *Proceedings of AP 2000 Millenium Conference on Antennas and Propagation,* Davos, Switzerland, 9–14 April.

Martín Polegre A. J., Roederer A., Caille G., and Boyer L. (2003), "Conformal Phased Array for ESA's GAIA Mission," in *Proceedings of 3rd European Workshop on Conformal Antennas,* Bonn, Germany, 22–23 October, pp. 99–102.

Martín Polegre A. J., Roederer A. G., Crone G. A. E., and de Maagt P. J. I (2001), "Applications of Conformal Array Antennas in Space Missions," in *Proceedings of 2nd European Workshop on Conformal Antennas,* The Hague, The Netherlands, 24–25 April.

Mathews C. P. and Zoltowski M. D. (1994), "Eigenstructure Techniques for 2–D Angle Estimation with Uniform Circular Arrays," *IEEE Transactions on Signal Processing,* Vol. 42, No. 9, pp. 2395–2407.

Neff H. P. and Tillman J. D. (1960), "An Electronically Scanned Circular Antenna Array," *IRE International Convention Record,* Vol. 8, pp. 41–47, March.

Nelson E. A. and Britt P. P. (1972), "Cylindrical Phased Arrays: Beam Scanning and Sidelobe Control," in *NAECON '72 Record,* pp. 189–196.

Nester W. (1968), "The Fast Fourier Transform and the Butler Matrix," *IEEE Transactions on Antennas and Propagation,* Vol. AP-16., No. 3, 360.

Nguyen C. T.-C., Katehi P. B., and Rebeiz G. M. (1998), "Micromachined Devices for Wireless Communications," *Proceedings of IEEE,* Vol. 86, No. 8, pp. 1756–1768.

Peeler G. D. M. and Archer D. H. (1953), "A Two-Dimensional Microwave Luneberg Lens," *IRE Transactions on Antennas and Propagation,* Vol. AP-1., No. 1, pp. 12–23.

Peeler G. D. M., Kelleher K. S., and Hibbs H. H. (1952), "An Organ-Pipe Scanner," *IRE Transactions on Antennas and Propagation,* Vol. 1, pp. 113–122.

Peeler G. D. M. and Gabriel W. F. (1955), "Volumetric Scanning GCA Antenna," *IRE National Convention Record,* Vol. 3, Part 1, pp. 20–27, December.

Pettersson L., Danestig M., and Sjöström U. (1997), "An Experimental S-band Digital Beamforming Antenna," *IEEE Aerospace and Electronics Systems Magazine,* Vol. 12, No. 1, pp. 19–27.

Provencher J. H. (1970), "A Survey of Circular Symmetric Arrays," in *Proceedings of Phased Array Symposium,* Polytechnic Institute of Brooklyn, pp. 292–300.

Raabe H. P. (1976), "Fast Beamforming with Circular Receiving Arrays," *IBM Journal of Research and Development,* pp. 398–408, July.

Raffaelli S. and Johansson M. (2005), "Cylindrical Array Antenna Demonstrator: Simultaneous Pencil and Omni-directional Beams," in *Proceedings of 4th European Workshop on Conformal Antennas,* Stockholm, Sweden, 23–24 May, pp. 93–96.

Rahim T., Guy J.R.F, and Davies D.E.N. (1981), "A Wideband UHF Circular Array," in *IEE Proceedings of 2nd International Conference on Antennas and Propagation,* Part 1, pp. 447–450, April.

Rai E., Nishimoto S., Katada T., and Watanabe H. (1996), "Historical Overview of Phased Array Antenna for Defense Application in Japan," in *Proceedings of IEEE International Symposium Phased Array Systems and Technology,* 15–18 October, pp. 217–221.

Rebeiz G. M. and Muldavin J. B. (2001), "RF MEMS Switches and Switch Circuits," *IEEE Microwave Magazine,* pp. 59–71, December.

Recio R. F., Sarkar T. K., and Kim K. (2002), "Elimination of Mutual Coupling in a Conformal Adaptive Array Antenna," in *Proceedings of IEEE International Symposium AP-S,* 16–21 June, pp. 106–109.

Rehnmark S. (1980a), "Instantaneous Bearing Discriminator with Omnidirectional Coverage and High Accuracy," *IEEE MTT-S International Symposium Digest,* pp. 120–122, May.

Rehnmark S. (1980b), "Improved Angular Discrimination for Digital ESM Systems," in *Proceedings of 2nd Military Microwaves Conference,* pp. 157–162, October.

Reindel J. (1974), "A Matrix Feed System for a Cylindrical Antenna Array," in *Proceedings of 4th European Microwave Conference on,* Montreux, 10–13 September, pp. 288–292.

Rotman W. and Turner R. F. (1963), "Wide-Angle Microwave Lens for Line Source Applications," *IEEE Transactions on Antennas and Propagation,* Vol. AP-11, No. 6, pp. 623–632. See also Leonakis G. L. (1986), "Correction to 'Wide-Angle Microwave Lens for Line Source Applications'," *IEEE Transactions on Antennas and Propagation,* Vol. AP-34, No. 8, p. 1067.

Schmidt R. O. (1979), "Multiple Emitter Location and Signal Parameter Estimation," in *Proceedings of RADC Spectrum Estimation Workshop,* October, 1979, pp. 243–258. (Reprinted in *IEEE Transactions on Antennas and Propagation,* Vol. AP-34., No. 3, pp. 276–280, March 1986.)

Schwartzman L. S. and Stangel J. (1975), "The Dome Antenna," *Microwave Journal,* pp. 31–34, October.

Sheleg B. (1968), "A Matrix Fed Circular Array for Continuous Scanning," *IEEE AP-S International Symposium Digest,* pp. 7–16, 9–11 September.

Sheleg B. (1973), "Butler Submatrix Feed Systems for Antenna Arrays," *IEEE Transactions on Antennas and Propagation,* Vol. AP-21, No.2, pp.228–229.

Shelton J. P. and Kelleher K. S. (1961), "Multiple Beams from Linear Arrays," *IRE Transactions on Antennas and Propagation,* Vol. AP-9, No. 2, pp.154–161.

Shestag L. N. (1974), "A Cylindrical Array for the TACAN System," *IEEE Transactions on Antennas and Propagation,* Vol. AP-22, No. 1, pp.17–25.

Sierra M. and Bernal J. C. (1989), "Multibeam Antenna Array," *IEEE AP-S International Symposium Digest,* pp. 150–153, 26–30 June.

Skahill G. and White W. D. (1975), "A New Technique for Feeding a Cylindrical Array," *IEEE Transactions on Antennas and Propagation,* Vol. AP-23, No. 2, pp. 253–256.

Skolnik M. (2003), "Attributes of the Ubiquitous Phased Array Radar," in *Proceedings of IEEE International Symposium on Phased Array Systems and Technology 2003,* Boston, USA, 14–17 October, pp. 101–105.

Stein S. (1962), "On Cross Coupling in Multiple-Beam Antennas," *IRE Transactions on Antennas and Propagation,* Vol. AP-10, No. 5, pp. 548–557.

Steyskal H., Hessel A., and Shmoys J. (1979), "On the Gain-versus-Scan Trade-Offs and the Phase-Gradient Synthesis for a Cylindrical Dome Antenna," *IEEE Transactions on Antennas and Propagation,* Vol. AP-27, No. 6, pp. 825–831.

Steyskal H. (1987), "Digital Beamforming Antennas—An Introduction," *Microwave Journal,* pp. 107–124, January.

Steyskal H. (1991), "Array Error Effects in Adaptive Beamforming," *Microwave Journal,* pp. 101–112, September.

Svantesson T. (2001), *Antennas and Propagation from a Signal Processing Perspective,* Ph.D. Thesis, Chalmers University of Technology, Dept. of Signals and Systems.

Thies W. H. (1973), "Omnidirectional Multibeam Array Antenna," U.S. Patent 3,754,270. 21 August.

Thomas D. T. (1979), "Design Studies of Wide Angle Array Fed Lens," *IEEE AP-S International Symposium Digest,* pp. 340–343.

Uhlman M. (1975a), "Möglichkeiten der Speisung für Phasengesteuerte Zylinder-Strahlgruppen," *NTZ,* Vol. 28, Nr. 9, pp. 299–305.

Uhlmann M. (1975b), "Cylindrical Phased Array with PIN-Diode Controlled Parallel-Plate Feeding System," in *Proceedings of 5th European Microwave Conference,* Hamburg, 1–4 September, pp. 458–465.

Valentino P. A., Rothenberg C., and Stangel J. J. (1980), "Design and Fabrication of Homogeneous Dielectric Lenses for Dome Antennas," *IEEE AP-S International Symposium Digest,* pp. 580–583.

White W. D. (1962), "Pattern Limitations in Multiple-Beam Antennas," *IRE Transactions on Antennas and Propagation,* Vol. AP-10, No. 4, pp. 430–436.

Wirth W. D. (1978), "Radar Signal Processing with an Active Receiving Array," in *Proceedings of Military Microwaves Conference,* London, pp. 420–431.

Wirth W. D. (1995), "Long Term Coherent Integration for a Floodlight Radar," in *Proceedings of IEEE International Radar Conference,* 8–11 May, pp. 698–703.

Wolfson R. I. and Cho C. F. (1979), "A Low-Inertia Commutating Feed for Circular Arrays," in *Proceedings of Antenna Application Symposium,* September.

Wolfson R. I., Cho C. F., and Charlton G. G. (1980), "Experimental Results of a Commutating Feed for Circular Arrays," in *Proceedings of Antenna Application Symposium,* September.

Vourch E., Caille C., Martin M. J., Mosig J. R., Martin A., and Iversen P. O. (1998), "Conformal Array Antenna for LEO Observation Platforms," *IEEE AP-S International Symposium Digest,* pp. 20–23.

CONFORMAL ARRAY
PATTERN SYNTHESIS

In this chapter, we will address the synthesis problem for conformal array antennas. Previous chapters have mostly dealt with analysis of various types and shapes of conformal antennas. The inverse problem—synthesis—is usually more difficult than the analysis. It is in many cases nonlinear, and there is usually no unique solution to the problem. And yet, this is the problem that faces the design engineer all the time.

The main focus here is on various methods for antenna pattern synthesis. However, synthesis of a conformal array antenna entails more than synthesizing a pattern. One should also include the problems of optimizing the shape, the distribution of radiating elements, optimizing the impedance characteristics, polarization, bandwidth, and so on. A coherent approach to all of this is not possible due to the difficulty of specifying all requirements in a rigorous way. We will, however, touch upon these aspects as well, and some related material is also found in the other chapters.

10.1 INTRODUCTION

Antenna pattern synthesis is a vast subject in the antenna literature. The amount of publications is extensive; see, for example [Schell and Ishimaru 1969, Bucci et al. 1994]. Most papers deal with linear array antennas; the results can then be carried over to planar antennas. It is in general not possible to use these results also for curved arrays, although projections of excitations from a planar case to a curved array have been attempted.

Conformal Array Antenna Theory and Design. By Lars Josefsson and Patrik Persson
© 2006 Institute of Electrical and Electronics Engineers, Inc.

Several of the classical methods used for linear arrays are both elegant and analytical. We have, for instance, the Woodward–Lawson synthesis, Dolph–Chebyshev synthesis, Fourier synthesis, and methods derived from the Taylor line-source synthesis, to mention just a few [Hansen 1992]. Since 1970, numerical synthesis methods have become increasingly popular. The objectives can be divided into (a) pencil beam/low sidelobe synthesis, (b) shaped beam synthesis, and (c) optimization of performance indices such as directivity and bandwidth.

Today, iterative methods based on optimization techniques are very powerful tools for pattern synthesis thanks to the modern computer. Arbitrarily shaped beam patterns can be synthesized with minimum effort (as long as the patterns are physically realizable). Some of the more general methods can also be applied to nonplanar array antennas, that is, conformal arrays. It may still be difficult to predict errors beforehand and to ensure that the numerical solution is not trapped in a local optimum.

Finding a radiation pattern that meets the required pattern shape is only part of the job. Next, we must find the corresponding excitations of the radiating elements. And, since elements must have a feeding structure, we need to find feeding voltages and currents that will result in the desired excitations in the presence of mismatch, mutual coupling, and so on. Finally, the robustness of the solution should be assessed, as well as the impact of frequency variations, error tolerances, polarization and so on. Maybe a more stable solution is required instead of the first one derived.

In conformal arrays, the radiating elements point in different directions. This makes it essential to include the element characteristics in the synthesis procedure. In planar arrays, the element factor can often be disregarded and the synthesis is concentrated on the array factor as if the elements were isotropic. However, it is still of interest to study the synthesis of arrays with isotropic radiators, since much fundamental knowledge can be gained this way. And some arrays do have isotropic radiators, at least in one plane. We will, therefore, present some cases with isotropic radiators as canonical cases of reference.

The polarization information is carried by the radiating elements. Since they do not combine to form a common element factor, as in planar arrays, the polarization characteristics become particularly significant for the conformal antenna synthesis problem. It is possible to include polarization in the synthesis; for example, to optimize for low sidelobes and low cross polarization at the same time.

Low sidelobe/narrow main beam synthesis will be discussed in relation to Dolph–Chebyshev synthesis for circular arrays. Both isotropic elements and directive elements will be treated. The method makes use of phase-mode theory [Taylor 1952, Davies 1981] and is based on the discussion of this topic in Chapter 2.

The pattern requirements are most often expressed in power (or decibels) and the pattern phase is usually not important. If this can be exploited in the synthesis, it adds degrees of freedom and gives better results than a field synthesis. Beam shaping, as in cosecant-squared patterns, is a typical example where phase is of no concern. Thus, more than one solution for the excitation to the synthesis problem may exist.

Adaptive array theory is a cross-disciplinary subject combining antenna arrays and signal processing. Signal processing algorithms in adaptive arrays make it possible to modify the radiation pattern according to changing environments by, for example, placing pattern nulls in directions where there is strong interference. The methods can also be applied in a more static fashion, that is, to find excitations for low sidelobes in specific angular regions.

All the essential aspects of antenna synthesis for conformal array antennas can not be covered in a limited space. We will, however, discuss principles and by examples illustrate some of the important points that we have mentioned here. The chapter concludes with a detailed presentation of a shaped-beam/low-sidelobes synthesis for a cylindrical array using the method of alternating projections. Least squares approximation is included and the effect of mutual coupling is illustrated.

Finally, an effort will be made to summarize the usefulness of the various synthesis methods in tabular form.

10.2   SHAPE OPTIMIZATION

The shape of the conformal antenna determines its radiation characteristics to a great extent. The shape is, therefore, an important design parameter. The shape is in many cases dictated by other requirements like aerodynamics in the case of an aircraft, but in other cases we may be free to choose a suitable antenna shape. Conformal antenna shapes were discussed in detail in Chapter 3 and also in Chapter 8 in relation to packaging considerations.

There is currently an interest in shape optimization for a variety of applications. For large structures, there will be a large number of design parameters. One would, therefore, need efficient optimization methods such as gradient methods, which require the calculation of the derivatives with respect to the design parameters. It has been shown that this calculation can be done very efficiently from a single solution of the adjoint problem [Pironneau 1984]. Shape optimization methods have been used successfully in aerodynamics [Weinerfelt and Enoksson 2001], but also for microwave devices [Chung et al. 2000] and for radar cross section reduction [Bondeson et al. 2004]. The technique could certainly also be applied to optimizing the shape of conformal antennas.

10.3   FOURIER METHODS FOR CIRCULAR RING ARRAYS

In Chapter 2, we analyzed circular arrays and their radiation properties. We demonstrated how the circular antenna excitation can be expanded in phase modes (for a discussion of phase modes, see Section 2.3). The excitation phase modes are related to a similar far field phase-mode expansion, which uniquely defines the radiation pattern. The inverse problem, i.e. synthesizing the excitation according to a required antenna pattern shape, follows directly. Let us take a circular array with isotropic radiators. By expanding the desired far field $E^{\text{des}}(\phi)$ in a Fourier series of phase modes we write

$$E^{\text{des}}(\phi) = \sum_{-M}^{M} A_m e^{jm\phi} \tag{10.1}$$

where the coefficients A_m, the far field phase-mode amplitudes, are obtained from the desired pattern according to the inverse transform,

$$A_m = \frac{1}{2\pi} \int_0^{2\pi} E^{\text{des}}(\phi) e^{-jm\phi} d\phi \tag{10.2}$$

The number of used far field modes has been limited to $\pm M$, hence, the synthesis is approximate. When isotropic radiators are used, the excitation phase-mode amplitudes C_m are found [cf. Eq. (2.25)] from

$$C_m = \frac{1}{j^m J_m(kR)} A_m \qquad (10.3)$$

The circular antenna excitation is then given by

$$V(\varphi) = \sum_{-M}^{M} C_m e^{jm\varphi} \qquad (10.4 = 2.18)$$

The number of modes M should not exceed kR, where R is the array radius, according to the discussion in Section 2.3. A few higher modes could possibly be added, but they will not radiate efficiently. If more modes are necessary to match the desired pattern shape, one will, therefore, have to choose a larger array diameter.

The Fourier method solution provides a match to the desired pattern in the least mean square sense. Drawbacks of the Fourier method are oscillations and overshoots (Gibbs' phenomenon) that appear when the desired pattern has rapid variations or discontinuities; see, for example [Schell and Ishimaru 1969].

The expression for the circular array radiation with $2M + 1$ phase modes [Eq. (10.1)], is similar to the corresponding expression for a linear array of $2N + 1$ isotropic elements:

$$E_{\text{lin}}(\phi) = \sum_{-N}^{N} V_n e^{jknd \sin \phi} \qquad (10.5)$$

Thus, if the far field phase-mode excitations A_m are set equal to the element excitations V_n, the same pattern functions will result. However, the pattern in the circular array case would be given in the ϕ space, whereas the pattern in the linear array case would be given in the $\sin \phi$ space. One can, for example, create $\sin x/x$ type patterns for both arrays in the two domains by equating far field phase-mode amplitudes and element amplitudes, respectively.

An obvious problem regarding synthesis arises if the denominator in Equation (10.3) vanishes for modes that are essential for the desired pattern. Another observation is that the method does not always give good results because of radiation of spurious modes. We will take a closer look at these effects and the role of the element pattern in the next section. As an example, we will take a Dolph–Chebyshev pattern for circular arrays using the Fourier phase-mode approach.

10.4 DOLPH–CHEBYSHEV PATTERN SYNTHESIS

10.4.1 Isotropic Elements

The array factor of a linear, equally spaced array antenna [Eq. (10.5)], can be identified as a polynomial by substituting $w = e^{jkd \sin \phi}$ [Schelkunoff 1943], thus obtaining

$$E_{\text{lin}}(\phi) = \sum_{-N}^{N} V_n w^n \qquad (10.6)$$

By making this a Chebyshev polynomial of the same order, the excitation coefficients V_n can be determined to obtain a pattern function with Chebyshev characteristics [Dolph 1946]. This is the basis for Chebyshev pattern synthesis for linear arrays [see, e.g., Balanis 1997, p. 297]. In a circular array, the corresponding expansion [Eq. (10.1)], is given in terms of phase modes with amplitudes A_m instead of the element excitations V_n. Thus, in the following example we have taken the element excitations of a synthesized linear array to represent the circular array phase-mode excitations. See Figure 10.1, which shows a nine-element linear –25 dB Dolph–Chebyshev array pattern and the corresponding nine-phase-mode, circular –25 dB Dolph–Chebyshev array pattern. The beamwidth is about 2.5 times larger in the circular array case than in the linear array case, corresponding approximately to the difference in array transverse width (length vs. diameter) for the two cases.

Having found the far field phase modes does not mean that we have solved the synthesis problem completely. The remaining problem is to find a realistic element excitation that will generate the phase modes. The transformation is given by Equation (10.3); Figure 10.2 presents both mode sets (A_m and C_m) for the circular array pattern of Figure 10.1 (b).

The next step is to find the element excitation $V(\varphi)$ expressed in the near field (excitation) modes according to Equation (10.4). Sampling the continuous excitation $V(\varphi)$ at the element positions φ_n gives the element excitation values [Figure 10.3 (a)].

Has the discrete excitation we found solved the problem? From Chapter 2 we know that additional spurious modes are generated (distortion terms) and we have to make sure that they are sufficiently attenuated. We notice from Figure 10.3 (b) that the pattern computed from the discrete element excitation has an increased side lobe level compared to the design goal of –25 dB, possibly due to spurious effects. Let us take a closer look at the relevant design criteria.

First of all, the number of modes used (we had nine) must be less than or equal to the number of elements (we had ten) in order to avoid aliasing of the phase-mode spectrum (cf. Table 2.1). Second, in order to radiate the useful unambiguous mode spectrum the number of phase modes must be less than about $2kR$ (cf. Figure 2.12). A commonly stated limit is $2kR + 1$ [Taylor 1952]; in our case we had $2kR + 1 = 9$. With 11 elements instead of 10, we would get $2kR + 1 = 9.8$, which is more on the safe side. Using this number, we get the pattern shown in Figure 10.4. Now we have obtained a result that is very close to the ideal Chebyshev pattern. However, this is a narrowband solution since kR cannot be allowed to vary much. We have the requirement

$$2M + 1 < 2kR + 1 < 2N + 1 \tag{10.7}$$

where $2M + 1$ is the number of phase modes and $2N + 1$ is the number of elements. For an increased bandwidth, the number of elements has to be increased; hence, element spacings are much closer in terms of wavelengths at the low-frequency end.

10.4.2 Directive Elements

We found in Chapter 2 that arrays with directive elements exhibit less frequency sensitivity (greater "mode stability") than arrays with isotropic elements. This was illustrated by a design with omnidirectional patterns. Let us now also look at the Chebyshev array case using directive elements. The critical factor is the expression with the Bessel function $A_m/C_m = j^m J_m(kR)$ appearing in the denominator of Equation (10.3). The corresponding expression for conversion from far field phase modes to excitation phase modes when directive elements are used was derived in Chapter 2 [see Eq. (2.34)]:

$$A_m/C_m = \left[\sum_{p=-\infty}^{\infty} D_p j^{m-p} J_{m-p}(kR) \right] \tag{10.8}$$

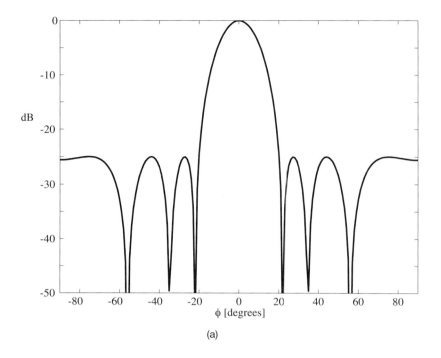

(a)

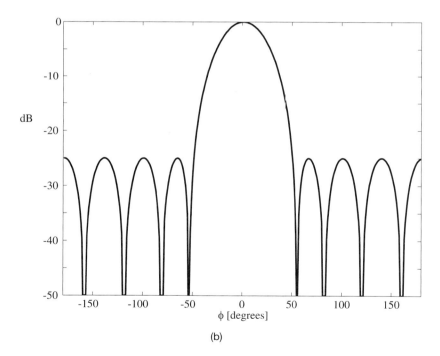

(b)

<u>Figure 10.1.</u> Chebyshev patterns: (a) Linear array with 9 elements, (b) Circular array with 9 phase modes (and 10 elements). Both cases have isotropic elements spaced 0.4 λ.

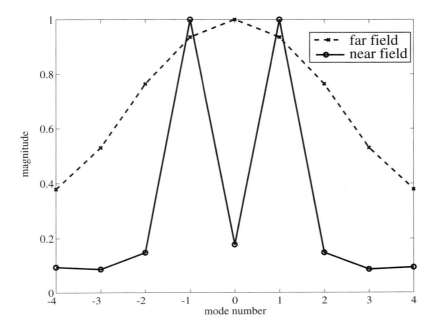

<u>Figure 10.2.</u> Far field mode (dashed line) and excitation mode (solid line) amplitudes for the circular array case in Figure 10.1 (b).

The coefficients D_p are the coefficients of a Fourier expansion of the element pattern. The summation in Equation (10.8) has typically much less variation with frequency than the single Bessel function; hence, it is more stable.

We repeat the Dolph–Chebyshev pattern synthesis as before, using 9 modes and 10 elements but for directive elements with $1 + \cos \phi$ shape. The resulting phase-mode amplitudes are shown in Figure 10.5 for two element spacings, 0.3 λ and 0.35 λ, representing a frequency shift of about 15%. Also shown are results for isotropic elements (dashed lines).

Notice the much larger variation of the excitation mode amplitudes for the array with isotropic elements compared to the array using directive elements. These variations need to be implemented in the feeding of the array for the required pattern shape to be maintained over this frequency range. Obviously, this is not an easy task. Special frequency filters in combination with digital beam forming could be one solution [Steyskal 1989].

10.5   AN APERTURE PROJECTION METHOD

This method starts from a planar aperture; see Figure 10.6. It is assumed that this planar antenna has been synthesized to fulfil the pattern requirements. The element locations of the curved array are projected onto the planar aperture, and the aperture excitation is sampled at the projected points. The sampled excitation is corrected for the element patterns and the sampling point density. A cophasal distribution is assumed, thus determining the element phases.

The method was tested experimentally [Chiba et al. 1989] for an array using 60 microstrip-patch elements on a ±46° segment of a half sphere with radius 3.2 λ. The element spacings in θ and ϕ were 0.65 λ. The starting planar aperture had a Taylor –25 dB excita-

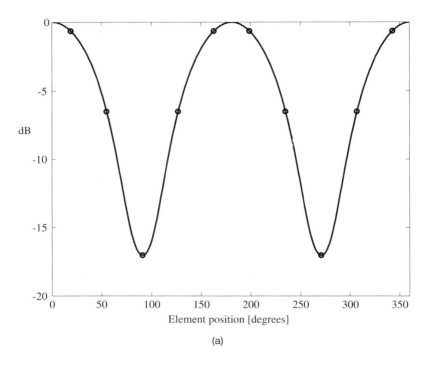

(a)

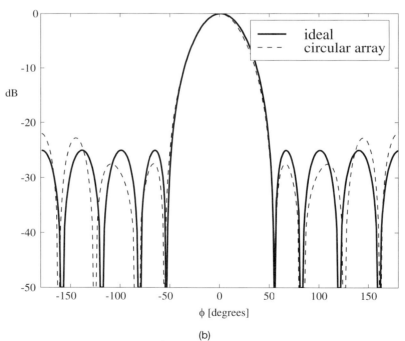

(b)

<u>Figure 10.3.</u> (a) Excitation function with samples at the element positions, (b) Radiation pattern (dashed line) compared to the ideal case (solid line) from Figure 10.1 (b). Ten radiating elements spaced 0.4 λ.

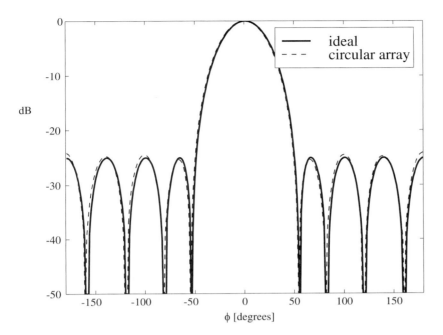

Figure 10.4. Pattern from an 11 element circular array, nine phase modes, and 0.4 λ element spacing (dashed line), compared with the desired Chebyshev pattern (solid line).

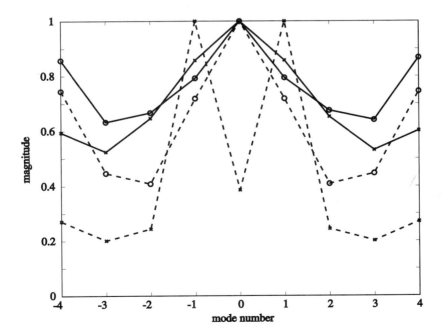

Figure 10.5. Circular array phase mode excitations for array with 10 elements and 9 phase modes. Legend: element spacings 0.3 λ (rings) and 0.35 λ (crosses). Dashed line: isotropic elements, solid line: directive $(1 + \cos \phi)$ elements.

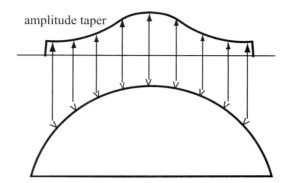

Figure 10.6. Projecting the element locations onto the planar aperture to sample excitation.

tion. Using measured element patterns in the synthesis, a –25 dB sidelobe pattern was also measured for the spherical array.

Thus, the projection method proved to be successful, at least when the pattern requirements are not too severe. In fact, we demonstrated in Chapter 8 that one can take the Taylor weighting directly on a curved (circular) array and achieve reasonable results. The increased projected element density toward the edges tends to be compensated for by the falloff of the element pattern. Theoretical results by Mailloux 1994 [Chapter 4] demonstrated this for –35 to –40 dB sidelobe levels using circular array radii down to about 10 λ. For still lower sidelobes, the mutual coupling has to be compensated for (and, in the practical case, careful calibration will probably be necessary). Kojima et al. [2003] included mutual coupling compensation based on measured embedded-element patterns. They demonstrated as low as –45 dB sidelobes for a 60 element, 22 λ radius sector array antenna.

A related projection method [Schuman 1994] starts also from a planar array solution that is projected onto a curved array. Here, however, the curved array is part of an enclosure that completely surrounds the planar array. Thus, the planar array illuminates the inside of the enclosure, setting up equivalent currents on the enclosure surface. These currents, according to the equivalence theorem [Balanis 1997, Chapter 12], radiate a field outside the enclosure that is identical to the field from the planar array in the absence of the enclosure. In this way, we have created a curved surface with a known current distribution that fulfils the pattern requirement.

The next steps are not so obvious. The current sheet must be replaced by real elements, the extent of the array on the surface must probably be limited, and the radiating element characteristics, cross polarization effects, shadowing, and so on will cause the pattern to deteriorate. The result reported [Schuman 1994] for a conical array was not very encouraging, but could perhaps be taken as a starting point for further optimizations by other methods. Rizk and coworkers [1985] also discuss the advantages and limitations of the method when applied to a conical surface.

10.6 THE METHOD OF ALTERNATING PROJECTIONS

The method of alternating projections (see, e.g., [Prasad 1980, Poulton 1986, Bucci et al. 1990]), is quite general and applicable to both planar and conformal antenna pattern syntheses. Two sets are defined, one (a) consisting of all realizable patterns, and the second (b) consisting of patterns meeting the given requirements. The requirements can, for example, be posed in terms of sidelobes not exceeding x dB and a main lobe meeting a certain shape within $\pm y$ dB. Thus, there are many possible solutions meeting the requirements, that is, many points in both sets. For a solution to exist, it is necessary that the two sets overlap somewhere, and the synthesis problem is to find at least one point in the overlapping region. The method of alternating projections can be visualized as successive projections between the two sets until the solution region is found; see Figure 10.7.

The method can be implemented in the following steps:

1. Select a starting pattern from the (A) domain.
2. Compare with the desired pattern. If satisfactory, then stop iteration. Otherwise, in angular regions where requirements are not met, replace values with the desired ones. This gives the (B) domain pattern.
3. Calculate excitation coefficients for a realizable pattern that matches the corrected pattern in the least-mean-square sense and, hence, belongs to (A). Calculate this pattern.
4. Go back to step 2.

There is a risk that the solution will stop at a local optimum, depending on the selection of the starting pattern. However, experience with the method, both for planar arrays and conformal arrays, is generally good. Steyskal [2002] used the method for synthesizing a low-sidelobe pattern for a conformal array of 116 microstrip-patch elements wrapped around a representative aircraft wing profile. A picture of this conformal wing array is found in Chapter 1 (Figure 1.3). Mutual coupling was included by using embedded-element patterns. Another detailed example is presented in Section 10.11 for a circular array pattern synthesis.

The method of alternating projections can be used for power pattern synthesis or field pattern synthesis, depending on the formulation. Field pattern synthesis consumes, of course, more degrees of freedom. It is also possible to include various other constraints. Vescovo [1995] used a starting pattern from a 32 element circular array of isotropic elements obtained by Fourier synthesis. He then introduced constraints on the excitation values and also imposed nulls [Vescovo 1997] at specific angles. The problem was solved using alternating projections.

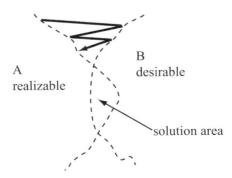

Figure 10.7. Method of alternating projections, illustrating the successive projections and the overlapping solution area.

10.7 ADAPTIVE ARRAY METHODS

Adaptive array antenna systems [Applebaum 1976, Hudson 1981, Gabriel 1992] make use of techniques for signal processing of digitized received signals. By deriving optimum weights on the element channels, based on the received signals, including noise and interference, the signal-to-noise-plus-interference ratio can be maximized. The result can be seen as creating nulls in the received pattern in the directions of interfering sources while preserving a high gain beam toward the useful signal. The mathematical principles can be carried over to antenna pattern synthesis, both for linear and conformal arrays [Brennan 1972, Sureau and Keeping 1982, Dufort 1989, Shi et al. 1997].

Imagine a scenario in which many uncorrelated noise sources are spaced all around the antenna except for a narrow sector where, instead, a signal source is located. After a successful adaptation, the element weights would produce a narrow beam toward the signal source and low sidelobes in the directions of the noise sources. By controlling the noise amplitudes, the shaping of the sidelobe region can be modified. The number of noise sources can be selected to be two to three times the number of elements in order to simulate a continuous noise distribution (see Figure 10.8).

Sureau and Keeping [1982] simulated a case with 372 noise sources evenly distributed on a circle around a 96 element circular array of which 32 elements were used. The embedded-element pattern was based on measured data (phase and amplitude), the same for all elements. With a noise-free window of about 10°, a pattern with a narrow beam and −30 dB sidelobes was achieved. They also demonstrated the synthesis of a difference pattern. The sidelobe taper could be modified by changing the amplitude distribution of the noise sources. In an early paper [Brennan 1972], the same technique was applied to the synthesis of a 40 element parabolic arc array using a uniformly distributed noise field.

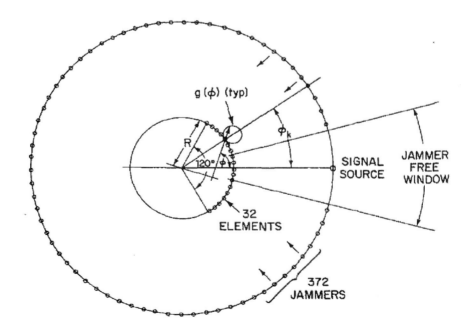

Figure 10.8. A jamming scenario according to [Sureau and Keeping 1982] for low sidelobe synthesis.

The adaptive array synthesis method is quite general, and applicable to both linear and conformal arrays. It is suited for designing specific sidelobe profiles and, of course, also for creating nulls in the patterns. Mutual coupling effects can be included by using embedded-element pattern data. The inversion of the covariance matrix (see Chapter 9) can be time-consuming for large arrays, and has to be repeated if, for example, the beam direction is changed. The method involves some experimentation; there are no simple design rules for achieving a particular side lobe level, and so on. However, iterative methods have been developed for successively adjusting the excitation so that an acceptable approximation to the desired pattern is obtained [Olen and Compton 1990, Shi et al. 1997].

10.8   LEAST-MEAN-SQUARES METHODS (LMS)

We can write the pattern function in matrix form:

$$E = A \cdot V \tag{10.9}$$

where E is the radiation pattern vector, A the system matrix, and V the excitation. The synthesis problem implies that the pattern $E(\phi_k)$ is given and we want to find the corresponding excitation V. For this inverse problem, that is, the synthesis problem, the solution can formally be written as

$$V = A^{-1} \cdot E \tag{10.10}$$

If the system of Equation (10.9) is over-determined, that is, the pattern $E(\phi_k)$ contains several more angles than there are elements in the array, only an approximate solution is possible. The system can be solved in the least-mean-squares (LMS) sense [Lindfield and Penny 1995]:

$$V = (A^H A)^{-1} A^H E \tag{10.11}$$

where the exponent H stands for complex conjugate transpose. (In MATLAB notation Equation (10.11) is simply written $V = A \backslash E$.)

Even if Equation (10.11) provides a best match to the desired pattern in the least-squares sense, it does not always provide the desired trade-off between various parts of the pattern, so some kind of iteration may be necessary.

Guy [1988] describes a method in which the difference E_{diff} between the desired pattern E_{des} and the first obtained LMS pattern E_0 is taken as a new desired pattern function to be synthesized in the least-squares sense. This is then added to E_0, and a new difference pattern is obtained to be synthesized, added, and so on. Only angles for which the pattern is outside the tolerance limits are included. The method converges rather quickly after 10–20 iterations in the cases studied. These include a parabolic cylindrical array with cosecant-squared main beam, and an ellipsoidal array with height/diameter = 1, synthesized for about –30 dB sidelobes. The method can be modified to handle phase-only synthesis (amplitudes are fixed) as well as amplitude-only synthesis (phases are fixed), but the convergence is much slower.

In many cases, it is important to choose a good starting pattern. It should reasonably well represent a realistic desired pattern, and thus ensure good convergence. Dinnichert

[2000] introduces a virtual planar array perpendicular to the desired main beam direction; see Figure 10.9. The planar array corresponds approximately to the projection of the conformal array in this direction. The virtual planar array can be synthesized by any known planar synthesis method to meet the conformal array requirements. A similar method is used by Schmid [2001].

Vaskelainen [1997a,b] describes a variation of the LMS method in which he introduces different weights for different directions, and also iteratively changes the weights. The solution [Eq. (10.11)] is modified to

$$V = (A^H W A)^{-1} A^H W E \qquad (10.12)$$

where W is a diagonal matrix with weights for each direction. In the iteration, the desired pattern is given the phase values obtained in each previous step, the weights are adjusted according to the relative errors in each point, a new approximate pattern is calculated, and so on. This method of using weights makes it possible to control the sidelobe profile and other parameters. The weight corrections, however, must be made carefully in order to ensure convergence. Among the examples shown by Vaskelainen [1997a] is a cosecant-squared shaped beam synthesized from a parabolic array. This is the same as that of Guy

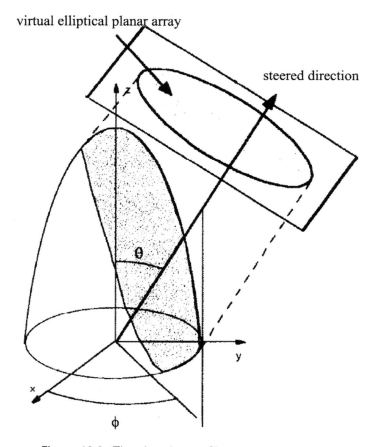

Figure 10.9. The virtual array [Dinnichert 2000].

[1988], and the two results agree well. Vaskelainen [1997b] also demonstrates results for a spherical array with fixed amplitude excitation, using phase synthesis only and the method of iterative weight correction. A number of nulls in the pattern could easily be added. However, when using phase synthesis only, the starting case must be selected carefully and the number of elements required is usually larger than in a phase and amplitude synthesis.

10.9   POLARIMETRIC PATTERN SYNTHESIS

Most of the discussion so far has dealt with the synthesis of a scalar pattern, with no regard for the polarization. One can extend the synthesis by considering both co- and cross-polarized components, or θ and ϕ components. We would then have an extended pattern vector:

$$E = \begin{bmatrix} E_\theta \\ E_\phi \end{bmatrix} \tag{10.13}$$

The same procedures as before in, for example, the LMS synthesis can be applied. We get

$$E = A \cdot V \tag{10.14}$$

where the system matrix also describes the coupling between polarizations:

$$A = \begin{bmatrix} A_{\theta\theta} & A_{\theta\phi} \\ A_{\phi\theta} & A_{\phi\phi} \end{bmatrix} \tag{10.15}$$

In this way, we can assign different requirements to the two components of the radiation pattern, typically the co- and cross-polarization patterns. The excitation must be applied to arrays with two-port radiators. It is, of course, essential to know the polarization characteristics of the element patterns. Mutual coupling can be included, either as coupling matrix or by the embedded-element patterns. It was shown by Heberling and Winterfeld [1992] that even for a cylidrical array with one row of two-port patch elements, the introduction of embedded-element patterns (instead of isolated patterns) in the synthesis resulted in significantly less cross polarization.

Schmid [2001] reported excellent polarization purity for two synthesized arrays, one cylindrical with elliptical cross section and one ellipsoidal. The very low cross polarization was mainly due to the assumption of ideal cross-polarization-free Huygens source elements consisting of a short magnetic dipole crossed with a short electric dipole.

By including two polarization components, the synthesis problem is essentially doubled. In the same way, one can extend vectors and matrices to include data for two or more frequencies. In this way, the synthesis can be made for a range of frequencies, also with different weights for different frequencies.

10.10   OTHER OPTIMIZATION METHODS

We have discussed several synthesis methods from an antenna perspective. There are, however, many types of general optimization methods that can be applied to antenna synthesis problems, for example, genetic algorithms and simulated annealing. The genetic algorithm method is based on an idea from Charles Darwin's theory of evolution. The parameter space (of "genes") is often discretized. Each set of parameters (the "chromosome") represents one candidate, which performs according to its value of the cost function. A population of several candidates is treated in each iteration step. Crossing genes and mutation form new generations of the population. Then, a selection of the best candidates (and rejection of the worst) is made, followed by a new crossover, and so on [Haupt 1995]. These steps reduce the risk of becoming trapped in local minima. However, the convergence is slow, especially for large problems. In spite of this the genetic algorithm has been used successfully for pattern synthesis problems, for example, a linear array on a cylinder [Allard et al. 2001] and linear and circular arrays [Yan and Lu 1997]. Phase and amplitude constraints can easily be included, and element locations can be perturbed.

Virga and Zhang [2000] combined the genetic algorithm method with the adaptive array method. Their application was a circular array with 60 Vivaldi type elements, 20 of which covered the active 120° arc. First, the genetic algorithm was used to maximize the directivity for a given –25 dB sidelobe level. Subsequently, the adaptive array method was used to insert several nulls into the pattern. A typical pattern with nine nulls imposed is shown in Figure 10.10.

Simulated annealing is another global search method in which the parameters are changed randomly. If the result is improved, the design is accepted; if not, it is rejected. However, occasionally a worse result is also accepted with a probability governed by the "temperature" parameter T, which is gradually reduced. After several steps the lower temperature makes it less likely that a worse result is accepted. Ferreira and Ares [1997, 2000] did shaped-beam synthesis using this method for cylindrical, spherical, and conical arrays. The acceptance of "worse" results in the beginning of the optimization reduces the risk of being trapped in a local optimum. The mutations in the genetic

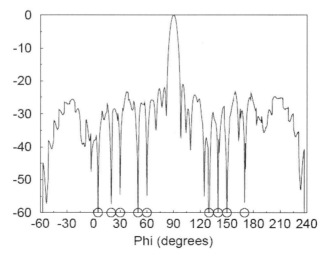

Figure 10.10. A –25 dB sidelobe pattern synthesized with the genetic algorithm method and nulls imposed by the adaptive array method [Virga and Zhang 2000].

algorithm method and, sometimes, restarting from new parameters, serve the same purpose.

10.11   A SYNTHESIS EXAMPLE INCLUDING MUTUAL COUPLING

We will conclude this chapter with a detailed presentation of pattern synthesis for a simple but typical application, and show results in terms of pattern performance and active impedance, including mutual coupling effects. We take as the basis for our example the cylindrical array antenna that we have referred to several times (see Figure 8.1a), using the method of alternate projections. The assumed requirements include both the sidelobe region and the shape of the main beam. We include the effect of the mutual coupling in the synthesis; neglecting the mutual coupling leads otherwise to poor results, especially in the shaped-beam region [Josefsson and Persson 1999].

For simplicity, we assume that the elements have a $\cos\phi$ isolated pattern and study one row of waveguide aperture elements spaced $0.53\,\lambda$ on an arc of about $120°$ on the 4.3 λ radius cylinder. We use calculated mutual coupling data for this row of 18 waveguide elements (see Chapter 6).

The radiation pattern $E(\theta_k)$ for a range of angles θ_k is a function of the complex element excitations $V(n)$. In matrix notation this can be written as

$$E = A_e \cdot V \tag{10.16}$$

The matrix A_e contains the geometrical and other factors that govern the relationship between pattern and excitation:

$$E(\phi_k) = \sum_n V_n \, EL(\phi_k - n\Delta\varphi)e^{jkR\,\cos(\phi_k - n\Delta\varphi)} \tag{10.17}$$

where $EL(\phi)$ is the element pattern, R is the cylinder radius, $k = 2\pi/\lambda$, and $\Delta\varphi$ is the angular spacing of elements around the cylinder.

We assume that the array is fed with an excitation vector $V^+(n)$ (waves from a feed network toward the elements). The mutual coupling among the elements will cause reflected waves $V^-(n)$ given by the mutual coupling scattering matrix S according to

$$V^- = S \cdot V^+ \tag{10.18}$$

Thus, the total excitation at the elements (in our case at the apertures) becomes

$$V^{\text{tot}} = (I + S) \cdot V^+ \tag{10.19}$$

where I is the identity matrix. Let us insert the last expression in Equation (10.16), yielding

$$E = A_e \cdot (I + S) \cdot V^+ = A \cdot V^+ \tag{10.20}$$

We see that Equation (10.20) is of the same form as Equation (10.16). Equation (10.20) provides a transformation from the excitation (from the feed) to the pattern, including the effect of the mutual coupling.

The synthesis problem will be solved by the method of alternating projections. The procedure is illustrated by the following example. A cosecant-squared main beam with unequal sidelobe envelopes will be synthesized. The specification is:

−180 to 0°: maximum −20 dB sidelobes
0 to 35°: cosec^2 within ±1dB
0 to 180°: maximum −30 dB sidelobes

Any pattern within the mask (dashed lines in Figure 10.11) is a solution. The Figure also shows the starting pattern for the iteration.

The pattern is successively modified in the regions in which the requirements are not met. For example, after three iterations shown in Figure 10.12, corrections are made around −100°, where −20 dB replaces the high sidelobe, but the phase values are retained. Similar corrections are done in the shaped beam region and around 50°. The next step is to find a realizable pattern that matches the corrected pattern in the LMS sense, and so on. A synthesized beam pattern after 100 iterations is shown in Figure 10.13.

The pattern is shown in polar form in Figure 10.14, where the locations of the 18 radiating elements are also indicated by small circles.

If we now repeat the synthesis but neglect the mutual coupling effect, we find that the pattern changes, Figure 10.15. Figure 10.15 (b) shows a close-up of the main beam analyzed with mutual coupling but synthesized without coupling. It is apparent that the mutual coupling plays an important role; it is essential to include the mutual coupling in the synthesis.

The change in excitation due to the mutual coupling is demonstrated in Figure 10.16.

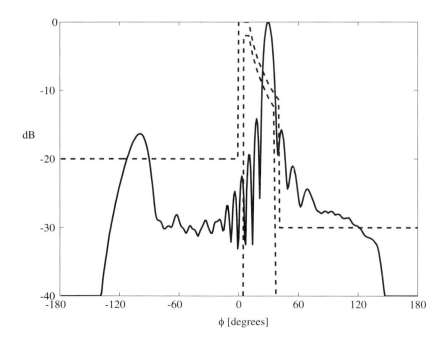

Figure 10.11. Starting case: Equal amplitude excitation with the main beam directed at 30 degrees. Dashed lines represent the requirement template.

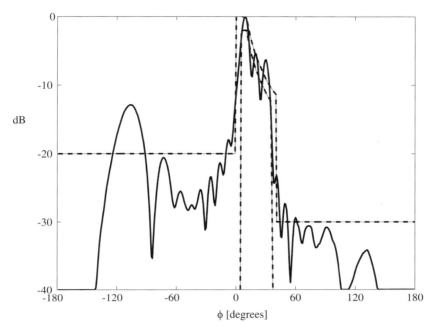

Figure 10.12. Result after three iterations.

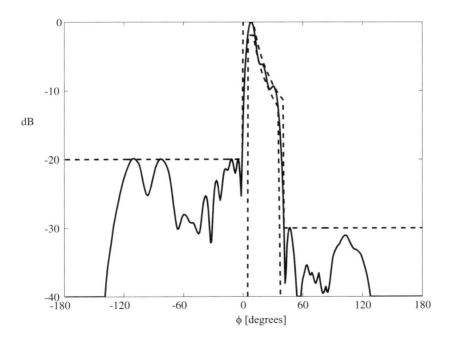

Figure 10.13. Synthesized pattern after 100 iterations.

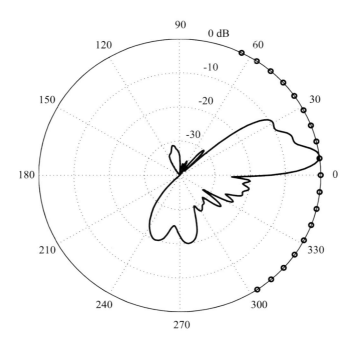

Figure 10.14. Synthesized pattern with coupling, also showing element locations.

The effect on the phase is shown next. Figure 10.17 shows the phase difference between the excitations without coupling and with coupling.

The active impedance expressed as the active reflection coefficient is presented in Figure 10.18 as a polar plot (corresponding to a Smith chart). The variations from element to element are significant.

By this example, we have illustrated a simple power-pattern synthesis for a circular array antenna. That it is essential to include the mutual coupling was shown by the impact of the coupling on the excitation amplitudes and the shaped beam pattern. The mutual coupling also modifies the active element impedance considerably and causes reflections on the feeding lines. These effects were not considered in the optimization. However, they could in principle be included by introducing additional parameters in the destination function, for example, the reflection coefficients, in addition to the shape of the radiation pattern.

10.12   A COMPARISON OF SYNTHESIS METHODS

Many optimization methods suitable for antenna array synthesis exist. We have mainly discussed those that have found use in connection with conformal antennas. There are many variations to most methods as well and new methods that are being developed. One example is the shape optimization method briefly mentioned in Section 10.2. Another is the particle swarm optimization method [Boeringer and Werner 2004] that resembles genetic algorithms, but is claimed to have simpler implementation. Furthermore, different performance indices can be chosen. We have given examples in which an average opti-

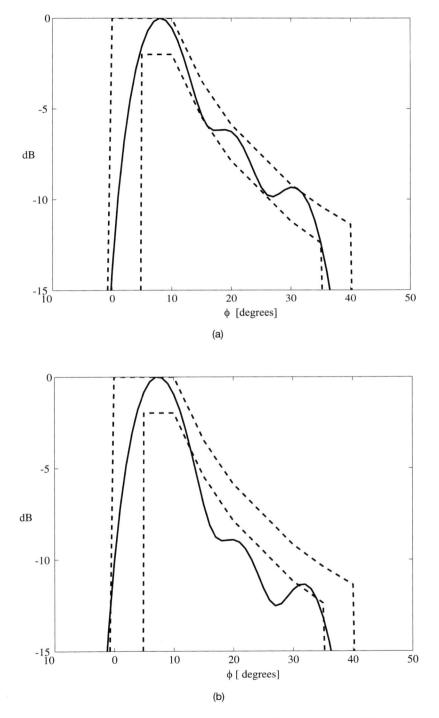

(a)

(b)

<u>Figure 10.15.</u> Pattern synthesis with coupling (a). The pattern in (b) demonstrates the effect of the mutual coupling if not included in the synthesis.

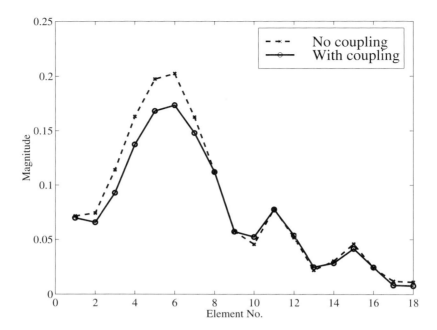

Figure 10.16. The mutual coupling effect on the synthesized excitation amplitude.

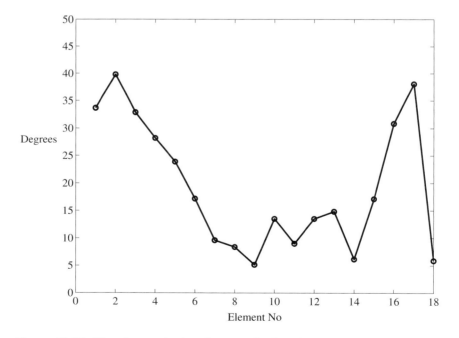

Figure 10.17. The change in the phase excitation due to the mutual coupling.

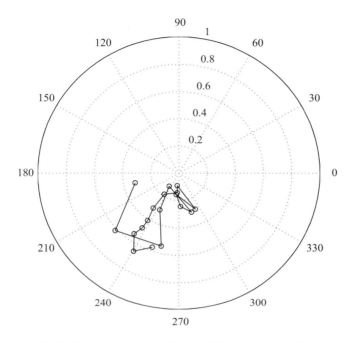

<u>Figure 10.18.</u> The active reflection coefficient for the 18 elements.

mum in the least-squares sense was sought. Other, more detailed criteria can be used instead, such as mini-max or equal ripple.

Thus, the suitability of the various methods for a specific application may depend on the detailed requirements, the available parameter space, previous experience, available computer codes, and simply the user's personal preferences. It is not possible to rank the methods, or even to include all possible methods in a comparison. However, an effort to summarize the main characteristics with conformal arrays in mind is presented in Table 10.1, partly based on [Guy 1999].

<u>Table 10.1.</u> Summary of synthesis methods for conformal arrays

Method	Characteristics
Array shape optimization (Section 10.2)	Not much done yet in this area.
Fourier methods (Section 10.3)	For cylindrical geometries. Can have problems with overshoots. Sometimes satisfactory as starting case for other methods.
Aperture projection (Section 10.5)	Three-dimensional, simple, not good for advanced requirements.
Alternating projections (Section 10.6)	General, effective, flexible.
Adaptive array (Section 10.7)	General, but computation intensive.
Iterative least-squares (Section 10.8)	Flexible, several variations possible.
Polarimetric synthesis (Section 10.9)	Can be included in other methods.
Genetic Algorithm (Section 10.10)	General, slow convergence.
Simulated annealing (Section 10.10)	General, slow convergence.

REFERENCES

Allard, R. J., Werner D. H., and Werner P. H. (2001), "A Radiation Pattern Synthesis Technique for Conformal Antenna Arrays Mounted on Truncated PEC Circular Cylinders," in *Proceedings of 17th Annual Review of Progress in Applied Computational Electromagnetics,* Monterey, CA.

Applebaum S. P. (1976), "Adaptive Arrays," *IEEE Transactions on Antennas and Propagation,* Vol. AP-24, No. 5, pp. 584–598.

Balanis C. A. (1997), *"Antenna Analysis, Theory, and Design,"* Wiley, New York.

Boeringer D. W. and Werner D. H. (2004), "Particle Swarm Optimization versus Genetic Algorithms for Phased Array Synthesis," *IEEE Transactions on Antennas and Propagation,* Vol. AP-52, No. 3, pp. 771–779.

Bondeson A., Yang Y., and Weinerfelt P. (2004), "Optimization of Radar Cross Section by a Gradient Method," *IEEE Transactions Magnetics,* Vol. 40, No. 2, pp. 1260–1263.

Brennan L. E. (1972), "Design and Adaptive Control of Conformal Arrays," in *Proceedings of Array Antenna Conference,* Naval Electronics Laboratory Center, San Diego CA, USA, paper 30, February.

Bucci O., Franceschetti G., Mazzarella G., and Panareillo G. (1990), "Intersection Approach to Array Pattern Synthesis," *IEE Proceedings, Part H,* Vol. 137, pp. 349–357.

Bucci O. M., D'Eila G., Mazzarella G., and Panariello G. (1994), "Antenna Pattern Synthesis: A New General Approach," *Proceedings of IEEE,* Vol. 82, No. 3, pp. 358–371.

Chiba I., Hariu K., Sato S., and Mano S. (1989), "A Projection Method Providing Low Sidelobe Pattern in Conformal Array Antennas," *IEEE AP-S International Symposium Digest,* pp. 130–133.

Chung Y. S., Cheon S., Park I. H., and Hahn S. Y. (2000), "Optimal Shape Design of Microwave Device using FDTD and Design Sensitivity Analysis," *IEEE Transactions Microwave Theory and Techniques,* Vol. 48, pp. 2289–2296, December.

Dinnichert M. (2000), "Full Polarimetric Pattern Synthesis for an Active Conformal Array," in *Proceedings 2000 IEEE International Conference on Phased Array Systems and Technology,* Dana Point, CA, USA, 21–25 May, pp. 415–419.

Dolph C. L. (1946), "A Current Distribution for Broadside Arrays which Optimizes the Relationship between Beam Width and Side Lobe Level," *Proceedings of IRE,* Vol. 34, pp. 335–348, June.

Dufort E. C. (1989), "Pattern Synthesis Based on Adaptive Array Theory," *IEEE Transactions on Antennas and Propagation,* Vol. AP-37, No. 8, pp. 1011–1018.

Ferreira J. A. and Ares F. (1997), "Pattern Synthesis of Conformal Arrays by the Simulated Annealing Technique," *Electronics Letters,* Vol. 33, No. 14, pp. 1187–1189, 3rd July.

Ferreira J. A. and Ares F. (2000), "Radiation Pattern Synthesis for Conformal Antenna Arrays," *Journal of Electromagnetic Waves and Applications,* Vol. 14, pp. 473–492.

Gabriel W. (1992), "Adaptive Processing Array Systems," *Proceedings of IEEE,* Vol. 80, No. 1, pp. 152–162.

Guy R. F. E. (1988), "General Radiation-Pattern Synthesis Technique for Array Antennas of Arbitrary Configuration and Element Type," *IEE Proceedings, Part H,* Vol. 135, pp. 241–248.

Guy R. F. E., Lewis R. A., and Tittensor, P. J. (1999), "Conformal Phased Arrays," in *Proceedings of 1st European Workshop on Conformal Antennas,* Karlsruhe, Germany, 29 October.

Haupt, R.L. (1995), "An Introduction to Genetic Algorithms for Electromagnetics," *IEEE Antennas and Propagation Magazine,* Vol. 37, No. 2, pp. 7–15.

Heberling D. and Winterfeld C. V. (1992), "Investigations on Mutual Coupling Effects in Polarimetric Conformal Patch Arrays," in *Proceedings of Radar 92,* 12–13 October 1992, pp. 118–121.

Hudson J. E. (1981), *Adaptive Array Principles,* Peter Peregrinus Ltd., London.

Josefsson L. and Persson P. (1999), "Conformal Array Synthesis Including Mutual Coupling," *Electronics Letters,* Vol. 35, No. 8, pp. 625–627.

Kojima N., Hariu K., and Chiba I. (2003), "Low Sidelobe Pattern Synthesis using Projection Method with Mutual Coupling Compensation," in *Proceedings of IEEE International Symposium on Phased Array Systems and Technology 2003,* Boston USA, 14–17 October, pp. 559–564.

Lindfield G. and Penny J. (1995), *Numerical Methods Using Matlab,* Ellis Horwood.

Mailloux R. J. (1994), *Phased Array Antenna Handbook,* Artech House.

Olen C. A. and Compton Jr. R. T. (1990), "A Numerical Pattern Synthesis Algorithm for Arrays" *IEEE Transactions on Antennas and Propagation,* Vol. AP-38, No. 10, pp. 1666–1676.

Pironneau O. (1984), *Optimal Shape Design for Elliptic Systems,* Springer-Verlag, New York.

Poulton, G. T. (1986), "Antenna Power Pattern Synthesis using Method of Successive Projections," *Electronics Letters,* Vol. 22, No. 20, pp. 1042–1043.

Prasad S. (1980), "Generalized Array Pattern Synthesis by the Method of Alternating Orthogonal Projections," *IEEE Transactions on Antennas and Propagation,* Vol. AP-28, No. 3, pp. 328–332.

Rizk M. S. A. S., Morris G., and Clifton M. P. (1985), "Projected Aperture synthesis Method for the Design of Conformal Array Antennas," in *Proceedings of 4th International Conference on Antennas and Propagation ICAP 85,* pp. 48–52.

Schelkunoff S. A. (1943), "A Mathematical Theory of Linear Arrays," *Bell System Technical Journal,* Vol. 22, pp. 80–107.

Schell, A. and Ishimaru A. (1969), "Antenna Pattern Synthesis," in *Antenna Theory,* Part I, R. Collin and F. Zucker, Eds., McGraw-Hill, New York.

Schmid O. (2001), "Pattern Properties of Singly and Doubly Curved Arrays," in *Proceedings of 2nd European Workshop on Conformal Antennas,* The Hague, The Netherlands, 24–25 April.

Schuman J. K. (1994), "Conformal Array Synthesis," *IEEE AP-S International Symposium Digest,* pp. 526–529.

Shi X., Yoo K.-S., Park J.-H., and Lee H.-J. (1997), "Pencil-Beam Pattern Synthesis for Arbitrary Arrays," *Electronics Letters,* Vol. 33, No. 12, pp. 1007–1008.

Steyskal H. (1989), "Circular Array with Frequency-Invariant Pattern," *IEEE AP-S International Symposium Digest,* San Jose, 26–30 June, pp. 1477–1480.

Steyskal H. (2002), "Pattern Synthesis for a Conformal Wing Array," in *Proceedings of IEEE Aerospace Conference,* Vol. 2, pp. 2-819–2-824.

Sureau J. C. and Keeping K. J. (1982), "Sidelobe Control in Cylindrical Arrays," *IEEE Transactions on Antennas and Propagation,* Vol. 30, No. 5, pp 1027–1031.

Taylor T.T. (1952), "A Synthesis Method for Circular and Cylindrical Antennas Composed of Discrete Elements," *IRE Transactions on Antennas and Propagation,* pp. 251–261, August.

Vaskelainen L. I. (1997a), "Iterative least-Squares Synthesis Methods for Conformal Array Antennas with Optimized Polarization and Frequency Properties," *IEEE Transactions on Antennas and Propagation,* Vol. 45, No. 7, pp 1179–1185.

Vaskelainen L. I. (1997b), "Phase Synthesis of Conformal Antenna Arrays," in *Proceedings of Antenn 97,* Göteborg, Sweden, 27–29 May, pp. 349–356.

Vescovo R. (1995), "Constrained and Unconstrained Synthesis of Array Factor for Circular Arrays," *IEEE Transactions on Antennas and Propagation,* Vol. 43, No. 12, pp 1405–1410.

Vescovo R. (1997), "Pattern Synthesis with Null and Excitation Constraints for Circular Arrays," in *Proceedings of ICEAA97,* pp. 453–456, 1997.

Virga K. L. and Zhang H. (2000), "Spatial Beamformer Weighting Sets for Circular Array STAP," in *Proceedings 2000 IEEE International Conference on Phased Array Systems and Technology,* 21–25 May, pp. 561–564.

Weinerfelt P. and Enoksson O. (2001), "Aerodynamic Shape Optimization and Parallel Computing Applied to Industrial Problems," in *Parallel Computational Fluid Dynamics,* Elsevier Science, Oxford.

Yan K-K. and Lu Y. (1997), "Sidelobe Reduction in Array-Pattern Synthesis Using Genetic Algorithm," *IEEE Transactions on Antennas and Propagation,* Vol. 45, No. 7, pp 1117–1122.

11

SCATTERING FROM CONFORMAL ARRAY ANTENNAS

In this chapter, we will discuss conformal antennas from a scattering point of view. This is an important issue, in particular for antenna installations on stealth vehicles where it is desired to minimize all scattering contributions. Efficient antennas are in principle also good scatterers, so the problem at hand is a difficult one. However, conformal antennas have some advantages over traditional planar antenna installations.

11.1 INTRODUCTION

An efficient receiving antenna is usually also an efficient scattering antenna. The antenna receives only part of the energy incident on the antenna; a significant part is also scattered. A portion of this scattered energy radiates back toward the illuminating transmitter. Thus, the antenna exhibits a radar scattering cross section (RCS). In fact, the antenna can be responsible for the larger part of the total RCS of, for example, an aircraft, especially if the aircraft is designed to have stealth characteristics. The antenna scattering can sometimes also be the source of EMC (electromagnetic compatibility) problems and cause interference with other systems on the same platform.

The general subject of electromagnetic scattering is extensive, see, for example, the numerous references in the *Proceedings of IEEE,* August 1965, Special Issue on Radar Reflectivity, where 30 pages (!) of literature references are listed. It has long been recognized that there exist relations between the transmitting and scattering properties of antennas [Sinclair et al. 1947, Stevenson 1948, Appel-Hansen 1979]. However, the scattering from practical antennas is not well covered in the literature. Several theoretical papers on scattering from basic antenna types like dipole, horn, slot, or similar elementary antenna elements have been published. For an overview, see among others [Collin and Zucker 1969, Williams 1986, Balanis 1997, Tai 2001, Bach Andersen 2003].

A number of authors have adopted a scattering matrix approach for single element antennas; see, for example, [Collin and Zucker 1969, Knott et al. 1993, Hansen 1989]. In [Harrington 1964], Harrington presents a formulation for the problem of scattering by an N-port loaded antenna where the representation is made in terms of open-circuited impedance parameters and short-circuited admittance parameters.

Some authors have treated the problem of scattering from planar array antennas, for example, [Tittensor and Newton 1989, Wills 1991, Chen and Jin 1995, McNamara et al. 1992, Fan and Jin 1999]. Rather few references deal with scattering from conformal array antennas. Ko and Mittra [1989] investigated the scattering properties of an array of patches placed on a parabolic cylindrical surface. Fan and [Jin 1997] analyzed the scattering from a slotted waveguide array bent into a cylindrical shape.

In this chapter, we will present some examples of scattering from a cylindrical conformal array antenna. This array has waveguide aperture elements, and we will investigate cases with and without a dielectric coating on the array. We will also give some examples of a trade-off between antenna performance and scattering characteristics by optimizing the antenna load.

The results for the cylindrical array are based on recent research by Dr. Björn Thors and coworkers; see [Thors 2003] and related references.

11.2 DEFINITIONS

The scattering characteristics of an object in terms of the radar cross section (RCS) is defined as

... an area that, when multiplied by the power flux density of the incident wave, would yield sufficient power that could produce, by isotropic radiation, the same radiation intensity as that in a given direction from the scattering object. (*IEEE Standard Definitions of Terms for Antennas,* IEEE Std 145-1993)

In equation form, the RCS can be written as

$$\sigma = \lim_{r \to \infty} 4\pi r^2 \frac{P^S}{P^i} = \lim_{r \to \infty} 4\pi r^2 \frac{\overline{H}^S \cdot \overline{H}^{S*}}{\overline{H}^i \cdot \overline{H}^{i*}} \qquad (11.1)$$

where P^S is the scattered power density and P^i is the incident power density. $\overline{H}^S$ is the scattered magnetic field and $\overline{H}^i$ is the magnetic field of the incident plane wave. Instead of the magnetic fields, we can use the corresponding electric field components—they are proportional. The definition is valid for a particular set of polarizations: that of the illuminating wave and that of the receiving antenna. A complete description of the scattering characteristics (at one angle and one frequency) is provided by a scattering matrix instead of just the scalar cross section σ. Let us break down the incident and scattered fields into two orthogonal components, the TE and TM wave components. Thus, four RCS components form the radar cross-section matrix:

$$[\sigma] = \begin{pmatrix} \sigma_{\text{TETE}} & \sigma_{\text{TETM}} \\ \sigma_{\text{TMTE}} & \sigma_{\text{TMTM}} \end{pmatrix} \tag{11.2}$$

where the subscripts TE/TM etc. indicate the polarizations of the incident/scattered fields.

The radar cross section is usually defined as a far field quantity, but can in principle also be defined in the near field. In this chapter we assume far field conditions with the definition as in Equations (11.1) and (11.2). We will analyze the co-polarized RCS—the matrix component σ_{TETE}—as well as the cross-polarized component, σ_{TETM}. When the illuminating antenna and the receiving antenna are at the same location, we have the monostatic radar cross section. If the two are separated in angle, we get the bistatic radar cross section. However, we will only consider the monostatic case here. The TE and TM field components are defined relative to the z axis, that is, the transversal (T) components are perpendicular to the z direction (cylinder axis in the cases to be studied).

The scattering cross section depends also on the antenna matching. A short-circuited antenna reradiates the energy that could otherwise have been absorbed in a matched termination. This part of the cross section is termed the antenna mode contribution; the amplitude and phase depend on the antenna load. The remaining part is the residual mode contribution; together they make up the total radar cross section of the antenna. Other definitions can be found in the literature, but the ones just described are chosen here for their simplicity. See also [Hansen 1989, Munk 2003]. We will assume that the cylindrical array that we are going to analyze has the feed lines terminated in matched loads. However, in Section 11.6 we will introduce a certain mismatch as a means for controlling the total scattering.

11.3 RADAR CROSS SECTION ANALYSIS

11.3.1 General

It is often claimed that an antenna scatters (at least) as much as it receives, even when it is matched. A large planar array antenna can in theory absorb all the incident energy from a fixed direction, thus becoming "invisible." However, on the rear side of the antenna it will cast a shadow. This can be seen as the result of the forward scattering canceling the incident field in that region [Williams 1986, Munk 2003]. In practice, we will have to account for some additional scattering due to contributions from the antenna/ground plane edges as well as scattered energy due to variations in the active matching as the direction of illumination is varied.

Most practical (not perfectly matched) planar antennas will give rise to a very large back-scattered field at normal incidence, since all parts of the antenna radiate back in phase. For a curved array, this will not happen; the phase compensation of the curvature in the antenna feed system will not work for scattering. Another advantage of conformal arrays is the conformal integration with the platform (aircraft, etc.), which reduces the effect of antenna edges and back structure.

We will present a detailed analysis of a canonical case: scattering from an array in a conducting cylinder. The cylinder itself will of course also scatter, but most of that is concentrated to the specular direction. In the monostatic case, this is significant only in the direction perpendicular to the cylinder axis. For other directions, there will be very little backscattering if the cylinder is large; the major contributions will come from the array.

11.3.2   Analysis Method for an Array on a Conducting Cylinder

The conformal array antenna with rectangular waveguide antenna elements in an infinitely long circular cylinder is shown in Figure 11.1 (*a*).

For the analysis of the scattering, we will use a hybrid of the method of moments (MoM) and one of the many different high-frequency methods discussed in Chapter 4. This approach is suitable for medium- to large-size arrays. The mutual coupling between the antenna elements can be included. We will present an outline of the analysis method; more details are found in Chapter 4 and in [Thors and Josefsson 2003] and other references.

The computational domain is divided into two regions as shown in Figure 11.1 (*b*). By using the field equivalence theorem [Balanis 1989], the fields in the exterior region are decoupled from the fields in the interior region by covering the apertures with a perfectly conducting surface and introducing equivalent magnetic currents on the surface. The magnetic field in the exterior region can now be expressed as the sum of the incident field in the presence of the conducting cylinder without apertures, $\overline{H}^{\text{inc}}$, and the magnetic field $\overline{H}^{\text{sc}}$ caused by the equivalent magnetic currents.

The electric field in the interior region is represented by rectangular waveguide modes. a^{ip} and b^{ip} (waveguide *i*, mode *p*) are the complex amplitudes of the forward and backward propagating modes, respectively.

The magnetic field in the interior region is denoted by $\overline{H}^{\text{int}}$. Matching the tangential magnetic fields in the aperture results in the following integral equation:

$$(\overline{H}^{\text{inc}} + \overline{H}^{\text{sc}})|_{\overline{r} \in S_{\text{ap}}^+} = \overline{H}^{\text{ext}}|_{\overline{r} \in S_{\text{ap}}^+} = \overline{H}^{\text{int}}|_{\overline{r} \in S_{\text{ap}}^-} \tag{11.3}$$

where S_{ap} is the set of all apertures.

The next step is to calculate the incident excitation field $\overline{H}^{\text{inc}}$ in the presence of the cylinder. Two different methods have been considered for this, the high-frequency method UTD, and a modal solution based on expansions of the fields in cylindrical wave functions (modes). Both methods produce practically identical results. We choose UTD since it turns out to be significantly faster.

The problem of finding the incident excitation is, by reciprocity, related to the radiation problem. If the solution to the radiation problem is known, reciprocity will automatically give the incident field $\overline{H}^{\text{inc}}$ on the cylinder surface. For a more detailed discussion see [Thors 2003] where a discussion concerning the UTD and the modal solutions for the incident and scattered fields can also be found. Now, consider two infinitesimal sources, $d\overline{p}_m^Q$ and $d\overline{p}_e^P$, magnetic and electric current moments, respectively, located at the points *Q* and *P*, as shown in Figure 11.2. The electric and magnetic fields generated by the sources are denoted by $\overline{E}^P$, $\overline{H}^P$ and $\overline{E}^Q$, $\overline{H}^Q$.

By making use of reciprocity [Harrington 1961], the following relation can be established:

$$-\overline{H}^P(\overline{r}_Q) \cdot d\overline{p}_m^Q(\overline{r}_Q) = \overline{E}^Q(\overline{r}_P) \cdot d\overline{p}_e^P(\overline{r}_P) \tag{11.4}$$

With *P* located in the far zone, the electric current moment can be used to generate an incident plane wave, where the strength of the current moment specifies the amplitude of

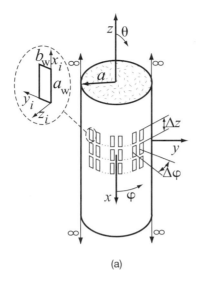

(a)

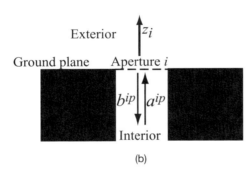

(b)

Figure 11.1. (a) The conformal array antenna. (b) The interior and exterior regions near an aperture.

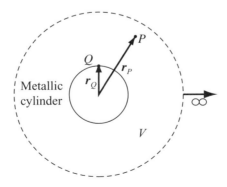

Figure 11.2. Location of sources at Q and P for reciprocity analysis.

the incident wave. Since $\bar{E}^Q(\bar{r}_P)$ can be expressed in terms of $d\bar{p}_m^Q(\bar{r}_Q)$ using the UTD, as shown by [Pathak et al. 1981], the incident magnetic field on the cylinder surface can be obtained from Equation (11.4).

The aperture fields can now be solved from Equation (11.3) using the method of moments with the waveguide mode functions as testing functions. The aperture fields then yield the scattered far field.

11.3.3 Analysis Method for an Array on a Conducting Cylinder with a Dielectric Coating

Many practical array antennas have a dielectric coating (radome) for mechanical protection. The dielectric layer will affect both the radiation and the scattering properties of the antenna. Figure 11.3 illustrates the problem for this case.

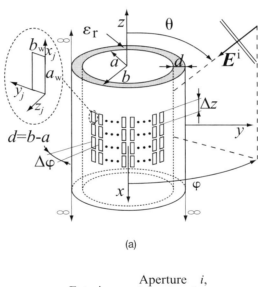

(a)

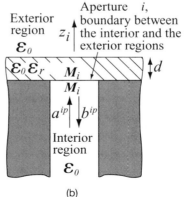

(b)

Figure 11.3. (a) The conformal array antenna coated with a dielectric layer with relative permittivity ε_r. (b) The exterior and the interior regions of aperture i.

The introduction of the dielectric coating makes the analysis problem much more diffi-cult compared to the noncoated surface discussed in Section 11.3.2. The techniques for calculating the excitation field $\overline{H}^{\text{inc}}$ and the field $\overline{H}^{\text{sc}}$ scattered from the apertures must in-clude the dielectric coating. For this purpose, special asymptotic techniques are used in combination with MoM [Thors and Rojas 2003]. These asymptotic techniques have pre-viously been used for radiation pattern and mutual coupling analyses; see Chapters 4 and 6. The following approximate rules illustrate when the asymptotic method is valid for coated surfaces (λ_0 is the free-space wavelength):

$$a > \lambda_0 \tag{11.5}$$

$$d < 0.15 \, \frac{\lambda_0}{\sqrt{\varepsilon_r}} \tag{11.6}$$

$$\theta_{\text{in}} > 10° \tag{11.7}$$

As mentioned in Chapters 4 and 6, the paraxial (near axis) region of the coated cylinder must be treated separately. Here, we use a planar approximation with a corresponding spectral domain technique [Croswell et. al. 1967, Thors et. al. 2002]. This approximation is also used for calculating the coupling in the azimuth direction between closely spaced elements. For large cylinders coated with thin dielectric layers, it has been found that the planar approximation is very accurate if the distance between the transmitting and receiv-ing apertures is not too large [Thors et al. 2002]. Also, note that $\overline{H}^{\text{inc}}$ is defined in the presence of the coated PEC cylinder, hence $\overline{H}^{\text{inc}} \neq \overline{H}^i$; compare Equation (11.1). More details are presented in [Thors et al. 2004].

For the numerical results presented in the following sections, the receiver and trans-mitter are collocated, that is, the values represent the monostatic RCS. Due to the large dynamic range, a logarithmic power scale is used with values in dBsm, that is, 0 dBsm corresponds to $\sigma_{\text{ref}} = 1 \text{ m}^2$. In section 11.4, theoretical results for the PEC cylinder, in-cluding some experimental data, are presented. The case with a dielectric coating is dis-cussed in 11.5. Finally, Section 11.6 addresses the trade-off between scattering and radia-tion efficiency.

11.4 CYLINDRICAL ARRAY

11.4.1 Analysis and Experiment—Rectangular Grid

The radar cross section has been analyzed for a cylindrical array antenna with the follow-ing dimensions:

Element waveguide apertures	39 mm × 16 mm
Element grid	Rectangular, 3 × 18 elements
Element spacing	37.1 mm in the azimuth
	41 mm in the axial direction
Cylinder radius	300 mm

The above dimensions agree with those of the experimental antenna shown in Figure 11.5. The polarization is horizontal. At the frequency selected, 6.8 GHz, the element spacings in the rectangular grid are 0.84 λ in the azimuth and 0.93 λ in the axial direction. We can, thus, expect grating lobe effects[1] in the scattering pattern. The frequency was dictated by the available measurement setup; otherwise, a lower frequency could have been chosen to reduce the grating lobe complication.

For the calculations, a simple phase correction was applied to the interior fields in the apertures. The rectangular waveguide modes, as defined in the waveguide, were transformed to the convex surface of the aperture by adding a small phase correction corresponding to the distance between the two surfaces. See also the discussion in Section 4.5.2.

We will first present calculated results and compare them with experimental values. This will verify the theoretical model. We will then perform a more detailed study of the number of waveguide modes required and the polarization effects. The results are presented in terms of the polarization scattering matrix elements; compare Equation (11.2).

Figure 11.4 shows the theoretical co-polarized scattering σ_{TETE} and its variation in the two principal planes. Only the dominant waveguide mode was used in this calculation. We note that for this case the results with and without mutual coupling are very similar.

The scattering pattern in the E plane exhibits a considerable ripple. The periodicity corresponds to the angular spacing of the array elements on the cylinder (7.09°). The large ripple amplitude is due to the large E-plane element spacings, 0.84 λ in this case. We have (roughly speaking) three major scattering centers. One is the contribution from an area close to normal incidence. In this area, several elements backscatter in phase. Two other in-phase areas are located near 36° from the normal incidence direction, on both sides, where a grating lobe radiates back to the source. These three most significant contributions interfere, causing the strong ripple. With a closer element spacing (see Section 11.6), the ripple is substantially reduced.

In order to verify the theoretical results, measurements were performed on the experimental cylindrical array antenna built by Ericsson Microwave Systems AB in Mölndal, Sweden (Figure 11.5). It has the same dimensions as were used in the calculation. The waveguides were terminated in matched loads designed for the dominant TE_{10} mode. The measurements took place at the indoor measurement range at the Swedish Defence Research Agency (FOI) in Linköping, Sweden.

Since we were interested in the scattering from the apertures in the cylinder and not from the cylinder itself, efforts were made to reduce the scattering contribution from the finite cylinder. Thus, absorbers were used to minimize the scattering from the cylinder edges;, see Figure 11.5. To avoid the strong specular reflection at normal incidence, the antenna was tilted a small angle (about 4°), corresponding to the first RCS minimum of the cylinder without apertures. See Figure 11.6, where the backscattered cross section from a finite cylinder without apertures, σ_{cyl}, is compared with the predicted aperture-scattered cross section σ_{TETE}. The backscattered field from the cylinder was computed by a physical optics (PO) approximation, which is accurate in the region near $\theta = 90°$. From these results, it was decided to make the measurements (and calculations) for an angle $\theta = 86.3°$.

After RCS data had been collected for all azimuth angles, the apertures were covered with conducting tape and the measurements were repeated, now representing a metallic

---

[1]There are no *distinct* grating lobes for a curved array. The curvature causes the grating lobes to be smeared out, but the effect is still significant.

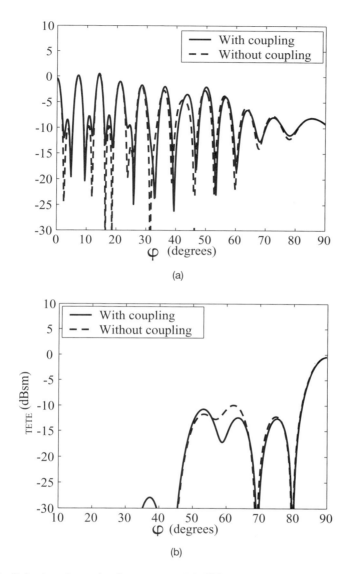

Figure 11.4. Calculated results for σ_{TETE} at 6.8 GHz with one waveguide mode, with mutual coupling (solid line) and without mutual coupling (dashed line). (a) *E* plane. (b) *H* plane.

cylinder without apertures. The array RCS was then obtained by subtracting the second results from the first. A comparison between the calculated and the measured results is presented in Figure 11.7.

Note that the measured curve is not perfectly symmetrical around $\varphi = 0°$. This is main-ly due to problems with aligning the antenna at the measurement range. The top of the supporting pylon was not exactly flat, causing the elevation angle θ to change slightly during azimuth rotation. Figure 11.6 shows that a small change in elevation angle from the desired $\theta = \theta_{min}$ corresponds to a large change in σ_{cyl}. This means that, for some az-

Figure 11.5. The experimental antenna prepared for RCS measurements. (Courtesy of Ericsson Microwave Systems AB, Göteborg, Sweden.)

imuth angles, two numbers of the same order of magnitude must be subtracted from each other, which results in a reduced accuracy. However, the calculated and experimental results seem to agree within the tolerances of the experimental setup.

11.4.2 Higher-Order Waveguide Modes

As shown in Figure 11.7, the single-mode approximation gives good results for the backscattered field in the E plane. However, as the elevation angle changes from normal incidence, the asymmetric illumination excites higher-order waveguide modes. Higher-

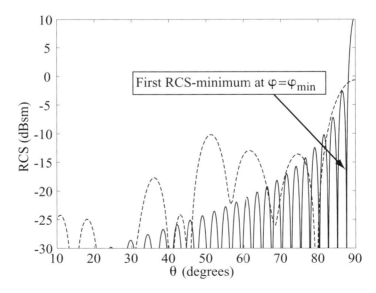

Figure 11.6. Comparison between the backscattered cross section from a finite cylinder without apertures (solid line) and the predicted array scattered cross section (dashed line) in the H plane.

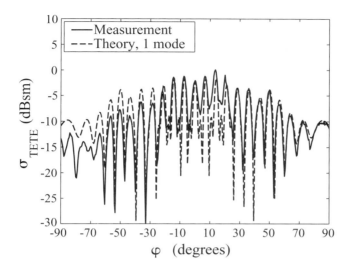

Figure 11.7. Comparison between the measured (solid line) and the calculated (dashed line) RCS at 6.8 GHz and $\theta = 86.3°$. The calculations included one waveguide mode and mutual coupling.

order modes must also be included for the case with cross-polarized illumination, that is, TM incidence. Figure 11.8 shows results for σ_{TETE} and σ_{TMTM} calculated with the three lowest-order waveguide modes (in cutoff order), that is, TE_{10}, TE_{20}, and TE_{01}.

As expected, the effect of the two higher-order modes is more pronounced in the H plane, the elevation plane (cf. the single-mode case in Figure 11.4). The dominant waveguide mode is not excited by an incident TM-polarized plane wave. However, the level of σ_{TETE} is only slightly higher than the levels of σ_{TMTM}.

Figure 11.9 shows the calculated radar cross section in the H plane with even more waveguide modes included. The E-plane results have already converged with three modes; in fact, the single-mode approximation appears to be sufficient for this plane.

For the H plane, a few additional modes are required in order to reach full convergence. Up to 40 modes have been tested [Thors 2003], but it was found that 15 modes are sufficient for convergence in the H plane (Figure 11.10). Not all modes are significant, however, and some modes are more important than other modes. Including the first five modes predicts the RCS quite well. In fact, the fifth mode, TM_{11}, is essential, as has previously been found in other studies [Bonnedal 1996, Persson and Josefsson 2001]. See also Section 6.2.3.3.

Figure 11.11 shows the RCS in a contour plot covering azimuth angles from $0°$ to $90°$ and elevation angles from $10°$ to $90°$ (normal incidence). Fifteen modes are used and mutual coupling is included. Figure 11.11 (a) shows the co-polarized case σ_{TETE} and (b) the cross-polarized case σ_{TETM}.

Due to symmetry, the radar cross section exhibits very low cross-polarization levels in the E and H planes. However, quite small angular changes will break the symmetry, resulting in an increased cross polarization. In some regions, the cross-polarization levels are of the same order of magnitude as the co-polarization levels. See also Figure 11.12, in which the components of the radar cross section matrix are compared at the elevation angle $\theta = 60°$.

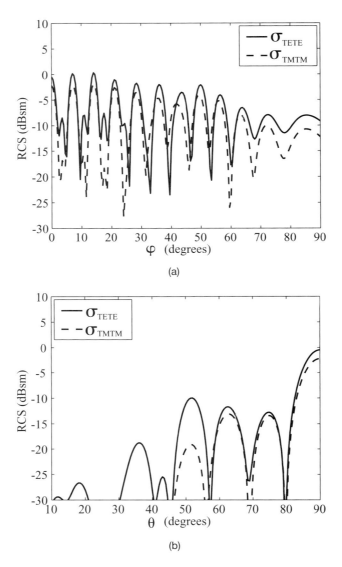

Figure 11.8. Calculated results for σ_{TETE} (solid line) and σ_{TMTM} (dashed line) with three waveguide modes at 6.8 GHz. (a) *E* plane. (b) *H* plane.

11.4.3 Triangular Grid

By staggering the middle row of the 3×18 elements array by half the azimuth spacing, we obtain a coarse triangular grid. By repeating some of the previous cases with this new grid, we can compare the results for the two grids.

Figure 11.13 shows how the *E*-plane co-polarization levels are modified when the elements are arranged in a triangular grid.

The reduction of the RCS levels for the triangular grid array compared to the rectangular grid is explained by the denser azimuth spacing of the triangular grid, which reduces the grating lobe contributions. Figure 11.14 shows the *H*-plane co-polarization levels. The

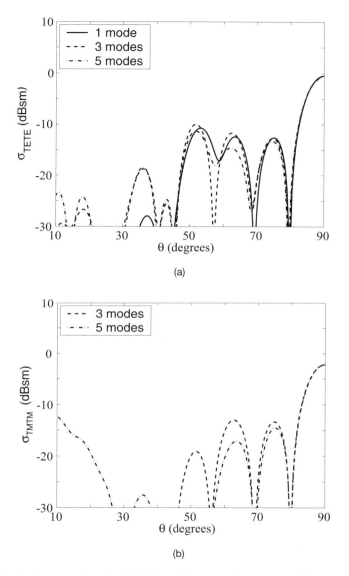

Figure 11.9. *H*-plane results for 6.8 GHz with up to five waveguide modes. (a) σ_{TETE}. (b) σ_{TMTM}.

maximum RCS is also reduced here. However, at some elevation angles the triangular grid array exhibits larger RCS.

Figure 11.15 compares the scattering properties in terms of σ_{TETE} for the triangular and rectangular grid arrays. Figure 11.16 shows the cross-polarization case.

11.4.4 Conclusions from the PEC Conformal Array Analysis

We conclude from the analysis so far that it is only for very accurate results that it is necessary to take the coupling between the antenna elements into account. The main features

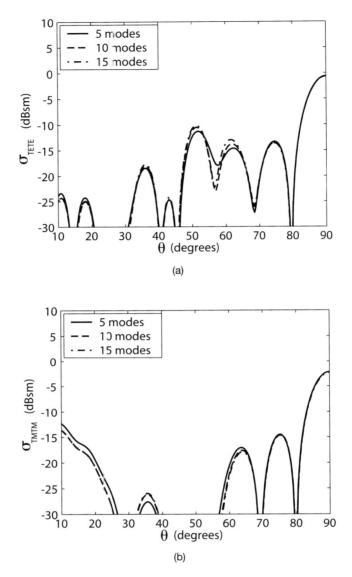

Figure 11.10. *H*-plane results for 6.8 GHz with up to 15 waveguide modes. (a) σ_{TETE}. (b) σ_{TMTM}.

of the scattering are obtained without introducing the mutual coupling (which simplifies the calculations).

Rather few waveguide modes are needed when the RCS is calculated in the *E* plane, whereas at least five modes are needed for the *H*-plane analysis. For best results, 15 modes should be used in the *H* plane.

It has been found that the antenna exhibits strong cross-polarization scattering outside the principal planes. It is in general important to include all components in the radar cross-section matrix when analyzing scattering from a conformal array antenna.

However, since the arrays have only three rows one cannot draw general conclusions

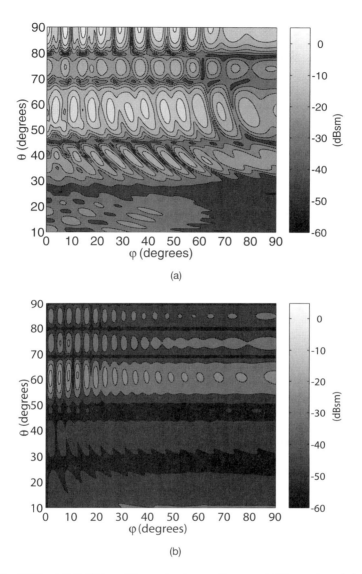

Figure 11.11. RCS at 6.8 GHz with mutual coupling and 15 modes. (a) σ_{TETE}, (b) σ_{TETM}.

regarding scattering from rectangular versus triangular grids from the simple example presented.

11.5   CYLINDRICAL ARRAY WITH DIELECTRIC COATING

The results for the coated array were obtained for an array configuration as defined below:

Element waveguide apertures	39 mm × 16 mm
Element grid	Rectangular, 4 × 32 elements

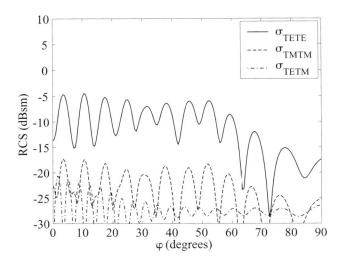

Figure 11.12. The level of cross-polarization σ_{TETM} (dash–dotted line) compared to the co-polarized cases σ_{TETE} (solid line) and σ_{TMTM} (dashed line). 6.8 GHz, 15 modes, and $\theta = 60°$.

Element spacing	20 mm in the azimuth
	41 mm in the axial direction
Cylinder radius	300 mm

Thus, the difference from the noncoated case is a denser grid and four rows of elements instead of three [Figure 11.3 (*a*)]. This will eliminate the grating lobe problem in the azimuth (at 6.8 Ghz, 20 mm corresponds to 0.45 λ). 32 elements cover about the same azimuth sector on the cylinder ($\approx 120°$) as before with 18 elements. The dielectric layer has a thickness of 2.4 mm. Permittivity values from 1.0 to 2.5 have been studied. Five waveguide modes were used in the calculations. We present results mainly for two frequencies, 4.3 GHz and 6.8 GHz.

11.5.1 Single Element with Dielectric Coating

In Figures 11.17 and 11.18, the effect of the dielectric layer on the *E*- and *H*-plane RCS is illustrated for a single-element antenna on the coated cylinder at 4.3 GHz.

By introducing the dielectric layer, the monostatic RCS decreases in the specular direction. With a coating, more energy propagates along the cylinder surface, which explains the higher RCS levels at $|\varphi| > 90°$. The oscillations result from the interference between the lit region geometrical optics ray and the rays that creep around the cylinder.

The corresponding results calculated at 6.8 GHz are presented in Figures 11.19 and 11.20.

The increased frequency makes the coating electrically thicker and more energy propagates along the cylinder surface. For elevation angles approaching the axial directions, the radius of curvature in the direction of the geodesics increases. Therefore, the surface rays shed less energy as they propagate along the surface. This effect is more pronounced

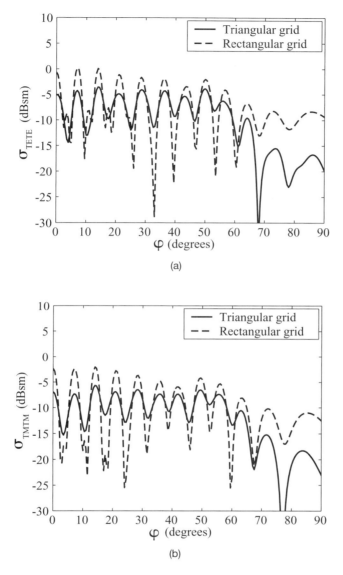

Figure 11.13. *E*-plane results for rectangular grid (dashed line) and triangular grid (solid line). (a) is σ_{TETE} and (b) is σ_{TMTM}. 6.8 GHz and 15 modes.

at the higher frequency. As before, there is interference between the lit region geometrical optics ray and the rays that creep around the cylinder.

11.5.2 Array with Dielectric Coating

Since we now have a relatively dense element grid, we do not experience a large ripple in the azimuth as we had before (cf. Figure 11.7). The *E*-plane RCS results shown in Figure

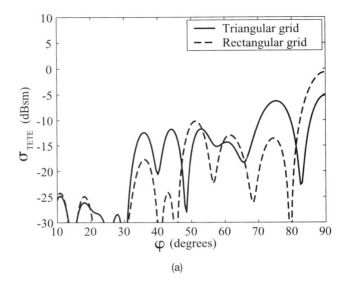

(a)

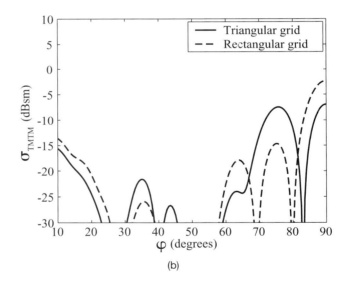

(b)

Figure 11.14. *H*-plane results for triangular grid (solid line) and rectangular grid (dashed line). (a) σ_{TETE}. (b) σ_{TMTM}. 6.8 GHz and 15 modes.

11.21 are fairly constant over the about 120° extent of the array. The effect of the dielectric layer on the RCS is not dramatic. The RCS patterns become more oscillatory when the dielectric coating is introduced. The corresponding *H*-plane results are presented in Figure 11.22.

At the lower frequency, the energy of the creeping waves are attenuated rather fast and the coating has a small effect on the backscattered field. The appearance of the RCS patterns is largely determined by the array factor. The peaks near 32 and 148 degrees in the *H* plane at 4.3 GHz [Figure 11.22 (*a*)] are due to grating lobes in the elevation

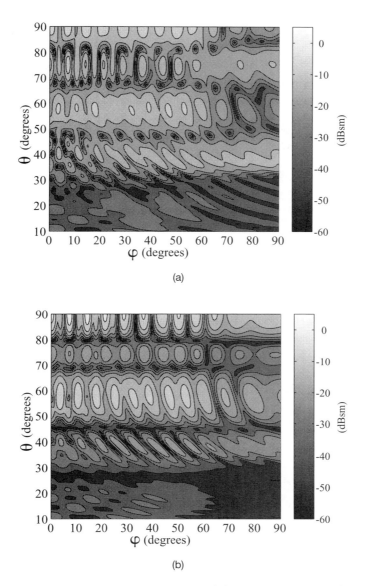

Figure 11.15. σ_{TETE} at 6.8 GHz with 15 modes. (a) Triangular grid. (b) Rectangular grid.

plane. At the higher frequency [Figure 11.22 (*b*)] the energy is guided much more efficiently along the cylinder surface and the dielectric coating has a larger impact on the backscattered field, especially in the paraxial regions where an increased RCS is obtained.

Figure 11.23 shows the RCS as a function of frequency for two different angles of incidence, $\theta = 90°$ and $\theta = 50°$, both with and without a dielectric coating. The resonance around 3.8 GHz in Figure 11.23 (*a*) corresponds to the cutoff frequency for the dominant waveguide mode ($f_{c,10}$). A similar resonance has also been observed for the isolated ele-

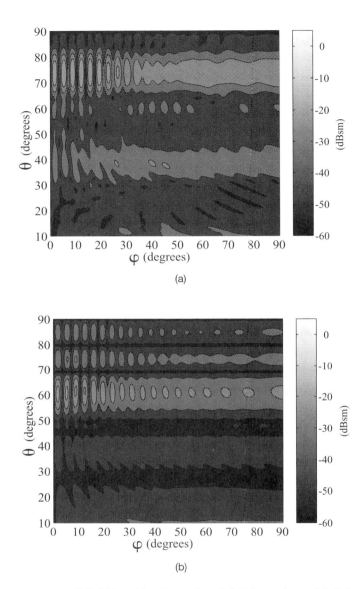

Figure 11.16. σ_{TETM} at 6.8 GHz with 15 modes. (a) Triangular grid, (b) rectangular grid.

ment [Figure 11.30 (*b*)]. For frequencies below $f_{c,10}$, no modes propagate in the waveguides and the RCS decreases quickly, as expected. The second mode in cutoff order is the TE$_{20}$ mode, with a cutoff frequency $f_{c,20} \approx 7.7$ GHz. Because of symmetry, this mode is not excited when $\theta = 90°$ and no resonance is observed around that frequency. However, when the elevation angle is changed from normal incidence the TE$_{20}$ mode becomes more important, as demonstrated in Figure 11.23 (*b*), where a resonance is observed around 7.7 GHz. The highest RCS in Figure 11.23 (*b*) occurs between 5 and 6 GHz, corresponding to a "grating lobe" that sweeps past as the frequency is changed.

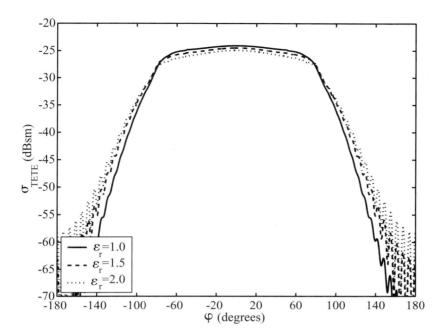

Figure 11.17. The *E*-plane RCS at 4.3 GHz for a single waveguide element on a cylinder with different dielectric coatings.

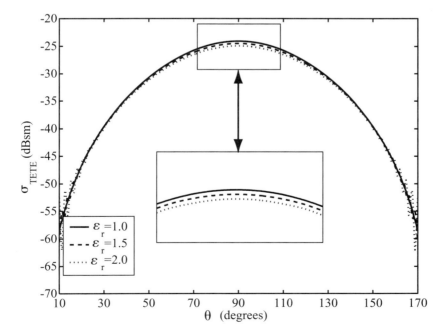

Figure 11.18. The *H*-plane RCS at 4.3 GHz for a single waveguide element on a cylinder with different dielectric coatings.

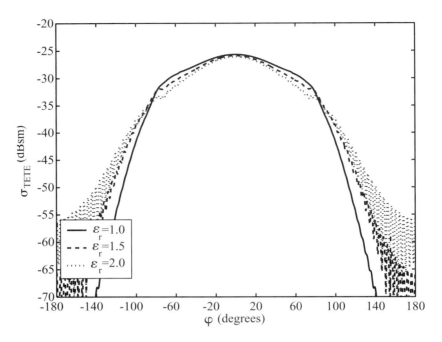

Figure 11.19. The *E*-plane RCS at 6.8 GHz for a single waveguide element on a cylinder with different dielectric coatings.

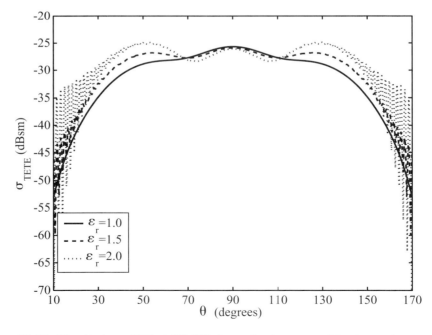

Figure 11.20. The *H*-plane RCS at 6.8 GHz for a single waveguide element on a cylinder with different dielectric coatings.

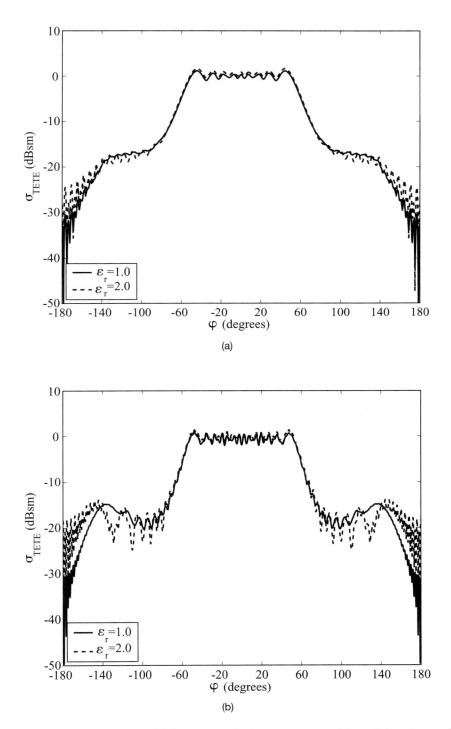

Figure 11.21. The *E*-plane RCS for a 4 × 32 element array with a dielectric coating (dashed line). The noncoated case is included as reference (solid line). (a) 4.3 GHz. (b) 6.8 GHz.

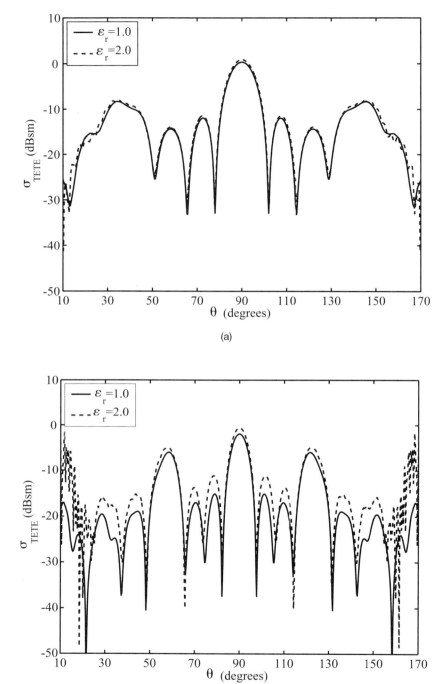

Figure 11.22. The *H*-plane RCS for a 4 × 32 element array with a dielectric coating (dashed line). The noncoated case is included as reference (solid line). (a) 4.3 GHz. (b) 6.8 GHz.

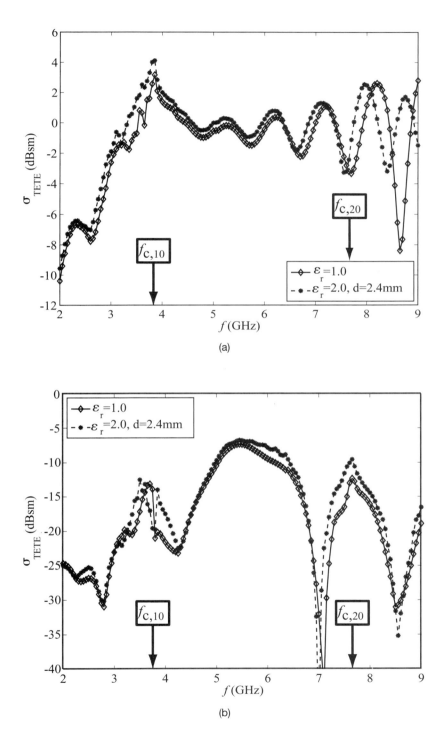

Figure 11.23. The RCS as a function of frequency for a 4 × 32 element array with dielectric coating, $\varepsilon_r = 2.0$ (dashed line), and without coating (solid line). (a): $\theta = 90°$, $\varphi = 0°$. (b): $\theta = 50°$, $\varphi = 0°$.

The effect of a changing cylinder radius has also been investigated. Keeping a constant aperture separation (in mm), results for different cylinder radii are shown in Figure 11.24. For the smallest cylinder radius investigated, the apertures cover an arc of approximately 237° on the cylinder. As shown in Figure 11.24 (a), this results in a large angular region with a rather constant RCS. As the cylinder radius is increased, the angular extent is reduced. As the radius a goes to infinity, the backscattering approaches the planar array case.

11.6 RADIATION AND SCATTERING TRADE-OFF

11.6.1 Introduction

A common technique to reduce a vehicle's radar cross section (RCS) is shaping, for example, using flat surfaces where the individual facets are oriented to deflect incoming electromagnetic energy away from the direction of the expected monostatic threat. By applying radar-absorbing materials to certain "hot spots" with strong reflections, the scattering can be further reduced. After these measures, the remaining dominant RCS contribution may very well be due to the antennas on the vehicle [Williams 1986, Wiesbeck and Heydrich 1992].

The backscattered field from an antenna can to some extent be controlled by adjusting the antenna load impedance. Such an adjustment will also affect the radiation properties of the antenna, and a trade-off between scattering and radiation properties is necessary. Several authors have reported work on this topic for simple antennas such as dipoles, circular loops, and horns; see, for example, [Garbacz 1962, Chen and Liepa 1964, Chen 1965, Lin and Chen 1968, Schneider 1996]. By a systematic variation of the element loads in an array antenna, it is possible to minimize the scattering in selected directions without too much penalty on the gain performance [Stjernman and Josefsson 2003].

In this section, we will present a trade-off between gain and RCS for waveguide aperture antennas mounted on a PEC circular cylinder. Figure 11.25 (a) shows the geometry. The geometrical parameters are the following:

Element waveguide apertures	39 mm × 8 mm
Element grid	Rectangular, 40 × 32 elements
Element spacings	20 mm in the azimuth
	41 mm in the axial direction
Cylinder radius	300 mm

A staggered triangular grid as shown in Figure 11.25 (d) is also analyzed for comparison. In all cases, the array covers an arc of about 120° on the cylinder. There is no dielectric coating.

The matching is controlled by placing reactive elements in the form of thin irises in the waveguides. Symmetrical inductive irises were used since it was found that inductive irises could be placed closer to the apertures than capacitive irises, thus providing a more broadband match [see Figure 11.25 (b)].

The waveguides are assumed to be connected to matched loads, that is, the modal amplitudes a_3^{ip} in Figure 11.25 (b) are zero for all apertures (numbered i) and modes (numbered p). Once the aperture electric field has been determined, the far-zone scattered

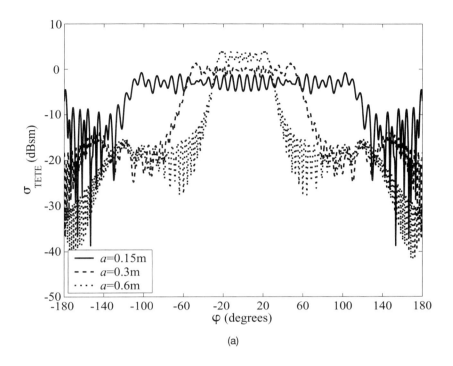

(a)

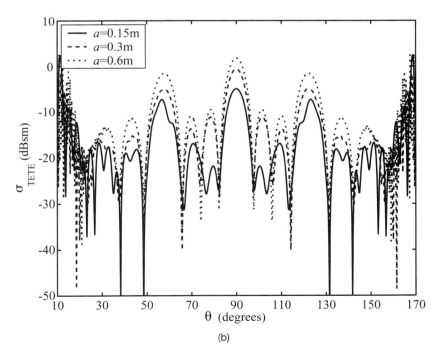

(b)

Figure 11.24. The effect of a changing cylinder radius for a 4 × 32 element array with $\varepsilon_r = 2.0$ coating at 6.8 GHz. The radius is 0.15 m (solid line), 0.3 m (dashed line), and 0.6 m (dash–dotted line). (a) *E* plane, (b) *H* plane.

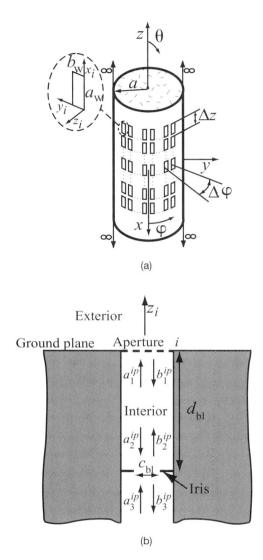

Figure 11.25. (a) The conformal array antenna. (b) The interior and exterior domains at aperture i. *(continued)*

fields are calculated using UTD [Pathak et al. 1981]. For the radiation problem, the array is fed from the waveguides, that is, $a_3^{i1} \neq 0$ and $\overline{H}^{\text{inc}} = 0$.

We will first present results for matched waveguide elements, both the single-element case and the array case. We will then modify the iris parameters in order to find a good compromise between gain and scattering cross section. The iris gap, c_{bl}, and position, d_{bl}, needed to obtain an impedance match, were determined using a single-mode transmission-line model based on approximate formulas for the susceptance values [Collin 1992]. For the arrays, the design was made for a center element with the array phased for the main beam pointing in the desired direction, also including the effects of the mutual coupling among the elements (active impedance match). The resulting iris was then used in

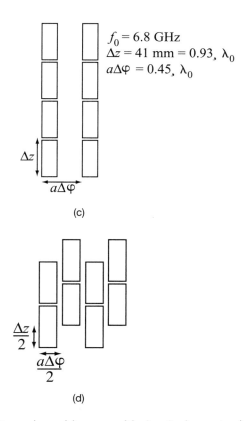

$f_0 = 6.8$ GHz
$\Delta z = 41$ mm $= 0.93, \lambda_0$
$a\Delta\varphi = 0.45, \lambda_0$

(c)

(d)

Figure 11.25. (*cont.*) (c) Rectangular grid array with 4 × 2 elements. (d) Triangular grid array with 4 × 2 elements.

all waveguides, leading to a situation of well-matched center elements and somewhat less well-matched edge elements.

11.6.2 Single-Element Results

In Section 11.4, it was found that the RCS in the E plane can be calculated with good accuracy using only the dominant waveguide mode. At least five waveguide modes are needed in the H plane; for best results, up to 15 modes should be included. Figure 11.26 shows the effect of higher-order modes (one mode vs. 15 modes), with and without iris, in the E and H planes for a single waveguide element situated at $\varphi = 0°$ on the cylinder. The frequency is 6.8 GHz.

The iris dimensions for the single-element match were found to depend slightly on the number of waveguide modes used in the optimization:

$c_{bl} = 20.6$ mm, $d_{bl} = 4.3$ mm as optimized with one waveguide mode
$c_{bl} = 20.7$ mm, $d_{bl} = 4.2$ mm as optimized with 15 waveguide modes

Note that the RCS level is considerably higher for the matched configurations compared to the unmatched cases. This is typical for many antennas and points to the difficult

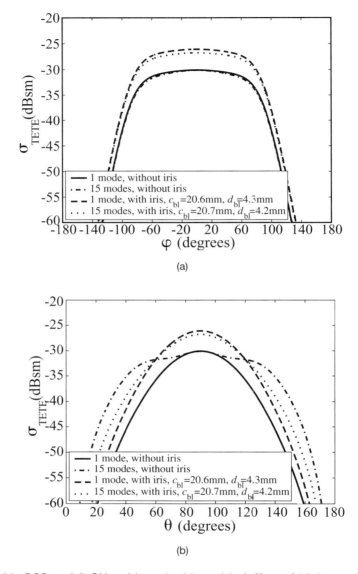

Figure 11.26. RCS at 6.8 GHz with and without iris (effect of higher-order modes). Solid line: one mode, no iris. Dash–dotted line: 15 modes, no iris. Dashed line: One mode, iris matched. Dotted line: 15 modes, iris matched. (a) *E* plane. (b) *H* plane.

problem of maximizing gain and at the same time minimizing the scattering. As expected, the fundamental mode is sufficient for the *E* plane where a very wide, almost flat response is obtained. When the elevation angle is changed from normal incidence, higher-order modes are excited and become important for the backscattered field as shown in Figure 11.26 (*b*). However, the single-mode approximation is reasonably accurate for the matched configuration. This is due to the position of the symmetrical inductive iris close to the aperture, causing the TE_{20} mode to become suppressed, thus increasing the importance of the TE_{10} mode. The effects of matching and higher-order modes on the antenna gain are shown in Figure 11.27.

For the antenna function, the excitation is symmetrical and the single-mode approximation predicts the gain rather well. Without matching, the gain decreases about 0.7 dB, but the RCS is reduced about 3.4 dB. Thus, increasing the mismatch reduces the RCS significantly without too much penalty on the antenna gain.

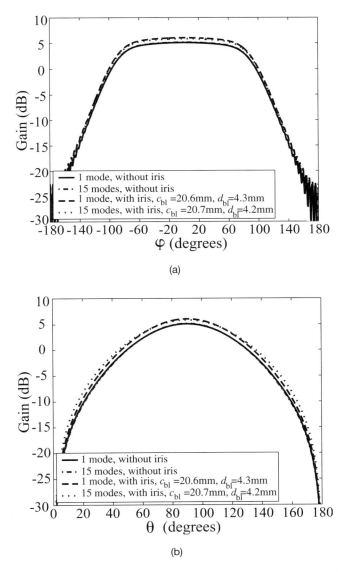

(a)

(b)

Figure 11.27. Gain at 6.8 GHz with and without iris and the effect of higher-order modes, *E* and *H* planes. Solid line: one mode, no iris. Dash–dotted line: 15 modes, no iris. Dashed line: One mode, iris matched. Dotted line: 15 modes, iris matched. (a) *E* plane. (b) *H* plane.

A systematic variation of the impedance loads (iris parameters) and the effects on reflection coefficient (gain) and RCS are shown in Figures 11.28 and Figure 11.29. Fifteen waveguide modes were used here. The dark areas have low reflection and low scattering, respectively.

The darkest region in Figure 11.28 (*a*) corresponds to values of c_{bl} and d_{bl}, where the antenna element is well matched. For a very small iris gap, the antenna is almost short-circuited, which obviously leads to a poor match, but very little backscattering [Figure 11.28 (*b*)].

From the results, we find that the dark regions in Figure 11.28 (*a*) and (*b*) do not coincide and that a trade-off between efficiency and RCS is necessary. A region of low RCS can be found in the lower-left corner for all frequencies, which indicates that an iris placed close to the aperture with a gap of $c_{bl}/a_w \approx 0.6$ may be a good choice [cf. Figure

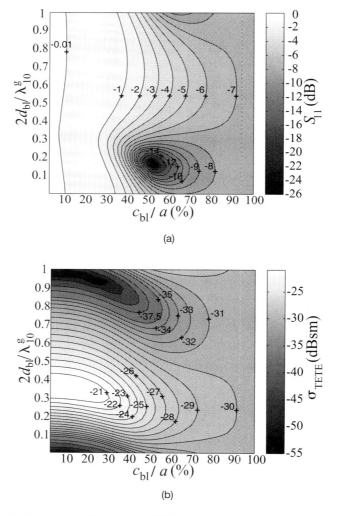

Figure 11.28. Reflection coefficient and RCS as a function of iris gap and iris position ($f = 6.8$ GHz). (a) S_{11}, (b) σ_{TETE} ($\theta = 90°$, $\varphi = 0°$).

11.28 (*a*)]. To illustrate this, the reflection coefficient and RCS are shown in Figure 11.30 (*a*) for an element with an iris placed 1 mm away from the aperture.

As the frequency increases from the working frequency band, it can be expected that higher-order modes will give a larger contribution to the scattered field, especially for *H*-plane scattering. This is illustrated in Figure 11.30 (*b*), where the RCS of a single element is given as a function of frequency in the direction ($\theta = 60°$, $\varphi = 0°$) for different numbers of modes. We notice in this Figure a sharp resonance around the cutoff frequency for the dominant mode ($f_{c,10} \approx 3.8$ GHz). For frequencies around and below $f_{c,10}$, the fundamental mode predicts the RCS very well and higher-order modes are of little consequence. For frequencies below $f_{c,10}$, no modes propagate in the waveguide and the RCS decreases quickly, as expected. However, as the frequency increases, the second mode in cutoff or-

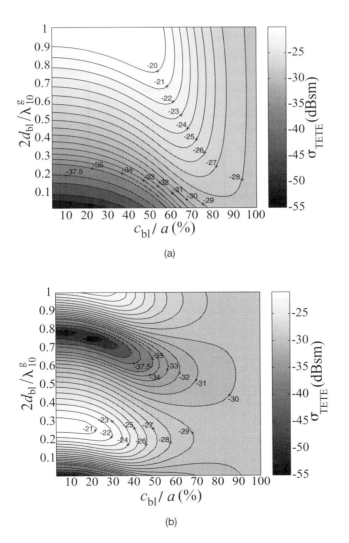

(a)

(b)

Figure 11.29. RCS as a function of iris gap and iris position ($\theta = 90°$, $\varphi = 0°$). (a) $f = 4.3$ GHz, (b) $f = 8.0$ GHz.

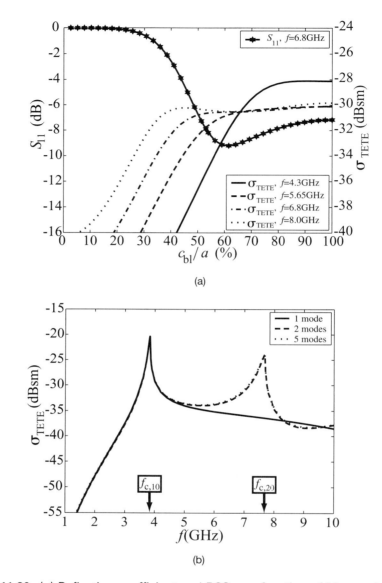

Figure 11.30. (a) Reflection coefficient and RCS as a function of iris gap for a single element with an iris placed 1 mm from the aperture (15 modes, $\theta = 60°$). (b) The RCS for a single element without iris as a function of frequency ($\theta = 60°$).

der (TE_{20}) becomes important. The second resonance around the cutoff frequency for the TE_{20} mode ($f_{c,20} \approx 7.7$ GHz) is not predicted by the single-mode approximation.

11.6.3 Array Results

The effects of mutual coupling have been investigated for a reduced array of 4×32 elements placed in a rectangular grid [cf. Figure 11.25 (a)]. It was found [Thors and Josefsson 2003] that the mutual coupling primarily affects the exact gain and RCS levels, whereas the general patterns can be obtained without coupling (within 0.5 dB). By neglecting mutual coupling, it is only necessary to consider the diagonal elements of the moment matrix elements. This reduces the memory requirements and the computational time significantly and makes it possible to analyze many cases and large arrays. In the following comparative study, we have neglected mutual coupling when investigating the effects of various impedance loads and element grid types.

The radiation and scattering properties of a rectangular and a triangular grid array consisting of 40×32 elements have been compared and the results are given in Figure 11.31. Due to the large number of elements, the RCS oscillates heavily, especially in the H plane. In order to simplify the comparisons, the RCS is presented as an arithmetic average (in power) calculated within a sliding window. The chosen window is $5°$ wide for the E-

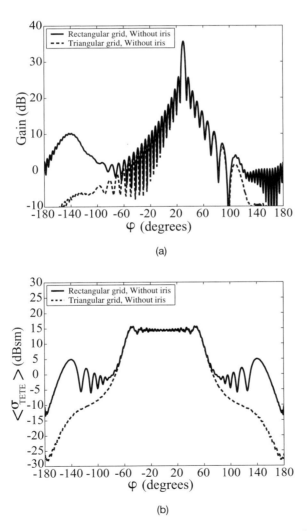

Figure 11.31. Comparison of rectangular (solid line) and triangular (dashed line) grid arrays with 40×32 elements at 6.8 GHz. (a) Pattern scanned to $\varphi_0 = 30°$ in the E plane, three modes. (b) The E-plane RCS ($5°$ sliding window average), one mode.

plane averaging and 2° for the *H* plane. The averaging makes the curves easier to read, but causes the narrow high peaks and large dips to appear reduced in amplitude.

Due to the rather large element spacing, the sidelobes increase in a region around $\varphi = -140°$ for the rectangular grid array, as shown in Figure 11.31 (*a*). This "grating lobe" effect is not observed for the triangular grid array, in which the effective element spacing is smaller. The larger element spacing of the rectangular grid array is also responsible for the higher RCS at azimuth angles outside $\varphi = \pm60°$ in Figure 11.31 (*b*).

Finally, the radiation and scattering properties of a cylindrical array antenna have been compared to the corresponding quantities for a planar array. The planar results were approximated by using a cylinder of radius $a = 100$ m in the calculations; the RCS results are presented in Figures 11.32 and 11.33.

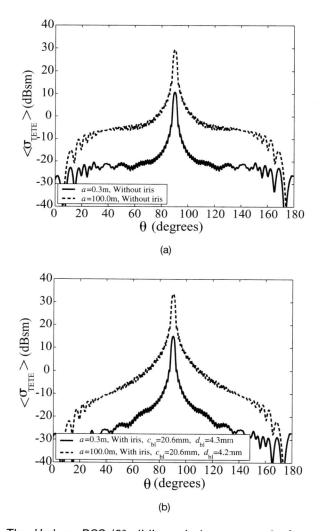

Figure 11.32. The *H*-plane RCS (2° sliding window average) of a cylindrical (solid line) and a planar (dashed line) array antenna (triangular grid, 40 × 32 elements, 6.8 GHz, 15 modes). (a) Without irises. (b) With irises (matched).

As shown in Figure 11.32, the *H*-plane RCS is generally much lower for the cylindrical array than for the planar array. This is a consequence of the curved geometry, which defocuses the backscattered field; compare with [Wills 1991]. Both unmatched arrays exhibit an increased RCS at elevation angles far from broadside but a lower peak value compared to the matched arrays (Figure 11.32). For the planar array, the *E*-plane RCS is similar to the *H*-plane RCS with a very high RCS response at normal incidence. The cylindrical array has an RCS response similar to the results presented in Figure 11.21 (*b*) with a moderately high RCS over a broad angular region where the elements are situated. Thus, the cylindrical array has a higher RCS in the *E* plane, which, however, coincides with the large structural scattering from the cylinder itself.

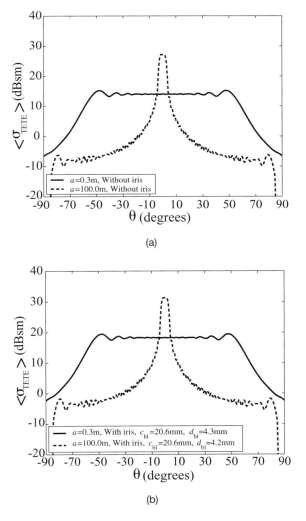

(a)

(b)

Figure 11.33. The *E*-plane RCS (5° sliding window average) of a cylindrical (solid line) and a planar (dashed line) array antenna (triangular grid, 40 × 32 elements, 6.8 GHz, 15 modes). (a) Without irises. (b) With irises (matched).

11.7 DISCUSSION

We have presented an analysis of scattering from a cylindrical array with waveguide aperture elements. The polarization is azimuthal, that is, horizontal with a vertical (z-directed) cylinder axis. Each aperture has a dominant magnetic current distribution in the z direction. Therefore, this configuration represents a realization of a canonical case with axial magnetic dipole elements. Like the dual case with axial electric dipoles, this array has low cross-polarized radiation, as discussed in Chapter 8.

Another canonical case is represented by an array with the waveguide apertures rotated to produce axial polarization. This case can be analyzed using the same methods as we have described here. The cross-polarization effects will be larger in that case. The effects of higher waveguide modes will also be different. Consider, for example, illumination in the azimuthal plane. Since the elements are oriented in different directions relative to the incoming wave, there will be different degrees of asymmetric illumination, causing different mode combinations in the individual elements. This effect will be small near the aperture resonance, but will increase for higher frequencies. An example is given in Figure 11.34, showing the electric field distribution along a narrow rectangular aperture (slot) backed by a waveguide. The illumination is 60° from the direction normal to the slot, in the H plane. The dashed line represents 9.375 GHz, which is close to slot resonance; here, the fundamental TE_{10} mode dominates. However, at 14 GHz, the slot is far from resonance, and higher modes are also excited; compare Figure 11.30 (b).

It is possible to improve the radiation function (increase the gain) and reduce the scattering (reduce the RCS) by loading the waveguide elements with a suitable reactive im-

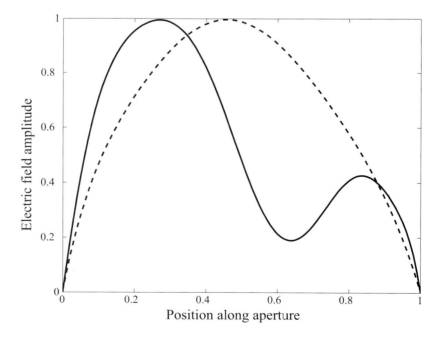

Figure 11.34. The electric field distribution along a narrow aperture (slot) illuminated at $\theta = 60°$ from normal incidence (H plane). Dashed line is near resonance (9.375 GHz) and solid line at 14 GHz. From [Josefsson 1990].

pedance. For the waveguide element investigated here, an optimum iris load was sought by calculating the reflection coefficient and RCS as a function of the iris gap, c_{bl}, and iris position, d_{bl}. A suitable compromise was obtained for an iris placed rather close to the aperture.

Grating lobes can be the source of high backscattering from arrays; therefore, close element spacings should be used. Hence, a triangular grid array is often to be preferred over a rectangular grid.

A general comparison of planar and cylindrical array antenna performance is a complicated issue. However, we could demonstrate that outside the plane $\theta = 90°$ the cylindrical array has much lower RCS compared to the planar array, since the cylindrical geometry defocuses the backscattered field. Just as the specular reflection at normal incidence should be avoided for a planar array, the cylindrical array antenna should be oriented in such a way that the cylinder normal points in directions where high RCS levels can be tolerated.

REFERENCES

Appel-Hansen J. (1979), "Accurate Determination of Gain and Radiation Patterns by Radar Cross Section Measurements," *IEEE Transactions on Antennas and Propagation,* Vol. AP-27, pp. 640–646, September.

Bach Andersen J. (2003), "Transmitting, Receiving, and Scattering Properties of Antennas," in *Proceedings of Antenn 03,* Kalmar, Sweden, May.

Balanis C. A. (1997), *Antenna Analysis, Theory, and Design,* (2nd ed.), pp. 90–97, Wiley.

Balanis C. A. (1989), *Advanced Engineering Electromagnetics,* Wiley.

Bonnedal M. (1996), "Coupling in a Rectangular Waveguide Array—Theory and Algorithms," Tech. Rep. SE/REP/0221/A, SAAB Ericsson Space, 1996.

Chen J. and Jin J. (1995), "Electromagnetic Scattering from Slot Antennas on Waveguides with Arbitrary Terminators," *Microwave and Optical Technology Letters,* Vol. 10, pp. 286–291.

Chen K. M. and Liepa V. (1964), "Minimization of the Backscattering of a Cylinder by Central Loading," *IEEE Transactions on Antennas and Propagation,* Vol. AP-12, pp. 576–582, September.

Chen K. M. (1965), "Minimization of the Backscattering of a Cylinder by Double Loading," *IEEE Transactions on Antennas and Propagation,* Vol. AP-13, pp. 262–270, March.

Collin R. E. (1992), *Foundations for Microwave Engineering,* pp. 340–341, McGraw-Hill.

Collin R. E. and Zucker F. J., eds. (1969), *Antenna Theory, Part 1,* pp. 123–131, McGraw-Hill.

Croswell W. F., Rudduck R. C., and Hatcher D. M. (1967), "The Admittance of a Rectangular Waveguide Radiating into a Dielectric Slab," *IEEE Transactions on Antennas and Propagation,* Vol. AP-15, pp. 627–633, September.

Fan G. X. and Jin J. M. (1997), "Scattering from a Cylindrically Conformal Slotted Waveguide Array Antenna," *IEEE Transactions on Antennas and Propagation,* Vol. 45, pp. 1150–1159, July.

Fan G. X. and Jin J. M. (1999), "Scattering from a Large Planar Slotted Waveguide Array Antenna," *Electromagnetics,* Vol. 19, pp. 109–130, February.

Garbacz R. J. (1962), "The Determination of Antenna Parameters by Scattering Cross-Section Measurements I-III," Tech. Rep., Antenna Laboratory, Ohio State University, November.

Hansen R. C. (1989), Relationships between Antenna as Scatterers and as Radiators," *Proceedings of IEEE,* Vol. 77, pp. 659–662, May.

Harrington R. F. (1961), *Time-Harmonic Electromagnetic Fields,* McGraw-Hill.

Harrington R. F. (1964), "Theory of Loaded Scatterers," *Proceedings of IEE,* Vol. 111, pp. 617–623, April.

Josefsson L. (1990), "Slot Coupling and Scattering," *IEEE AP-S International Symposium Digest,* pp. 942-945.

Knott E. F., Shaeffer J. F., and Tuley M. T. (1993), *Radar Cross Section,* pp. 410–416, Artech House, Norwood, MA, USA.

Ko W. L. and Mittra R. (1989), "Scattering by a Conformal Array of Metallic Patches," *Microwave and Optical Technology Letters,* Vol. 2, pp. 145–149, April.

Lin J. L. and Chen K. M. (1968), "Minimization of the Backscattering of a Loop by Impedance Loading—Theory and Experiment," *IEEE Transactions on Antennas and Propagation,* Vol. AP-16, pp. 299–304, May.

McNamara D. A., Malherbe J. A. G., and Pistorius C. W. I. (1992), "External Contribution to Electromagnetic Scattering from Planar Antenna Array of Open Rectangular Waveguides," *Electronics Letters,* Vol. 28, pp. 1415–1417, July.

Munk B. A. (2003), *Finite Antenna Arrays and FSS,* Wiley, 2003.

Pathak P. H., Wang N., Burnside W. D., and Kouyoumjian R. G. (1981), "A Uniform GTD Solution for the Radiation from Sources on a Convex Surface," *IEEE Transactions on Antennas and Propagation,* Vol. AP-29, pp. 609–622, July.

Persson P. and Josefsson L. (2001), "Calculating The Mutual Coupling between Apertures on a Convex Circular Cylinder Using a Hybrid UTD-MoM Method," *IEEE Transactions on Antennas and Propagation,* Vol. 49, No. 4, pp. 672–677, April.

Schneider R. K. (1996), "A Re-look at Antenna In-band RCSR via Load Mismatching," in *Proceedings of Antennas and Propagation Society International Symposium,* Vol. 2, pp. 1398–1401, July.

Sinclair G., Jordan E. C., and Vaughan E. W. (1947), "Measurement of Aircraft-Antenna Patterns Using Models," *Proceedings of IRE,* Vol. 35, pp. 1451–1467, December.

Stevenson A. F. (1948), "Relations Between the Transmitting and Receiving Properties of Antennas," *Quarterly of Applied Mathematics,* Vol. V, pp. 369–384, January.

Stjernman A. and Josefsson L. (2003), "Reduction of Radar Cross Section (RCS) of Antenna Arrays by Systematic Variation of Impedance Load," in *Proceedings of Antenn 03,* pp. 255–260, Kalmar, Sweden, May.

Tai C. T. (2001), "A Scattering Theory of Receiving Antennas," Techn. Rpt. RL 999, Radiation Laboratory, University of Michigan, USA, November.

Thors B. and Josefsson L. (2001), "The Radar Cross Section of a Conformal Array Antenna," in *Proceedings of 2nd European Workshop on Conformal Antennas,* The Hague, The Netherlands, April.

Thors B., Josefsson L., and Norgren M. (2002), "Self- and Mutual Admittance Calculations for Rectangular Waveguides Radiating into a Grounded Dielectric Slab," Techn. Rpt. TRITA-TET 02-07, Division of Electromagnetic Theory, Royal Institute of Technology, Stockholm, Sweden, November.

Thors B. (2003), *Radiation and Scattering from Conformal Array Antennas,* Ph.D. Thesis, Royal Institute of Technology.

Thors B. and Josefsson L. (2003), "Radiation and Scattering Trade-Off Design for Conformal Arrays," *IEEE Transactions on Antennas and Propagation,* Vol. AP-51, No. 5, pp. 1069–1076, May.

Thors B. and Rojas R.G. (2003), "Uniform Asymptotic Solution for the Radiation from a Magnetic Source on a Large Dielectric Coated Circular Cylinder, *Radio Science,* Vol. 38, No. 5, pp. 4-1–4-12.

Thors B., Josefsson L., and Rojas R. G. (2004), "The RCS of a Cylindrical Array Antenna Coated

With a Dielectric Layer," *IEEE Transactions on Antennas and Propagation,* Vol. AP-52, No. 7, pp. 1851–1858, July.

Tittensor P. J. and Newton M. L. (1989), "Prediction of the Radar Cross Section of an Array Antenna," in *Proceedings of Sixth International Conference on Antennas and propagation (ICAP 89),* April, pp. 258–262.

Wiesbeck W. and Heidrich E. (1992), "Influence of Antennas on the Radar Cross Section of Camouflaged Aircraft," in *Proceedings of Radar 92 International Conference,* Brighton UK, 12–13 October, pp. 122–125.

Williams N. (1986), "The Radar Cross Section of Antennas—an Appraisal," in *Proceedings of Military Microwave Symposium, MM-86,* pp. 502–508.

Wills R. W. (1991), "Calculation of Radar Cross Section for an Active Array," in *Proceedings of IEE Colloquium on Antenna Radar Cross-Section,* London UK, 7 May, pp. 8/1–8/4.

INDEX

ABOUT THE AUTHORS

Lars Josefsson was born in Norrköping, Sweden, in 1939. He graduated with an M.Sc.Eng. from The Royal Institute of Technology in Stockholm in 1962 and got his Ph.D. from Chalmers University of Technology in Göteborg in 1978. Thanks to a grant from the Marcus Wallenberg Foundation in 1981 and a Fulbright Scholarship in 1982, he spent 1982–1983 at the University of California, Los Angeles, as a visiting scientist. He was Adjunct Professor in Antenna Technology from 1983 to 1986 at Chalmers and held the same position at the Royal Institute of Technology in Stockholm from 1996 until 2003. Now he is back at Chalmers as Adjunct Professor in the Department of Signals and Systems.

Lars Josefsson was with Ericsson Microwave Systems AB in Mölndal, Sweden, for more than 40 years. He held a staff position responsible for the introduction of new antenna technology and systems, internal R&D projects, and internal education relating to antennas. In 2001, he was appointed Senior Expert, Antenna Systems. In 2004, he received the Thulin Silver Medal from the Swedish Aeronautical Society for his contributions to several generations of airborne radar antenna developments. He was elected Fellow of the IEEE in 1999. He is an Associate Editor of the *IEEE Transactions on Antennas and Propagation*.

Patrik Persson was born in Stockholm, Sweden, in 1973. He received the M.Sc.Eng. and Ph.D. degrees in electromagnetic theory from the Royal Institute of Technology, Stockholm, in 1997 and 2002, respectively. He is currently employed as a research associate in the division of Electromagnetic Theory, Royal Institute of Technology. His main research interests are development of analysis and design tools for conformal arrays.

Conformal Array Antenna Theory and Design. By Lars Josefsson and Patrik Persson
© 2006 Institute of Electrical and Electronics Engineers, Inc.

Patrik Persson was the cochairman of the 3rd European Workshop on Conformal Antennas (2003) and chairman and organizer of the 4th European Workshop on Conformal Antennas (2005). He was elected as a partner of ACE (Antenna Centre of Excellence) in the European Commission 6th Framework Program, in particular in the program for conformal antennas.

He spent 6 months of the academic year 2000–2001 at the ElectroScience Laboratory, Ohio State University, Columbus, Ohio as a visiting scholar, working on the development of high-frequency models for conformal antennas. He has also been a visiting scientist (part time, Feb. 2002–Apr. 2004) at the department of Physics and Measurement Technology, Division of Theoretical Physics, Linköping University in Sweden. During this period, he was involved with work on faceted conformal antennas.

Patrik Persson received the 2002 R. W. P. King Prize Paper Award given by IEEE.